Bioplastics and Biocomposites
A Practical Introduction

Green Chemistry Series

Editor-in-chief:
James H. Clark, *Department of Chemistry, University of York, UK*

Series editors:
Catherine Birch, *AgriFood X Limited, UK*
Graham Bonwick, *AgriFood X Limited, UK*
George A. Kraus, *Iowa State University, USA*
Andrzej Stankiewicz, *Delft University of Technology, The Netherlands*

Titles in the series:
1: The Future of Glycerol: New Uses of a Versatile Raw Material
2: Alternative Solvents for Green Chemistry
3: Eco-Friendly Synthesis of Fine Chemicals
4: Sustainable Solutions for Modern Economies
5: Chemical Reactions and Processes under Flow Conditions
6: Radical Reactions in Aqueous Media
7: Aqueous Microwave Chemistry
8: The Future of Glycerol: 2nd Edition
9: Transportation Biofuels: Novel Pathways for the Production of Ethanol, Biogas and Biodiesel
10: Alternatives to Conventional Food Processing
11: Green Trends in Insect Control
12: A Handbook of Applied Biopolymer Technology: Synthesis, Degradation and Applications
13: Challenges in Green Analytical Chemistry
14: Advanced Oil Crop Biorefineries
15: Enantioselective Homogeneous Supported Catalysis
16: Natural Polymers Volume 1: Composites
17: Natural Polymers Volume 2: Nanocomposites
18: Integrated Forest Biorefineries
19: Sustainable Preparation of Metal Nanoparticles: Methods and Applications
20: Alternative Solvents for Green Chemistry: 2nd Edition
21: Natural Product Extraction: Principles and Applications
22: Element Recovery and Sustainability
23: Green Materials for Sustainable Water Remediation and Treatment
24: The Economic Utilisation of Food Co-Products
25: Biomass for Sustainable Applications: Pollution Remediation and Energy
26: From C–H to C–C Bonds: Cross-Dehydrogenative-Coupling
27: Renewable Resources for Biorefineries
28: Transition Metal Catalysis in Aerobic Alcohol Oxidation
29: Green Materials from Plant Oils

30: Polyhydroxyalkanoates (PHAs) Based Blends, Composites and Nanocomposites
31: Ball Milling Towards Green Synthesis: Applications, Projects, Challenges
32: Porous Carbon Materials from Sustainable Precursors
33: Heterogeneous Catalysis for Today's Challenges: Synthesis, Characterization and Applications
34: Chemical Biotechnology and Bioengineering
35: Microwave-Assisted Polymerization
36: Ionic Liquids in the Biorefinery Concept: Challenges and Perspectives
37: Starch-based Blends, Composites and Nanocomposites
38: Sustainable Catalysis: With Non-endangered Metals, Part 1
39: Sustainable Catalysis: With Non-endangered Metals, Part 2
40: Sustainable Catalysis: Without Metals or Other Endangered Elements, Part 1
41: Sustainable Catalysis: Without Metals or Other Endangered Elements, Part 2
42: Green Photo-active Nanomaterials
43: Commercializing Biobased Products: Opportunities, Challenges, Benefits, and Risks
44: Biomass Sugars for Non-Fuel Applications
45: White Biotechnology for Sustainable Chemistry
46: Green and Sustainable Medicinal Chemistry: Methods, Tools and Strategies for the 21st Century Pharmaceutical Industry
47: Alternative Energy Sources for Green Chemistry
48: High Pressure Technologies in Biomass Conversion
49: Sustainable Solvents: Perspectives from Research, Business and International Policy
50: Fast Pyrolysis of Biomass: Advances in Science and Technology
51: Catalyst-free Organic Synthesis
52: Hazardous Reagent Substitution: A Pharmaceutical Perspective
53: Alternatives to Conventional Food Processing: 2nd Edition
54: Sustainable Synthesis of Pharmaceuticals: Using Transition Metal Complexes as Catalysts
55: Intensification of Biobased Processes
56: Sustainable Catalysis for Biorefineries
57: Supercritical and Other High-pressure Solvent Systems: For Extraction, Reaction and Material Processing
58: Biobased Aerogels: Polysaccharide and Protein-based Materials
59: Rubber Recycling: Challenges and Developments
60: Green Chemistry for Surface Coatings, Inks and Adhesives: Sustainable Applications
61: Green Synthetic Processes and Procedures
62: Resource Recovery from Wastes: Towards a Circular Economy
63: Flow Chemistry: Integrated Approaches for Practical Applications
64: Transition Towards a Sustainable Biobased Economy

65: Transportation Biofuels: Pathways for Production: 2nd Edition
66: Challenges in Green Analytical Chemistry: 2nd Edition
67: CO_2-switchable Materials: Solvents, Surfactants, Solutes and Solids
68: Green Toxicology: Making Chemicals Benign by Design
69: Sustainable and Functional Redox Chemistry
70: Waste Valorisation Using Ionic Liquids
71: Natural Product Extraction: Principles and Applications: 2nd Edition
72: Renewable Resources for Surface Coatings, Inks and Adhesives
73: Green Corrosion Inhibition: Fundamentals, Design, Synthesis and Applications
74: Chemical Valorisation of Carbon Dioxide
75: Green Sample Preparation Techniques: Concepts, Novel Materials and Solvents, and Applications
76: Methane Conversion Routes: Status and Prospects
77: Green Gasoline: A Green Spark Transportation Fuel
78: Agri-food Waste Valorisation
79: Bioplastics and Biocomposites: A Practical Introduction

How to obtain future titles on publication:
A standing order plan is available for this series. A standing order will bring delivery of each new volume immediately on publication.

For further information please contact:
Book Sales Department, Royal Society of Chemistry, Thomas Graham House, Science Park, Milton Road, Cambridge, CB4 0WF, UK
Telephone: +44 (0)1223 420066, Fax: +44 (0)1223 420247
Email: booksales@rsc.org
Visit our website at books.rsc.org

Bioplastics and Biocomposites
A Practical Introduction

Edited by

David Grewell
Northern Illinois University, USA
Email: dgrewell@niu.edu

ROYAL SOCIETY OF **CHEMISTRY**

Green Chemistry Series No. 79

Print ISBN: 978-1-78262-656-5
PDF ISBN: 978-1-78801-008-5
EPUB ISBN: 978-1-83916-020-2
Print ISSN: 1757-7039
Electronic ISSN: 1757-7047

A catalogue record for this book is available from the British Library

The Royal Society of Chemistry is a charity, registered in England and Wales, Number 207890, and a company incorporated in England by Royal Charter (Registered No. RC000524), registered office: Burlington House, Piccadilly, London W1J 0BA, UK, Telephone: +44 (0)20 7437 8656.

For further information see our website at www.rsc.org

Preface

Welcome to the fascinating world of bioplastics and biocomposites! This book serves as a comprehensive guide, offering a concise yet thorough summary of the key aspects of these remarkable materials. Whether you are a novice seeking an introduction to this rapidly growing industry or an expert looking to deepen your understanding, the authors have crafted this resource to provide valuable perspectives and insights.

In recent years, there has been a significant shift towards sustainable and environmentally friendly alternatives to conventional plastics. Bioplastics and biocomposites have emerged as promising solutions, offering a range of advantages such as reduced reliance on fossil fuels, lower carbon footprint, and enhanced biodegradability. As the demand for these materials continues to grow, it becomes essential to have a comprehensive understanding of their characteristics, synthesis, performance, processing techniques, applications, and end-of-life treatment options.

This book begins with an overview that sets the stage for the subsequent chapters. It provides a concise yet informative exploration of the fundamental concepts, historical developments, and current trends in the field of bioplastics and biocomposites. By laying this foundation, we aim to ensure that readers of all backgrounds can grasp the essence of these materials and their significance in the context of sustainability.

Moving forward, the book delves into the synthesis of bioplastics and biocomposites. It presents an in-depth analysis of the various manufacturing techniques, raw materials, and processing conditions involved in their production. By understanding the intricacies of synthesis, readers can gain valuable insights into the factors influencing material properties and tailor their application-specific requirements.

Performance evaluation is another crucial aspect covered in this book. Through comprehensive discussions, readers will explore the mechanical,

Green Chemistry Series No. 79
Bioplastics and Biocomposites: A Practical Introduction
Edited by David Grewell

Published by the Royal Society of Chemistry, www.rsc.org

thermal, and chemical properties of bioplastics and biocomposites. The authors provide a comprehensive overview of the characterization methods employed to assess these materials, allowing readers to understand their behavior under different conditions and make informed decisions for specific applications.

Additionally, the book explores the processing techniques employed for shaping bioplastics and biocomposites into useful products. Whether it be extrusion, injection molding, 3D printing, or other advanced methods, readers will gain insights into the challenges, advantages, and considerations associated with each process. The authors also discuss the latest advancements and emerging technologies, providing a glimpse into the future possibilities of these materials.

The diverse applications of bioplastics and biocomposites are highlighted in a dedicated chapter, covering a wide range of industries such as packaging, automotive, construction, electronics, and more. By showcasing real-world examples and case studies, this book aims to inspire readers and encourage the adoption of these sustainable materials across various sectors.

Finally, the book addresses the crucial topic of end-of-life treatment options for bioplastics and biocomposites. Understanding how these materials can be properly disposed of or recycled is vital for achieving a circular economy and minimizing environmental impact. The authors delve into the existing technologies, challenges, and opportunities in this area, shedding light on the importance of responsible waste management.

Throughout this book, the authors have drawn from their extensive expertise and research in the field of bioplastics and biocomposites. By synergizing their knowledge and providing a comprehensive overview, they aim to equip readers with valuable insights, enabling them to navigate the intricacies of this dynamic industry.

Whether you are a scientist, engineer, entrepreneur, student, or simply an enthusiast seeking to explore the world of sustainable materials, this book will serve as an invaluable resource. We hope that it not only expands your understanding of bioplastics and biocomposites but also inspires you to contribute to the ongoing efforts towards a greener and more sustainable future.

David Grewell

Contents

Green Chemistry Series No. 79
Bioplastics and Biocomposites: A Practical Introduction
Edited by David Grewell

Published by the Royal Society of Chemistry, www.rsc.org

CHAPTER 1

Plastics and Bioplastics Overview

D. GREWELL

College of Engineering and Engineering Technology, Northern Illinois University, 590 Garden Rd, Dekalb, IL 60115, USA
Email: Dgrewell@niu.edu

1.1 Plastics and Polymers

Consumers tend to see plastics as a low-cost, low-quality material for high-volume, low-performance applications such as trash bags and packaging. Yet, plastics display properties that range from soft and flexible to strong and rigid and are used in many high-performance applications. In automotive interiors, for example, the plastics often have a soft and flexible "feel", while plastics "under the hood" are able to support high structural loads at relatively high temperatures, such as intake manifolds. Bullet-proof vests, which can absorb a relatively large amount of energy, are made from a plastic material (Kevlar). Plastics are able to provide such a wide range of available thermal-mechanical properties because of their molecular structure.

Plastics are man-made polymers. The name "polymers" is of Greek origin and refers to long-chain molecules: many (poly) repeating units (mers). Scheme 1.1 shows a polyethylene molecule (a common plastic for trash bags, milk jugs, and other high-volume items) that consists of many repeating ethylene molecule units. A single ethylene unit is called a monomer. A polymer chain containing relatively few repeating units is called an oligomer. Once the chain length is even longer, the molecule is defined as a "polymer".

Relatively strong primary covalent bonds hold together the backbone of the long polymer chain; in the case of polyethylene, these are carbon–carbon

Green Chemistry Series No. 79
Bioplastics and Biocomposites: A Practical Introduction
Edited by David Grewell

Published by the Royal Society of Chemistry, www.rsc.org

bonds. The side groups, hydrogen in the case of polyethylene, are also attached by covalent bonds, making the molecule strong. The molecule must be exposed to enough energy (either through heat or mechanical loading) to break these primary bonds.

Large-scale industrial production of man-made plastics only began in the 1940s, and nature has no immediate pathway to break down or degrade them. When plastics are exposed to the environment, they require a long time to degrade. However, some plastics were specifically designed to degrade once exposed to the natural environment, although they are not part of the majority of plastics used ubiquitously today.

Natural, non-man-made polymers that can be found in nature include proteins, starches, and DNA and are readily degradable. Bioplastics (plastics derived from renewable feedstock such as corn, vegetable oil, soybean, or sugar cane) are often, but not always, degraded by nature.

1.2 Plastics *versus* Polymers

We will define plastics as man-made polymers based on petrochemical feedstock. Other polymers are based on renewable feedstock, including many natural polymers, such as proteins and keratin (polymerized amino-acids), and starches and chitin (polymerized sugars). Thus, while all plastics are polymers, not all polymers are plastics, see Figure 1.1.

We will use the term bioplastics for plastics derived from renewable feedstock. While they are also man-made, they are based on natural/renewable

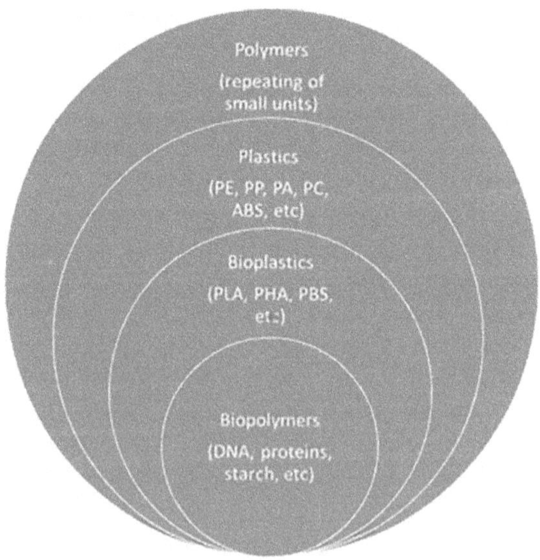

Figure 1.1 Illustration of the classes of polymers. Adapted from ref. 2 with permission from The Authors, Copyright © 2016 James A. Schrader, Heidi A. Kratsch, and William R. Graves.

feedstock such as corn, soybeans, cane sugar, and other plant products. Historically, the first plastics used commercially were bioplastics. Alexander Parkes produced Parkesine from cellulose and in 1856 used chloroform and castor oil to make a bioplastic.[3–5] In 1869, John Wesley replaced ivory with nitrocellulose acetate for the manufacture of billiard balls.[4] In 1865, Paul Schützenberger produced cellulose acetate butyrate for films and coatings. In 1899, Adolf Spitteler and W. Krische used milk protein (casein) to make coatings and a replacement for ivory based on work performed by ancient Egyptians.[4] The petrochemical industry started developing higher-value applications for petroleum shortly after WWII and developed new, less expensive plastics with superior performance. These products included synthetic rubber for tires, nylon for high temperature applications, and polythene for films, leading to the wide acceptance of petrochemical plastics in a variety of industries. Today, less than 5% of fossil fuels are used to produce more than 300 million metric tons of plastic material, corresponding to a revenue stream that compares to the energy sector. However, environmental concerns, price and availability volatility of non-sustainable feedstock, and governmental regulations, such as the USDA Bio-preferred program, paved the way for the bioplastics industry to grow by 20% annually, and it is projected to continue this growth for many years.[6] Although there are a number of companies producing bioplastics,[7] they have offset less than 1% of petrochemical plastics. For example, the production of polylactic acid (PLA) from corn and sugar feedstocks corresponds to approximately 19% of the total bioplastics market. Bio-based PET (polyethylene terephthalate), the plastic for producing carbonated beverage bottles, corresponds to approximately 6.2% with a total global production of approximately 4.7 million metric tonnes in 2022.[1] Table 1.1 provides a list of common bioplastics.

1.3 Molecular Structures

The length and configuration of polymer chains determine the thermal-mechanical properties of a material. We distinguish between three distinct morphologies: amorphous, semi-crystalline, and crosslinked, see Figure 1.2, each with its own molecular configuration. Non-crosslinked chains form thermoplastic plastics, *i.e.*, when heated to a sufficiently high temperature, the material will soften and can "flow". Typically, thermoplastics do not flow like water but rather like a viscoelastic material, similar to Silly-Putty®. This process is often described as "melting", but it is more accurate to use the term "plasticizing".[8] Thermoplastics can typically be recycled because they can be heated and cooled repeatedly to reshape them into new products.

Common industrial processes, such as injection molding and extrusion, use thermoplastics because they can be easily formed under heat and pressure. In amorphous materials, the polymer chains have no order or orientation, similar to a bowl of cooked spaghetti where the strands are all randomly oriented. Semi-crystalline plastics are formed by polymer chains that are folded onto each other, creating regions of parallel chains

Table 1.1 Common commercially available bioplastics. Reproduced from ref. 2 with permission from The Authors, Copyright © 2016 James A. Schrader, Heidi A. Kratsch, and William R. Graves.

Grade	Material	Full name	Tensile strength, yield/MPa	Tensile elongation/%	Notched Izod impact/J m^{-1}	Flexural modulus/MPa	Specific gravity/g cm^{-3}	Mold shrinkage/in in^{-1}	Melt temp./°C	Clarity
NatureWorks Ingeo 3001D PLA	PLA	Polylactic acid	62	3.5	16	3600	1.24	0.004	200	Transparent
Metabolix Mirel F1006 PHA	PHA	Polyhydroxyalkanoate	24	7	31	1300	1.3	0.014	165	Opaque
Ecoflex F Blend C 1200 PBAT	PBAT	Polybutyrate	44	710	12 000	89	1.26	n/a	120	Opaque
Unreinforced PET Base Resin	PET	Polyethylene terephthalate	160	5	5000	2500	0.75	0.015	230	Transparent
RTP 700 A Base Resin	LDPE	Low density polyethylene	7	10	347	138	0.92	0.0275	210	Translucent
RTP 700 UV Stabilized	HDPE	High density polyethylene	21	10	64	900	0.95	0.025	210	Opaque
RTP 200 Nylon 6/6 (PA)	PA66	Polyamide (Nylon) 6/6	83	10	53	2750	1.14	0.015	290	Opaque
RTP 600 ABS Base Resin	ABS	Acrylonitrile butadiene styrene	46	9.5	267	2400	1.06	0.006	225	Translucent
RTP 100 PP Base Resin	PP	Polypropylene	32	10	53	1380	0.91	0.0175	210	Translucent
RTP 400 PS Base Rein	PS	Polystyrene	53	2.5	21	3180	1.04	0.005	230	Transparent
Bay Plastics U-PVC	U-PVC	Unplasticized polyvinyl chloride (rigid)	55	3	3000	3000	1.36	0.004	250	Opaque

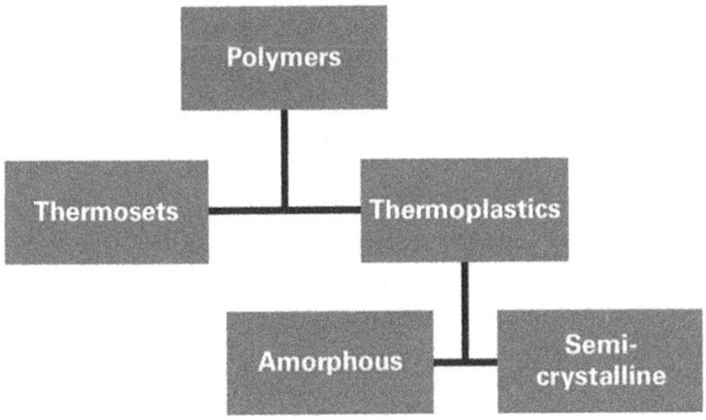

Figure 1.2 Types of plastics. Reproduced from ref. 2 with permission from The Authors, Copyright © 2016 James A. Schrader, Heidi A. Kratsch, and William R. Graves.

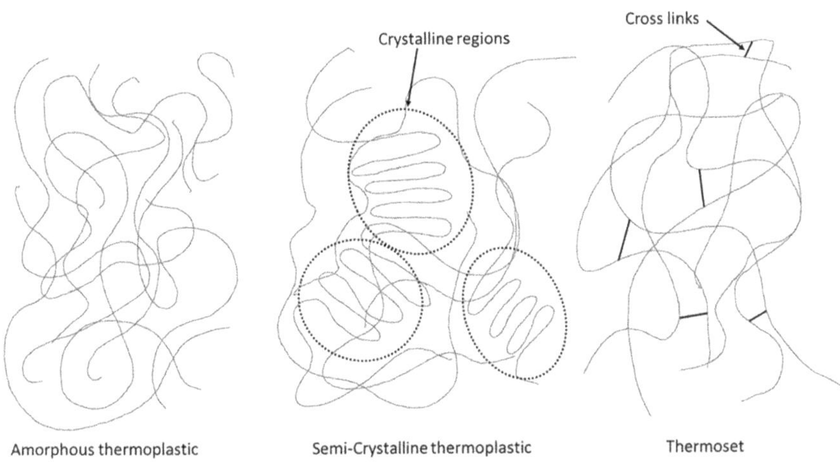

Figure 1.3 Possible molecular orientation (configuration) of polymers and resulting morphologies. Reproduced from ref. 2 with permission from The Authors, Copyright © 2016 James A. Schrader, Heidi A. Kratsch, and William R. Graves.

resembling uncooked spaghetti in a box. It is not common for plastics to be fully crystalline because of the length of the molecules, and therefore, they are commonly called "semi-crystalline." Figure 1.3 shows these various morphologies.

On the other hand, a thermoset is a plastic in which the molecular chains are "linked" to other chains by covalent bonds. These relatively strong, chemical bonds are formed during the synthesis process. Linked chains are not free to slide past each other and cannot flow, even when heated. These materials are called "thermally stable" or thermosets. At elevated temperatures these materials degrade (burn) rather than become pliable. Once a thermoset is formed through a chemical reaction, often in a mold or die, the material retains its shape and is not affected by heating.

Depending on their specific morphology, plastics react differently not only to heat, but also to solvents. Between the polymer chains in amorphous materials there is more free volume, and they tend to be softened or plasticized more easily by solvents compared to semi-crystalline thermoplastics or thermosets. Solvents do not cause chemical reactions. Rather, the solvent (usually a relatively small-molecule) is absorbed on an exposed surface and then diffuses into the bulk of the plastic. Because the small molecules insert themselves between the polymer chains, the bulk volume of the plastic is increased and thus the material "swells". The inserted solvent molecules also weaken the secondary bonds between the chains which allows the bulk plastic to swell further. Both the swelling and the weakening of the secondary bonds between the polymer chains reduce the bulk strength of the plastic. As a result, it is easier for the polymer chains to slide past each other under load after having been exposed to a solvent, resulting in deformation and eventual failure of the material. While the relatively low free volume in the crystalline regions of semi-crystalline plastics makes them less susceptible to solvent attack, they are not immune to solvent attack. Their susceptibility depends on the plastic and its processing conditions as well as on the solvent. Because of their covalent chemical crosslinked structure thermoset plastics are most resistant to solvents. While they may swell, their strength is often not significantly reduced. It is important to note that it does not take a "solvent" to cause these effects: nylon can experience solvent attack by water, and many cleaning solutions can promote solvent attacks in a wide range of plastics.[9]

Plastic materials come with a broad range of properties depending on their molecular structure and composition. Figure 1.4 shows that some plastics, such as non-impact modified polystyrene, can be relatively strong but brittle. Here, the Young's modulus (defined as the slope of the stress/strain curve in Figure 1.4) is high and material failure occurs with little strain, *i.e.*, no stretch. Other plastics, such as rubbers, are able to withstand

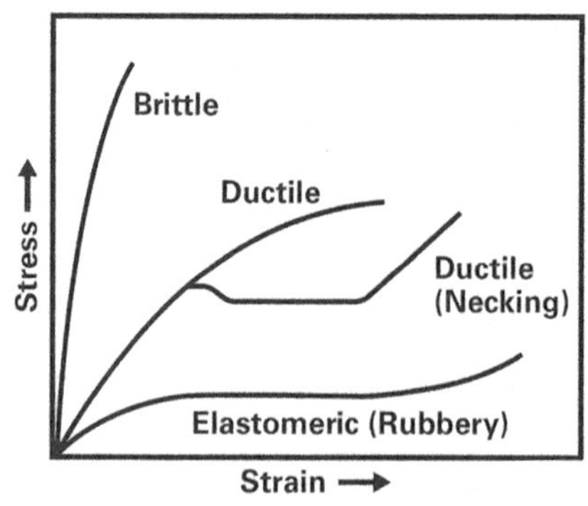

Figure 1.4 Typical stress/strain curve for plastics. Reproduced from ref. 2 with permission from The Authors, Copyright © 2016 James A. Schrader, Heidi A. Kratsch, and William R. Graves.

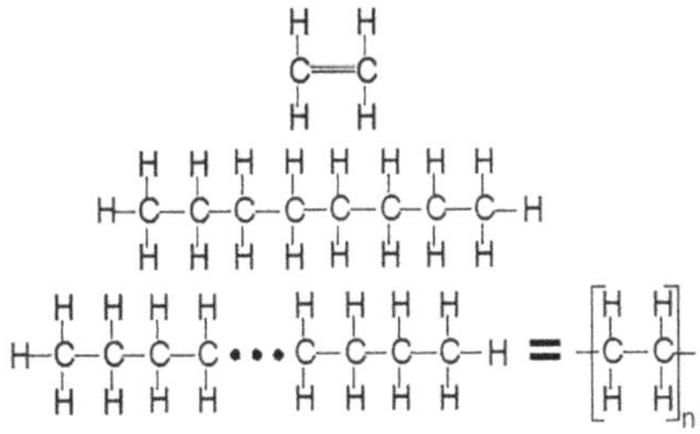

Scheme 1.1 Ethylene monomer (top), oligomer (middle), and polymer (bottom).

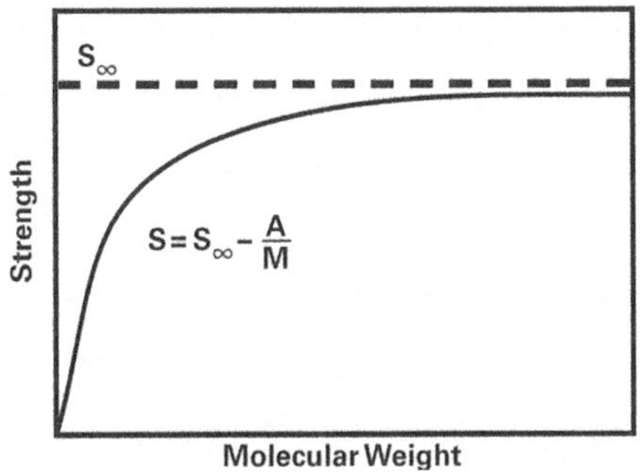

Figure 1.5 Typical relationship between ultimate strength as a function of molecular weight. Reproduced from ref. 2 with permission from The Authors, Copyright © 2016 James A. Schrader, Heidi A. Kratsch, and William R. Graves.

relatively high strain before failure; however, they typically exhibit limited tensile strength. The wide range of different plastics can be tailored to exhibit behaviors between the two extremes, brittle and rubbery.

Stiffness (Young's modulus) and ultimate strength can be derived from a tensile stress/strain curve. The material's toughness, described as a material's ability to absorb energy without experiencing failure, can also be determined by integrating the stress (σ)/strain (ε) curve between no strain to strain at failure (ε_f), which is the area under the stress/strain curve. A material's toughness (U_T) can therefore be determined by eqn (1.1).[10]

$$U_T = \int_0^{\varepsilon_f} \sigma \mathrm{d}\varepsilon \tag{1.1}$$

The most dominant factor affecting the thermo-mechanical properties of plastic materials is the material's molecular weight. There are different methods to determine the molecular weight of plastics, but considering the average molecular weight is the simplest approach. In commercial plastic, not all molecules have the same length/weight; therefore, the molecular weight of a plastic can be determined only as an average number. In general, the molecular weight of an individual molecule is the sum of the weight of its elements. For example, the molecular weight of polyethylene with a degree of polymerization of 100 (n = number of repeating ethylene units) is 2802 g mol^{-1}, as calculated in eqn (1.2).

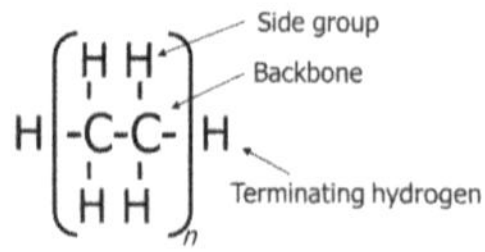

$$M_W = 100\left(\underbrace{(2.12)}_{\text{2 carbons}} + \underbrace{(4.1)}_{\text{4 hydrogens}}\right) + \underbrace{1+1}_{\text{terminating hydrogens}} = 2802\ \text{gram mole}^{-1} \tag{1.2}$$

Eqn (1.3) defines the average molecular weight for a system of molecules with varying degrees of polymerization and corresponding molecular weights. The calculation is based on the number of molecules, where M_w is the molecular weight of a molecule of the plastic with a total population of i molecules, with a corresponding weight fraction of f_n (portion of the population) of molecules with a molecular weight of M_w.

$$M_w = \sum_{1}^{n} M_n f_n \tag{1.3}$$

The increase in molecular weight of a plastic material typically causes its strength to increase asymptotically, as seen in Figure 1.5. This increase is the result of molecular entanglement as more energy and force is required to allow the molecules to slide past each other. The strength asymptotically approaches a maximum value as chain entanglement increases to a point where further entanglement no longer promotes an increase in strength, and failure occurs only because the primary chemical bonds along the backbone are broken.

The average molecular weight can also be calculated based on the number fraction (N) of molecules instead of weight, as shown in eqn (1.4).

$$M_n = \sum_{1}^{n} M_n N_n \tag{1.4}$$

Gel permeation chromatography is used to determine the molecular weight and its distrubution in polymer systems. Typically, a narrow distrubution is desirable – if possible, even a monodispersion, meaning all molecules have the same molecular weight, *i.e.*, $M_n = M_w$. The dispersity is formally defined by a ratio of the average molecular weight by weight and the average molecular weight by number, see eqn (1.5).

$$D = \frac{M_w}{M_n} \tag{1.5}$$

Unlike metals, plastics are rate- and temperature-sensitive: plastics typically exhibit higher strengths and lower strain to failure at higher loading rates (faster pulling). At higher loading rates, the molecules do not have enough time to slide past each other and as the load increases quickly, the stress quickly increases until the polymer chains fail along the covalent backbones. Because covalent bond failure requires high energy, bulk failure (breakage) of the plastic requires high energy and occurs at a relatively high rate, typically with limited strain. In contrast, at low strain rates (slow pulling), the polymer chains can slide past each other, resulting in relatively high strain (stretching) before failure. Given sufficient strain, plastics break at relatively low load forces because only a limited number of primary bonds are broken; this type of failure is exemplified by creep.[11] Under moderate, constant load, a plastic sample will slowly stretch until it fails or stretches to a maximum length [see Figure 1.6(a)]. Depending on the load amount, failure can occur quickly or can take as long as years. Plastics can also exhibit stress relaxation. Under constant strain or stretch, the level of stress on a plastic can decrease as a function of time [see Figure 1.6(b)]. Once the initial strain is applied, the stress is relieved over time as the polymer chains slide past each other, relieving the stress. Elevated temperatures amplify these effects as the molecular freedom is proportional to temperature. The effect of temperature on plastic is often more noticeable compared to metals or other materials.

1.4 Co-polymers and Blends

We distinguish between homo-polymers, where all monomers repeat and are of the same type, and co-polymers, where more than one monomer type is polymerized into a network of polymer chains. There are two general subgroups of co-polymers, branched and block co-polymers, see Figure 1.7.

The tailored synthesis of co-polymers allows for the targeted creation of many thermo-mechanical properties to meet specific application requirements. By combining two or more monomers into one plastic molecule, the final plastic will have the properties of both monomer types, resulting in a "hybrid" of properties.

It is also possible to mix two plastics to form a blend. Here, it is critical that the plastics being blended are compatible in terms of surface energies and processing temperatures. The addition of compatibilizers enhances the proper mixing of both blend partners, making it possible to produce a plastic blend that has the properties of both homo-polymers with a "hybrid" of thermo-mechanical properties.

1.5 Composites

A composite is a system of two or more dissimilar materials that retain a corresponding number of phases after being mixed. Composites are often used for lightweight, high-strength applications, such as the use of graphite fibers in the aerospace industry or the use of glass fibers in the production of skis. Strength and stiffness of the composite are proportional to the length of

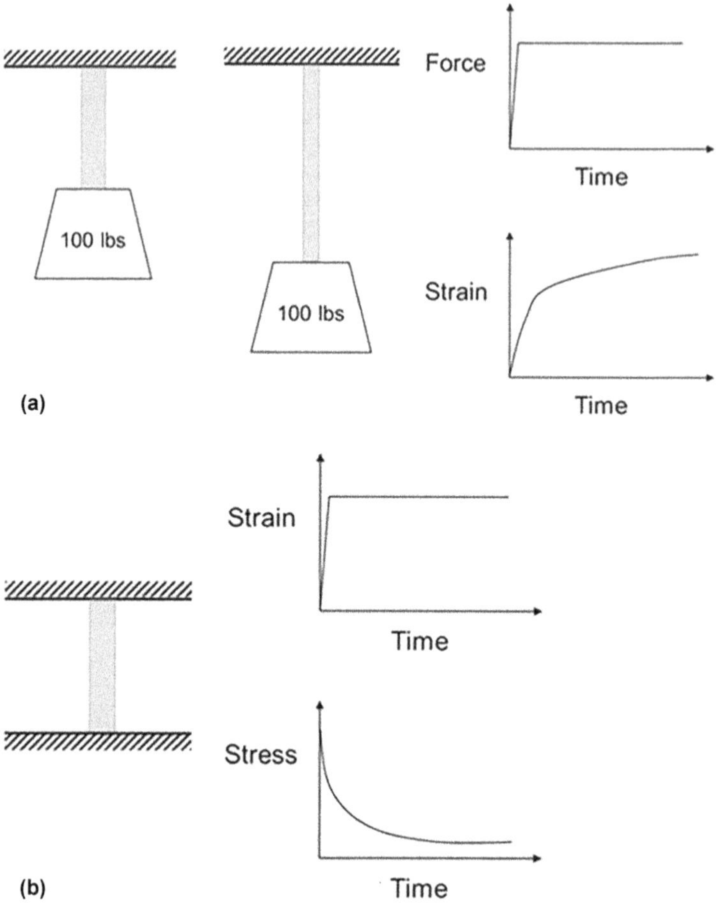

Figure 1.6 (a) Illustration of creep and (b) illustration of stress relaxation. Reproduced from ref. 2 with permission from The Authors, Copyright © 2016 James A. Schrader, Heidi A. Kratsch, and William R. Graves.

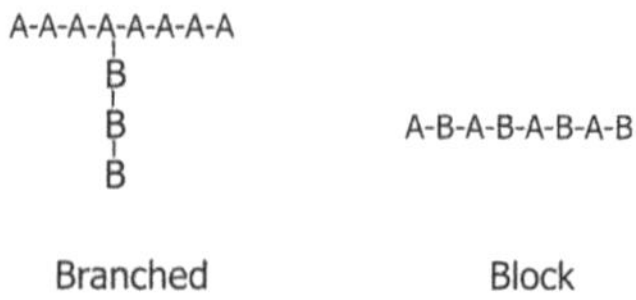

Figure 1.7 Illustration of branched and block co-polymers. Reproduced from ref. 2 with permission from The Authors, Copyright © 2016 James A. Schrader, Heidi A. Kratsch, and William R. Graves.

the fibers and the level of filler in the composite. The aerospace industry often uses graphite fibers, some of the strongest industry fibers, that are continuous. Making these relatively strong and stiff products with continuous fibers is labor-intensive as the fibers must be laid in place, typically in a mold, either manually or using semi-automated processes. Because most components are comprised of multiple layers of fibers, this process must be repeated several times. A vacuum-oven process is used to cure the component into a consolidated product once all layers are in place. To make this arduous process more efficient, the fibers are often received as a "pre-preg", *i.e.*, a tape of fibers with a thermosetting resin coating on the surface to serve as a binding agent.

While short fiber composites (fiber lengths of 2–3 mm) are typically not as strong as continuous fiber composites, they are relatively easy to process. The base resin (also called the matrix) can be a thermoplastic and thus produced using common thermoplastic processes such as extrusion and injection molding. These processes are further detailed in Chapter 4. The orientation of the fibers is random, and thus the strength of the components is relatively isotropic, meaning not dependent on load orientation.

Another common type of fiber-based composite is manufactured using long (from a few mm to longer than 100 mm), randomly orientated fibers. Unlike continuous fibers, where the individual layers of fiber have a distinctive direction, randomly oriented long fibers are placed into a mold in the form of mats. Here, typically, thermosets are used as the resins for these composites.

Other types of composites are manufactured using particles or plates/sheets. An example of particle-filled composites is talc-filled plastics. Often talc, a solid phase, is added to plastics to reduce the thermal expansion and to increase the toughness of the material. Like short fiber composites, particle-based composites can be processed with standard polymer processing equipment, and thus manufacturing with these materials is relatively easy.

1.6 End-of-life of Plastics

The introduction of plastic materials has changed modern life: whether consumer goods or electronics that are widely accessible or improving health through a wide range of medical devices. However, the positive impact on our quality of life comes with many social and environmental concerns: concerns regarding plastics taking up space in landfills for hundreds or even thousands of years and the increasingly dire consequences of their negative impact on water bodies large and small. While there is no single end-of-life treatment that is widely accepted for plastics or for bioplastics, the growing concerns over the negative impact on the environment are promoting the development of new technologies. It is important to note that not all bioplastics are degradable in nature and there are petrochemical plastics that are degradable in nature; thus, there is no clear distinction for best practices for end-of-life treatment for plastics and bioplastics.

1.6.1 Recycling

Recycling of plastics is a widely accepted solution, particularly in Europe where more than 40% of PET bottles are recycled.[6] However, the recycling rate globally is only 14%. In the U.S., plastics in general have a recycling rate of less than 5%. Among the factors that limit the growth of recycling rates are social reasons, lack of infrastructure, and costs. With the current rise in production of natural gas through hydro-fracking technology, the cost of new or virgin plastic materials is relatively low making it harder for recycled plastic materials to compete. In Europe, land usage has a higher social and environmental priority, therefore many countries have a long history of sustaining and updating recycling infrastructure.

While the acceptance of recycled plastic is growing slowly, the limiting force is and will be its cost. On the other hand, the cost for petroleum-based plastics is determined by increased feedstock prices, price volatility, and social and environmental costs. The notion that recycling can reduce greenhouse gas production by reducing the energy required to make products from virgin plastics needs careful review. It is critical to complete a full life cycle assessment for the process to assure that energy, water usage, *etc.*, are in fact lower than those required to produce virgin material. By using bioplastics, the amount of greenhouse gas release can be even further reduced because bioplastics are derived from biomass. The plants/organisms used for the production of bioplastics have captured carbon from the atmosphere, and the same carbon now serves as the chemical backbone for the polymer (bioplastic). It is important to note that complete life cycle analyses must be completed on bioplastics and their recycling to fully understand the costs, environmental impacts, and energy/water usage of these technologies.

1.6.2 Incineration

While it may be counterintuitive to consider burning plastics and bioplastics as a viable option, this approach has a benefit: the production of energy. Worldwide, there are many power plants that provide supplemental energy by burning municipal waste. Here, the plastic waste content, which is mostly carbon and hydrogen, enhances the energy content. During the production of plastics and bioplastics energy is stored in their chemical structure. When incinerated, this energy is released and can be converted to heat or electricity for consumer use. In the case of bioplastics, the carbon that was captured by the plant/organism during its life cycle is simply released back into the environment. It is important that the incineration of plastics must be completed at high temperatures with special exhaust systems to prevent the release of dangerous by-products. If incineration is properly implemented, similar to the infrastructure in Denmark, it is possible to use waste, including plastics, effectively as a source of energy.[12]

1.6.3 Degradation

The following sections describe the three general natural degradation mechanisms.

1.6.3.1 Hydrolysis

When plastics are polymerized by the removal of water (condensation polymerization) they can be de-polymerized through the addition of water, called hydrolysis. Exposing polylactic acid (PLA) to water at the proper pH level and temperature will cause its depolymerization. It is important to note that these conditions are not prevalent in nature. During depolymerization, short-chain PLA molecules (oligomers) can be created as intermediates, but eventually only lactic acid will remain and can be metabolized by a range of organisms in the environment.[12,13]

1.6.3.2 Ultraviolet Degradation

The ionizing radiation of direct sunlight will depolymerize most plastics over time. The radiation excites the chemical primary bonds which causes them to break. With sufficient time, the polymer chains become shorter and the plastic becomes brittle and the environment can degrade the plastic into small particles (micro plastics). It is critical to note that often these small particles remain in the environment for a very long time where they can have adverse effects, such as concentrating hydrophilic toxins.

1.6.3.3 Biological Degradation

Biodegradability is a very broad term and not well defined or understood, yet some bioplastics currently on the market are advertised as such. Here, we will define "biodegradable" as a material that can be consumed by living organisms (microbes) as a source of nutrition. Both anaerobic and aerobic environments facilitate biodegradability: anaerobic degradation occurs under exclusion of oxygen and typically results in the conversion of plastics to methane and water. If the methane is captured as a source of energy, this type of degradation can have a positive environmental impact. However, if the methane is released into the atmosphere, the negative greenhouse gas impact of methane is 21 times higher than that of carbon dioxide. In aerobic biodegradation, oxygen is present, and the plastic is converted into carbon dioxide and water. To make this method efficient within a relatively short time of less than two to three months, the environment must be relatively warm (35 °C) and at a relatively high moisture level (100% RH). These conditions are referred to as "commercial composting environments" and maintaining them typically requires special equipment.

Abbreviations

PLA	Polylactic acid
PET	Polyethylene terephthalate
U_T	Material's toughness
UV	Ultraviolet

References

1. A. Surendren, A. K. Mohanty, Q. Liuc and M. Misra, A review of biodegradable thermoplastic starches, their blends and composites: recent developments and opportunities for single-use plastic packaging alternatives, *Green Chem.*, 2022, DOI: 10.1039/d2gc02169b.
2. J. A. Schrader, H. A. Kratsch and W. R. Graves, *Bioplastic Container, Cropping Systems: Green Technology for the Green Industry*, The Sustainable Horticulture Research Consortium, 2016, ISBN: 978-1541324794.
3. A. Morgon, The Centenary of Casein, *Gloucestershire Soc. Ind. Archaeol. J.*, 1999, 44–50.
4. L. Phillips, Bioplastics Boom, *Am. Sci.*, 2008, **96**(2), 109.
5. Ellen Macarther Foundation, The New Plastics Economy-Rethinking the Future of Plastics, 2014.
6. Bioplastics Council, Bioplastics Industry Overview Guide, The Society of the Plastics Industry Bioplastics Council, 2012.
7. A. Tadmor and C. Gogos, *Principle of Polymer Processing*, John Wiley & Son, 2nd edn, 2006.
8. G. Ehrenstein and S. Pongratz, *Resistance and Stability of Polymers*, Hanser Publications, 2013.
9. W. Brostow and R. Corneliussen, *Failure of Plastics*, Hanser Publications, 1986.
10. W. Callister, *Materials Science and Engineering, An Introduction*, John Wiley & Sons, Inc, 1985.
11. Danish Environmental Protection Agency, Waste in Denmark, 2004.
12. H. Endres and A. Siebert-Rath, *Engineering Bioploymers, Markets, Manufacturing, Properties and Applications*, Hanser Publications, 2011.
13. Z. Kuruppalil, Plastics packaging: The challenge of going green, Accepted for publishing in The First International Conference on Green and Sustainable Technology conference proceedings, University of North Carolina A & T, 2010.

CHAPTER 2

Polymerization and Synthesis from Biobased Feedstocks

MICHAEL J. FORRESTER,† FANG-YI LIN† AND ERIC W. COCHRAN*

Department of Chemical and Biological Engineering, Iowa State University, 618 Bissell Road, Ames, Iowa 50011, USA
*E-mail: ecochran@iastate.edu

2.1 Introduction

Given the increasing importance of finding biorenewable alternatives to petroleum-based polymers, an abundance of time and resources have been put toward finding affordable, quality replacements. In general, there are two strategies for realizing these, which we will refer to as the "bioadvantaged" and "bioreplacement" approaches. A "bioreplacement" is simply the coaxing of biochemical processes such as fermentation and catalysis to produce petrochemical monomers from feedstocks such as sugars, oils, and lignocellulose, ideally at viable cost and quality levels.[1] Because the same building block chemicals (such as monomers) are produced by a renewable pathway rather than the non-renewable petrochemical pathway, renewable polymers can be produced within established infrastructure and markets. The limitation with this approach is its value proposition, which rests solely on preferable social and environmental impacts rather than potential for cost savings. Accordingly, to date there are only a few examples of real success (*e.g.*, the ethylene glycol used in the Plant Bottle by the Coca-Cola Company).

†These authors contributed equally to this work.

Green Chemistry Series No. 79
Bioplastics and Biocomposites: A Practical Introduction
Edited by David Grewell

Published by the Royal Society of Chemistry, www.rsc.org

The bioadvantaged concept is typically more challenging in that it embodies the use of new monomers that are derived from nature with no existing petrochemical counterpart. A polymer produced from a bioadvantaged monomer may have similar physical properties to an incumbent petrochemical plastic, but it may have vastly different chemical and/or mechanical properties. If these differentiated properties add sufficient value to the end-use application, a "bioadvantaged" product can add significantly to its economic viability. The consideration of bioadvantaged molecules introduces a wide variety of new materials with properties that are difficult or impossible to achieve from petrochemical polymers. While bioadvantaged polymers offer great potential to provide real competition with petrochemicals, there remains a significant need for the development of markets and supporting infrastructure. Thus, this chapter will focus more on bioadvantaged materials, with a brief overview of bioreplacement polymers as they have relevance, especially when looking at step-growth polymerization.

Because there are a wide range of polymerization pathways that are possible and each has complex chemistry synthesis and governing conditions, this chapter will provide a general overview of some of the more common methods that are currently (or potentially could be) used in industrial practice to produce (biobased) plastics. Table 2.1 summarizes the techniques that will be reviewed in this chapter.

2.2 Step-growth Polymerization

2.2.1 Introduction to Step-growth Polymerization

Step-growth polymerization is one of the most widely used and studied polymerization techniques. The most common type of step-growth polymerization is condensation polymerization, in which two smaller molecules react together to form a larger molecule and release a small molecule "condensate" as a byproduct.

Step-growth polymerization has yielded some of the most important polymeric materials. For example, polyamides such as Nylon™ can be used in a wide variety of industries from clothing, to fuel tanks, to zip ties and many more applications. Polyesters such as polyethylene terephthalate (PET) have been used in a wide range of different industries but are most commonly known for their use in water and soda bottles.

This section will examine various step-growth polymers, with an introduction to the basic concept and some of the relevant equations that are involved for this polymerization technique. Afterwards, a few examples of well-known petroleum-based step-growth polymers will be discussed to give an overview of the type of materials that are desirable, followed by a discussion of biopolymers produced from this technique.

Table 2.1 Summary of polymerization pathways detailed in this chapter.

Name	Monomer species	Advantages	Limitations	Examples of traditional plastics	Examples of biobased plastics
Condensation polymerization	Diacids, diols, diamines, and acid anhydrides	Common method to produce polyesters, polyamides, and polyanhydrides.	Requirement of elevated temperature and/or reduced pressure to boil off the eliminated species.	Polyethylene terephthalate (PET), Nylon 6,6	Bioreplaced PET, polyadvantaged nylon, and polyadvantaged polyurethane
Anionic polymerization	Alkenes, epoxies, and lactones	Method to produce products of high molecular weight, low dispersity with a high reaction efficiency.	Monomers cannot carry any acidic or electrophilic groups which are commonly seen in bio-derived monomers.	Polystyrene, polyisoprene, and poly(meth)acrylates	Poly(lactic acid), poly((*S*)-β-butyrolactone) and poly(lactam) derived from menthone
Cationic polymerization	Alkenes, epoxies, and lactones	Less sensitive to common functional groups in bio-derived monomers compared to anionic polymerization.	Limited control of the molecular weight and dispersity of products. Incapable of producing thermoplastics from multifunctional bio-derived monomers.	Polystyrene	Poly(anethole), and poly(epoxidized caster oil)
Acyclic diene/triene metathesis polymerization	Acyclic dienes and trienes	Able to polymerize unsaturated plant oils directly.	Requirement of elevated temperature and/or reduced pressure to boil off the eliminated species. Limited ability to produce high molecular weight products.	Polyethylene	Poly(jojoba oil), poly(allylated lignin) and caster oil derived polymers

Table 2.1 (*Continued*)

Name	Monomer species	Advantages	Limitations	Examples of traditional plastics	Examples of biobased plastics
Ring opening metathesis polymerization	Cyclic alkenes	Good control of molecular weight and dispersity with a high reaction efficiency.	Stricter requirement of monomer structures.	Polynorbornene, poly(cyclooctene), and poly(dicyclopentadiene)	Poly(apopinene) and polymers functionalized by plant oils
Free radical polymerization	Alkenes	Facile approach to polymerize alkenes. Able to be activated by heat or light.	Limited control of the molecular weight and dispersity of products.	Polystyrene, and poly(meth)acrylates, and polyacrylamide.	Poly(soybean oil), modified ferulic acid, lignin, terpenes, plant oils polymers.
Reversible addition–fragmentation chain transfer polymerization	Alkenes	Facile approach to polymerize alkenes compared to ionic and other controlled radical polymerizations.	Polymerization efficiency highly dependent on chain transfer agent species. Prone to decompose on chain transfer agent sites.	Polystyrene, poly(meth)-acrylates, and polyacrylamide.	Poly(acrylated epoxidized soybean oil), poly(4-vinylguaiacol), and β-methylstyrene copolymers.
Atom transfer radical polymerization	Alkenes	Easy approach to design product structures through different initiators. Able to conduct surface-initiated polymerization.	Controllability of polymerization relied on the catalyst stability. Limited applications in pharmacy, biology, and packaging due to the use of metal catalysts.	Poly(meth)acrylates	*N*-Phenyl itaconimide copolymer, rosin acid derived polymers, and glycerol derived polymers.

2.2.1.1 Basics and Equations of Step-growth Polymerization Systems

The essence of a step-growth polymerization is the series of reactions that occurs between two complementary functional groups "A" and "B", with alcohol and acid being a typical example. Once reacted, the reactants are covalently joined and can be abstractly represented as "ab"; in the exemplary acid/alcohol pair, "ab" represents the ester formed after elimination of water as a condensation product. These simple reactions lead to macromolecules when monomers bearing "AB" functionality or monomer pairs bearing "AA" and "BB" functionality are provided. As described by Hiemenz and Lodge,[2] each step of the polymerization yields incrementally larger chains following simple combinatorial rules, as illustrated for a small system in Table 2.2.

These pathways provide insight into the fundamental mechanisms and lead to governing equations that define reaction rates, yields, and the expected molecular weight distribution. A detailed derivation of these governing equations is presented in Hiemenz and Lodge.[2] First, it is important to understand the relationship between the degree of polymerization and the extent of reaction. This is shown in eqn (2.1) and defines the number average degree of polymerization (N_n) as a function of extent of reaction, p. As the extent of reaction approaches 1, the degree of polymerization approaches infinity.

$$N_n = \frac{1}{1-p} \tag{2.1}$$

The dispersity, $Đ$, also commonly referred to as the polydispersity index, is formally the second moment of the chain length distribution function, a statistical measure of the breadth of the molecular weight distribution. It assumes a minimal value of 1 for systems in which all molecules are precisely the same size, with values greater than unity for nearly all synthetic polymers and many biologically produced polymers. For step-growth polymerization, $Đ$ is determined by the extent of reaction p. According to eqn (2.2), a maximum dispersity of 2 is obtained when all functionality is consumed:

$$Đ \equiv \frac{M_w}{M_n} = 1 + p \tag{2.2}$$

Table 2.2 Table showing a possible way in which a step-growth polymerization can proceed.

Reaction step	Molecular species present
1	AB AB AB AB AB AB AB AB AB AB
2	AbaB AbaB AbaB AbaB AB AB
3	AbababaB AbaB AbabaB AB
4	AbababaB AbabababaB AB
5	AbababaB AbababababaB
6	AbababababababababaB

These two equations allow the prediction of molecular weight/dispersity or extent of reaction as long as one of the other variables is known. If degree of polymerization is known, eqn (2.1) can be rearranged to solve for extent of reaction. If dispersity is known, eqn (2.2) can be used to solve for degree of polymerization, and then, the extent of reaction can be determined. Likewise, if the extent of reaction is known, eqn (2.1) can be used to determine the molecular weight. This is a useful relation as estimates of molecular weight can be determined by several different characterization techniques. Thus, if access to size exclusion chromatography (SEC) is unavailable, molecular weight can be estimated by knowing the conversion or extent of reaction. In addition, if SEC is available, the extent of reaction can be calculated based on the molecular weight.

Also of importance to step-growth polymerization is the equation for the kinetic rate. In the first of two possible examples of step-growth polymerization, eqn (2.3) shows the polymerization of a monomer that has both A and B functionality, while eqn (2.4) shows the polymerization of a monomer that is a two monomer system, one which is a difunctional A monomer and one which is a difunctional B monomer.

$$4\mathrm{AB} \rightarrow \mathrm{AbababaB} \tag{2.3}$$

$$2\mathrm{AA} + 2\mathrm{BB} \rightarrow \mathrm{AabbaabB} \tag{2.4}$$

In eqn (2.3), the initial concentrations of A and B groups, $[\mathrm{A}]_0$ and $[\mathrm{B}]_0$, are strictly equal, and eqn (2.5) is the resultant rate expression. In eqn (2.4), however, the initial concentrations of A and B are not necessarily equivalent and eqn (2.6) results.

$$\frac{1}{[\mathrm{A}]} - \frac{1}{[\mathrm{A}]_0} = k_c t \tag{2.5}$$

$$\frac{1}{[\mathrm{A}]_0 - [\mathrm{B}]_0} \ln\left(\frac{[\mathrm{A}][\mathrm{B}]_0}{[\mathrm{A}]_0[\mathrm{B}]}\right) = k_c t \tag{2.6}$$

There is a wide range of possible pathways for performing condensation polymerization; however, a typical procedure is described by Colonna *et al.* Monomers are mixed with a titanium catalyst and heated to 90 °C for a few hours. The temperature is then raised to 215 °C and then to 280 °C. Finally, a vacuum is applied in order to remove the last traces of water. At the end of this process, high molecular weight PET polymers are produced.[9] This is a fairly common polymerization method and is used to allow the monomers to first oligomerize before allowing them to fully polymerize under high temperature and vacuum.

2.2.1.2 Common Petroleum Step-growth Polymers

There are a wide range of step-growth polymers. Plastics such as polyethylene terephthalate – or PET – are used widely and produced by the condensation reaction of ethylene glycol and terephthalic acid or its derivatives. PET is commonly used in the food and beverage industry.[3] Roughly 30% of the global production of PET is used to produce bottles for food applications.[4] Additionally, PET is used in the production of fibers, accounting for approximately 60% of global PET consumption.[4] PET fibers are used in applications such as carpets and clothing. More recently, PET has been considered for generating films for flexible electronic applications.[5] PET is also being considered as an additive to asphalt to reduce the permanent deformation of the asphalt.[6] PET is probably the most important polyester commercially used and accounts for a major portion of step-growth polymers.

Polyamides are another type of step-growth polymer that has wide commercial applications. The most common of these polyamides is Nylon. Nylon can be processed in a variety of ways, such as being spun into fiber or being molded into parts for applications where rigidity and structural integrity are needed. The method of processing can determine the potential applications; for example, when spun to fibers it can be used widely in clothing. It can also be molded, allowing it to be used in applications such as automotive applications. Additionally, Nylon films are commonly used for food packaging. The total production of Nylon products is approximately 5 million tons per year.[7] Nylon 6,6 and Nylon 6 comprise over 90% of nylon usage.[7] Nylon can be injection molded, and this accounts for over 60% of the total produced Nylon. Approximately 30% is extruded into films, fibers, pipes, *etc.*, and the remaining 10% is split between blow molding, rotomolding, reaction injection molding, and powder coating.[7] Nylon is widely used in the automotive industry as it makes up parts of a car such as the underhood, seat belt components, and wheel covers. Nylon is also used widely in electrical components and is present in devices such as cable ties, plugs, or switches.

2.3 Biopolymer Step-growth Polymers

2.3.1 Bioreplacement Polymers

Terephthalic acid, which is half of the PET polymer structure, can be produced by a biological pathway by oxidizing 5-hydroxymethylfurfural (HMF). HMF is a major product of biomass processing and is a building block for a wide range of potential chemicals. Scheme 2.1 shows the pathway suggested by Pacheco and Davis to produce terephthalic acid.[8] This approach takes advantage of a Diels–Alder reaction in order to form an aromatic ring, releasing water and generating a material that can be oxidized to produce the desired terephthalic acid or a terephthalic ester.

Another pathway to producing terephthalic acid is to use limonene as a starting material. It is then reacted with ethylenediamine, iron chloride, and sodium in order to produce *p*-cymene. The *p*-cymene is then reacted with nitric acid and water and then potassium permanganate in the presence of sodium hydroxide to yield terephthalic acid.[9] This pathway is shown in Scheme 2.2.

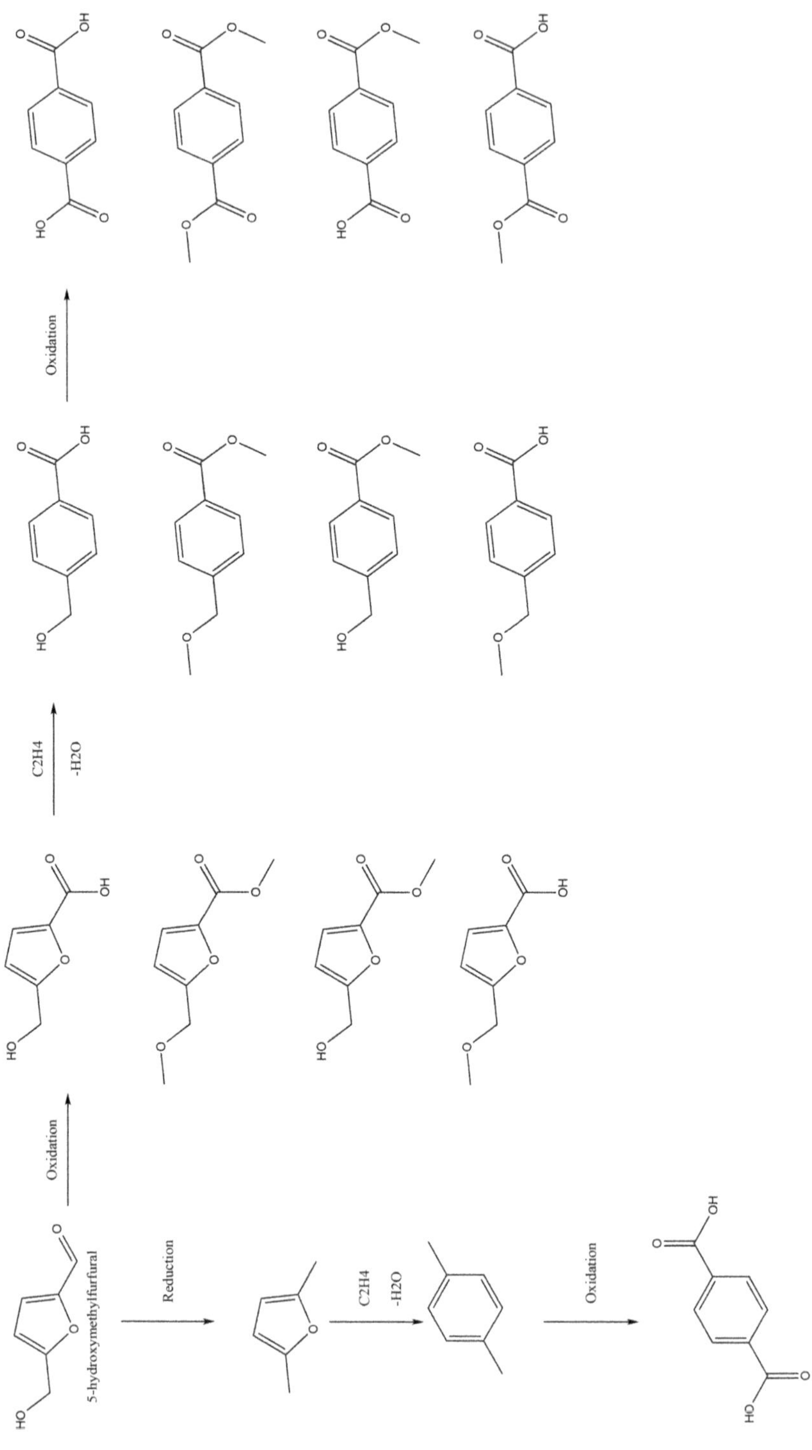

Scheme 2.1 Scheme showing a terephthalic acid bioreplacement pathway from HMF.

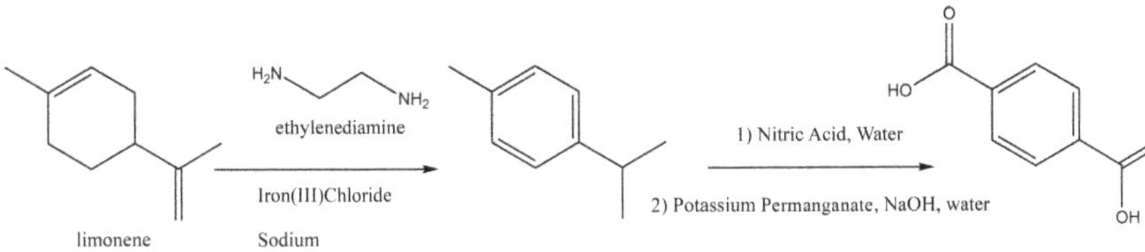

Scheme 2.2 Scheme showing a terephthalic acid bioreplacement pathway from limonene.

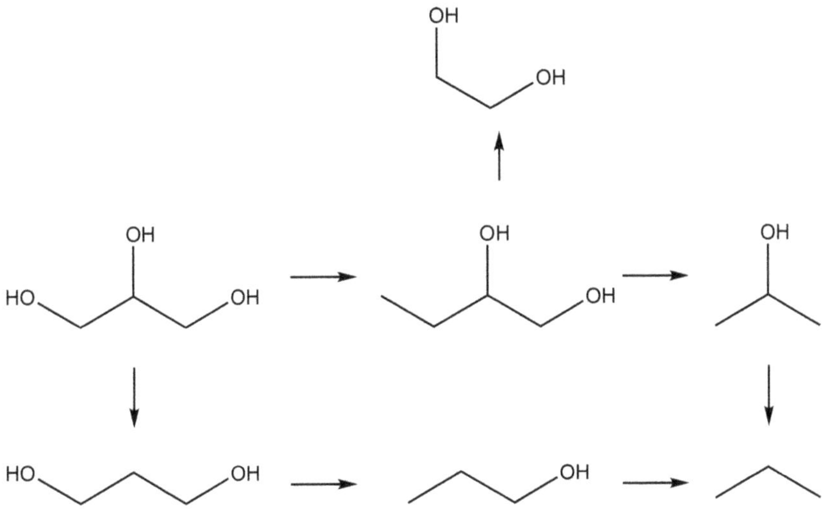

Scheme 2.3 Scheme showing a pathway for producing ethylene glycol from selective hydrogenation of glycerol.

These methods show that it is possible to produce a bioreplacement terephthalic acid, although there may be need for improvements to make these routes commercially viable. Ethylene glycol, the other half of PET, can be produced using a biorenewable alternative. For example, glycerol can be used to produce ethylene glycol. In the work by Zhou *et al.*,[10] glycerol is selectively dehydrated to convert first to 1,2 propanediol and then to ethylene glycol, see Scheme 2.3. Genetically engineered *Escherichia coli* is also capable of digesting D-xylose through multiple steps, as shown in Scheme 2.4, to yield ethylene glycol.[11]

Another potentially important biobased chemistry for biobased plastic is adipic acid. There are several methods for producing adipic acid, the second component for the synthesis of Nylon 6,6, using a biologically derived molecule. In work done by Polen *et al.*, they describe a process, as outlined in Scheme 2.5, where fermentation is used in conjunction with either glucose or benzoate to produce either glucaric acid or *cis,cis*-muconic acid. This is then further reacted to produce the desired adipic acid.[12] In the work done

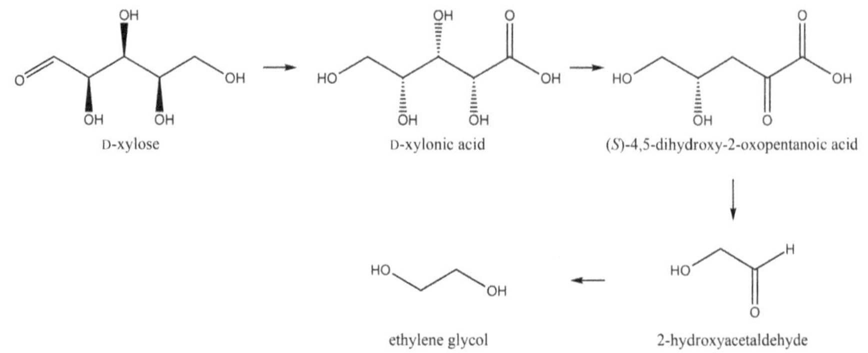

Scheme 2.4 Scheme showing bio-conversion of D-xylose to ethylene glycol.

by Beardslee and Picataggio, they demonstrate that it is possible to take fatty acids from oils such as coconut oil and convert them to C6 adipic acid with very high efficiencies (90%).[13]

This can then be polymerized using methods detailed in the work done by Suastegui *et al.*, where the adipic acid and hexamethylenediamine are mixed at a 1:1 molar ratio in methanol and then heated to 60 °C.[14] This pre-polymer is then precipitated and then heated to 250 °C in order to produce the desired Nylon 6,6 polymer. This is a relatively standard method, as it allows the formation of oligomers or pre-polymers. By performing an initial oligomerization step, unwanted volatilization of monomers is prevented. By using biological pathways to produce polymers that are typically petroleum-derived, it is possible to move towards a sustainable future in bio-based polymeric materials. These methods produce well defined and studied chemicals that will have exactly, or nearly exactly, the same properties as petroleum derived chemicals. It is also possible to consider bioadvantaged chemicals, chemicals that have some degree of difference from their petroleum derived counterparts. These differences can be both advantage and limitation, but recently researchers have made an effort to leverage the advantages, as detailed in the following sections.

2.3.1.1 *Bioadvantaged Polymers*

Bioadvantaged monomers that are further polymerized into bioadvantaged polymers have seen a dramatic increase in interest in recent years. These materials have the potential to have properties different from petrochemicals that give them an advantage compared to traditional petrochemical plastics. While making cost-effective direct replacements for petrochemicals may be an easier challenge for industry, it is sometimes more viable to produce plastics with superior properties or characteristics. This section will review some bioadvantaged polyesters and polyurethanes.

Polyesters are widely used in many industries. One example of a bio-advantaged polyester is shown in the work done by Qi *et al.* In this work, they

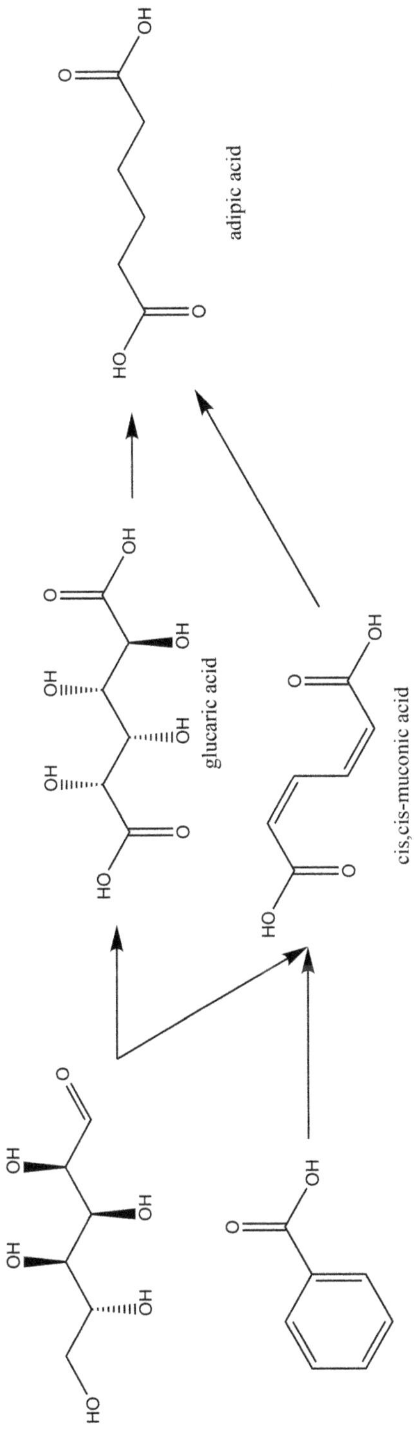

Scheme 2.5 Scheme showing the biomass conversion of glucose and benzoate to adipic acid.

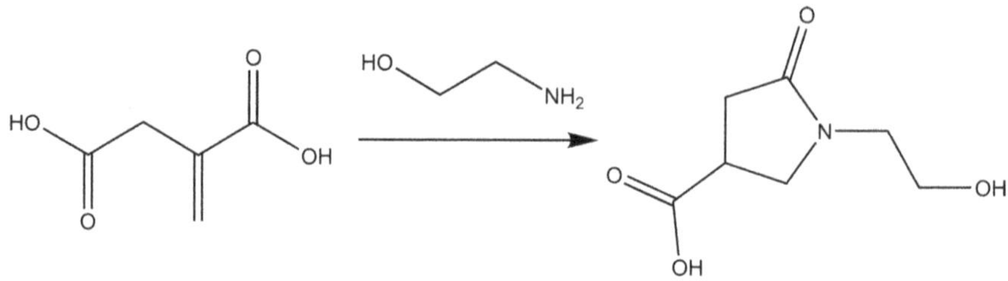

Scheme 2.6 Scheme showing the production of PEPC monomer.

Table 2.3 Table summarizing the work done on the itaconic polyester polymers. Reproduced from ref. 13 with permission from the Royal Society of Chemistry.

Polymer	Diacid : diol ratio	Molecular weight/kDa	Glass transition/°C
PEPC		20.8	60
C2 diol	1 : 1	10	56
C2 diol	1 : 1.2	23.2	62
C3 diol	1 : 1.1	19.5	50
C4 diol	1 : 1.1	24.9	39
C5 diol	1 : 1.05	21.2	29
C6 diol	1 : 1.05	23.4	24

used biomass-derived itaconic acid, which was converted to a condensation monomer by reacting with ethanolamine at reflux temperature according to Scheme 2.6. The monomers were then polymerized by adding catalytic amounts of antimony trioxide and heating to 180 °C for 4 hours and then to 240 °C under dynamic vacuum for an additional 8 hours. Upon completion of this polymerization, it is possible to recover poly(ethylene pyrrolidone carboxylate) or PEPC. It was also shown that it is possible to produce an AA type monomer by reacting itaconic acid with ethylenediamine at reflux temperature. This monomer can then be polymerized with C2–C6 end functionalized diols under the same conditions in order to produce the desired polyester. The polymers that were produced were between 10 kDa to 25 kDa. The glass transition and molecular weights of these materials are shown in Table 2.3.[15]

There are two characteristics that make this material potentially "bioadvantaged" compared to commodity polyesters like PET or polylactic acid (PLA). While PET has a glass transition of approximately 80 °C, PEPC and the C2-diol have glass transition temperatures of approximately 60 °C. Additionally, these materials are non-crystalline. This makes these materials easier to process with less shrinkage and warpage, and for many consumer applications, such as disposable water bottles, this is a temperature that is not likely to be achieved under normal circumstances. In contrast, PLA has a glass transition of approximately 52 °C, and conditions in which this temperature is met or is exceeded slightly are far more likely to occur. This can lead to a substantial decrease in performance of the material. Additionally, PET suffers in that it is not biodegradable. PEPC and some of the diol-itaconate diesters are more prone to degradation and after one year are almost completely degraded in many environments. For high volume, high waste, materials such as plastic bottles, this decreases their impact on the environment.[15]

Another polymer is produced from lignin product vanillin. In the work done by Mialon *et al.*, they produce polyesters from dihydroferulic acid synthesized from vanillin, which is a component of lignin. This produces an AB type monomer. The polymerization is then carried out with either antimony trioxide or zinc diacetate dihydrate under vacuum to remove the produced acetic acid, as shown in Scheme 2.7. This method produces polymers that have a T_g between 65 and 74 °C. This is very similar to that of PET, and in addition, this material has a melting temperature of approximately 30 °C lower than that of PET.[16] This material should have properties that are more similar to PET while having a lower processing temperature.

There are numerous examples of polyesters produced from bioadvantaged molecules; however, they follow a similar theme. A carboxylic acid or an ester is mixed with an alcohol or an ester, heated to relatively high temperatures $\geq$120 °C, and usually subjected to vacuum to remove water, alcohol, or carboxylic acids. This is all done in the presence of a catalyst, usually a Lewis acid such as antimony trioxide or zinc diacetate dihydrate. Upon completion of the polymerization, the polymers are usually dissolved in solvent and then precipitated. While these methods can change from paper to paper, the emphasis is on the removal of water.

Another possible family of bioadvantaged plastics are novel polyamides. In the work by Suastegui *et al.*, 3-hexenedioic was mixed with hexamethylenediamine at a 1 : 1 molar ratio in methanol according to Scheme 2.8.[14] The solution was then heated to 60 °C and allowed to stir. The precipitated pre-polymer was collected and dried. It was then slowly heated to 250 °C. The polymer was then collected and GPC and DSC were conducted to determine the molecular weight and thermal properties of the polymer. Molecular weights of 12.2 kDa were achieved, and interestingly, the glass transition was much lower: 60 °C for 100% unsaturated bionylon compared to 250 °C for commercial Nylon 6,6. By copolymerization, the T_g and crystallinity may be tuned, while also introducing functional sites capable of undergoing chemical modification. By reacting with the double bond the material can become stronger and tougher and have additional flame resistance and higher thermal conductivity.

2.3.2 Step-growth Conclusions

While there are various monomers that can be used in step-growth polymerization, the methods used to produce these polymers tend to be very similar. In addition, step-growth polymerization is usually completed at elevated temperatures and at reduced pressure in order to eliminate

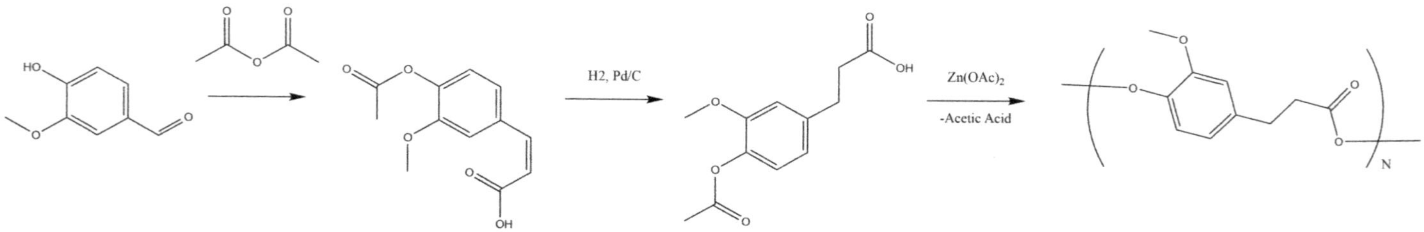

Scheme 2.7 Scheme showing the production of vanillin-based polyester.

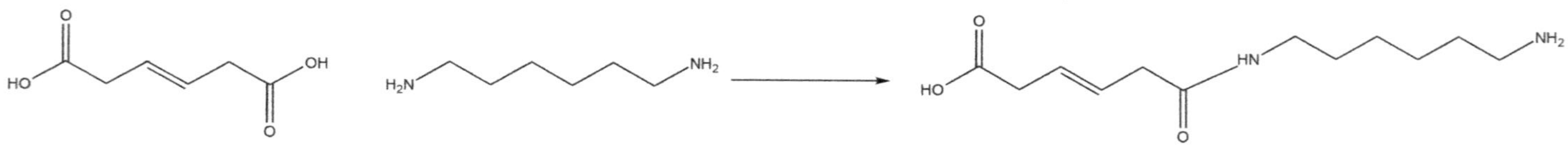

Scheme 2.8 Scheme showing the polymerization of 3-hexenedioic acid and hexamethylenediamine to yield a "bioadvantaged" variant of Nylon-6,6.

("boil off") small molecule byproducts. Bioadvantaged polymers can have a range of properties from soft rubbers and adhesives, to tough polymers, to brittle materials as well as strong. These are dependent upon composition, conditions of reaction, and several other factors. The use of these materials can produce biorenewable polymers that have a wide array of tunable properties.

2.4 Ionic Polymerization

2.4.1 Introduction to Ionic Polymerization

Ionic polymerization is a method of polymerizing monomers through either a cationic or an anionic active center. These ions are capable of polymerizing several different functional groups including alkenes, epoxides, lactones, and others. Both cationic and anionic polymerization belong to a class of polymerizations known as addition or also chain growth polymerizations. This means that they polymerize by adding to the end of the polymer chain rather than polymers and monomers linking together in a step-wise fashion as occurs in step-growth polymerization. Advantages are that no small molecules are produced as a byproduct as occurs in condensation polymerization and superior control over molecular weight and dispersity can be achieved. A limitation is that the reactive species is essentially restricted to vinyl-containing molecules for considerably less variety than afforded by step-growth polymerization. This section will review ionic polymerization in general with a discussion of the governing equations. In addition, some traditional petrochemical examples of anionic and cationic polymers will be reviewed with a continuation of some of the biopolymers produced using this polymerization method.

2.4.2 Basics and Equations of Anionic Polymerization

In Scheme 2.9, one of the most common examples of an anionic polymerization is seen. Styrene is activated with an organometallic, usually *sec*-butyllithium, which attacks the double bond to form the adduct which bears a stable carbanion terminus; the carbanion active center will further attach other styrene molecules to continue the polymerization. There are different anions that are capable of initiating this reaction, and the particular initiator is chosen according to the monomer or reaction conditions. Commonly used initiators include *sec*-butyllithium, *n*-butyllithium, *t*-butyllithium, potassium naphthalenide, or benzyllithium.

The rate of polymerization is:

$$R_{\mathrm{p}} = k_{\mathrm{app}}[\mathrm{M}][P^{*}]^{\frac{1}{4}} \tag{2.7}$$

where P^* is the concentration of active chain ends, and the k_{app} takes into account the propagation rate and the equilibrium between associated and free chains. When this polymerization is conducted in a hydrocarbon

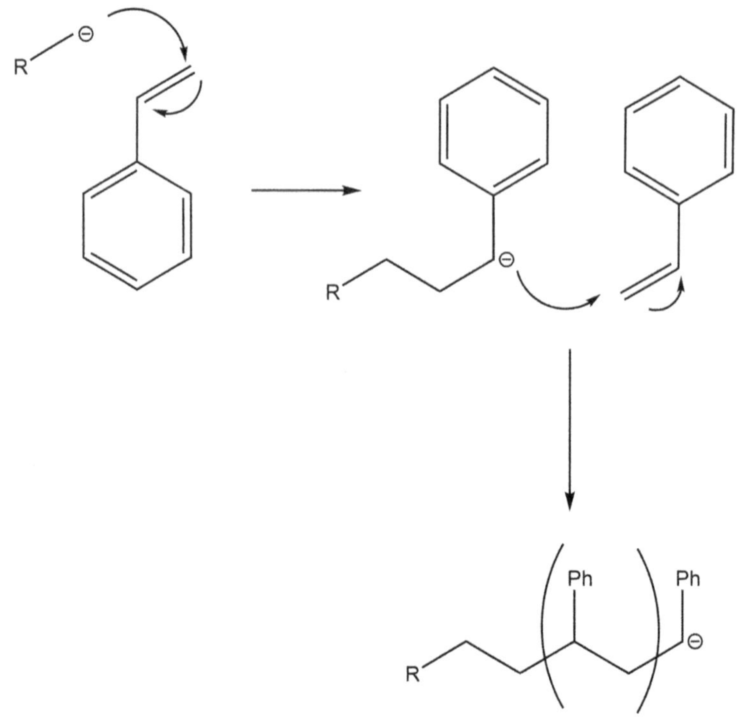

Scheme 2.9 Scheme showing an example of anionic polymerization of styrene.

solvent, the chain ends tend to associate significantly because of poor solubility of the ions; however, with a more polar solvent, the chains associate less. When the ions pair, they severely retard the polymerization at those sites, and thus the bulk of the polymerization occurs at free chain ends. This ratio of associated and free changes will significantly change the rate of polymerization and is thus an important design consideration.

If anionic polymerization is conducted under the correct conditions, it does not have a natural termination route. This means that the chain length at any given point is governed by eqn (2.8)

$$\nu_t = \frac{[\mathrm{M}]_0}{[\mathrm{I}]_0} p_t \tag{2.8}$$

where ν_t is the average chain length and corresponds to the moles of repeat units in a single polymer chain at a given time, p_t is the conversion at that given time, $[\mathrm{M}]_0$ and $[\mathrm{I}]_0$ are the initial monomer and initiator concentration respectively. It is worth noting that anionic initiators, and usually their propagating species, are some of the most sensitive compounds used to synthesize polymers. Typically chain end stoichiometric impurities, nearly anything acidic or electrophilic, will cause a complete failure of polymerization.

2.4.3 Basics and Equations of Cationic Polymerization

In cationic polymerization, cations are used as the initiator species. The monomer reacts with the cation to form the beginnings of a propagating polymer species. In general, cationic initiators tend to be Lewis acids such as boron trifluoride, aluminium chloride, or titanium chloride. Cationic polymerization also requires the use of a co-initiator. Scheme 2.10 shows cationic polymerization of styrene. Unlike anionic polymerization, cationic polymerization does have a termination possibility in which the cation combines with the anionic fragment to form a terminated chain. In addition to this termination mechanism, cationic polymerizations are also susceptible to chain transfer reactions in which a hydrogen can transfer to the anionic counterion or be abstracted by a monomer unit. These termination and chain transfer events, coupled with the required inconvenience of a co-initiator render the cationic system challenging to control, and often molecular weight distributions are broad and molecular weight control is limited.

The rate of polymerization is shown in eqn (2.9)

$$R_p = k_{co}\frac{k_i k_p}{k_t}[M][Cat][Co] \tag{2.9}$$

where [M], [Cat], and [Co] are the monomer, catalyst, and cocatalyst concentrations respectively. k_p, k_i, k_t, and k_{co} are rate constants for the

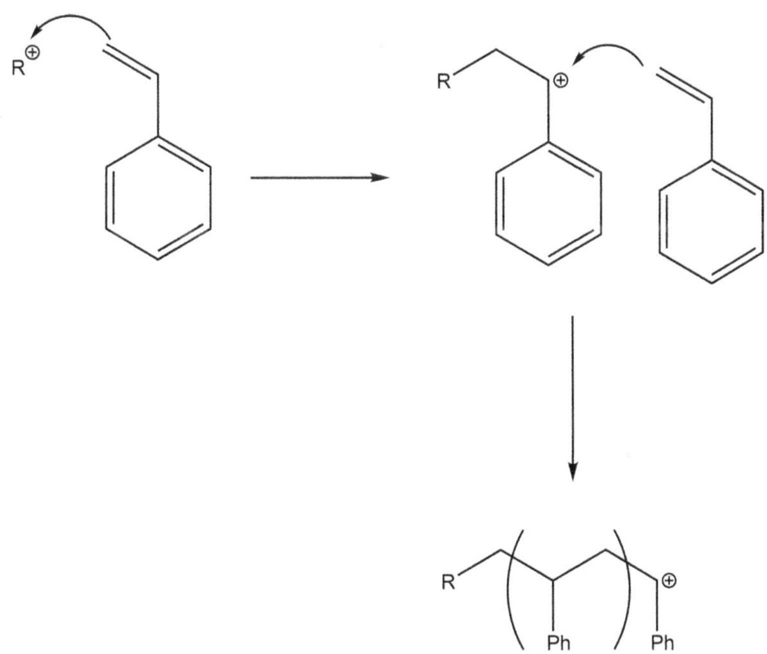

Scheme 2.10 Scheme showing an example of cationic polymerization of styrene.

propagation, initiation, termination, and formation of the catalyst/co-catalyst complex. The number-average degree of polymerization if the termination dominates over chain transfer is

$$N_\mathrm{n} = \frac{k_\mathrm{p}}{k_\mathrm{t}}[\mathrm{M}] \tag{2.10}$$

When chain transfer dominates over termination this becomes

$$N_\mathrm{n} = \frac{k_\mathrm{p}}{k_\mathrm{tr}} \tag{2.11}$$

where k_tr is the rate constant for chain transfer.

2.4.4 Common Petroleum Anionic Polymerization Products and Methods

Industrially, anionic polymerization is most widely used to produce styrenic copolymers and polydienes. Styrenic block copolymers comprise a large segment of the thermoplastic elastomers industry, and styrene/diene statistical copolymers are a common type of synthetic rubber used in the tire industry. Additionally, there is a wide variety of polymers that can be produced *via* this polymerization technique that are either of more academic or specialty interest.

2.4.4.1 Styrenic/Diene Polymers/Copolymers

Styrenic and diene polymers or copolymers are probably the most well-known anionic polymers. They are used in everything including tires, paints and adhesives, asphalt modification, and a wide variety of other areas. By considering the size of the asphalt industry, it can be seen why polymers are useful and, in most cases, necessary. Because of the fact that more and more is being expected of roads in terms of total loading and number of load cycles, the requirement for paving material to perform well continually increases. Pure asphalt tends to be prone to cracking at low temperatures and to creep and rutting at elevated temperature. The addition of select polymers can reduce many of the issues with bulk asphalt, and as such, accounts for an extremely high portion of the asphalt paving industry.[17]

These styrenic/diene block copolymers are produced in a relatively straightforward way. First, an initiator such as *sec*-butyllithium is dissolved in a non-polar solvent, most commonly cyclohexane or a mixture of cyclohexane and pentane/hexane. Polar aprotic solvents including diethyl ether and tetrahydrofuran can be used, although side reactions necessitate a decrease in temperature and, because they dissociate the living chain ends, the reaction rate increases dramatically. This tends not to be very scalable and as such is less common industrially because of the high cost associated with extreme conditions and more stringent exotherm management requirements. Benzene was a common choice used until its health risks became widely understood. In industrial practice, the monomer is added to the initiator solution slowly at room temperature. The reaction is exothermic (typically near 45 $kJ\,mol^{-1}$) and is kept below 60 °C using standard plant cooling water.

Once the styrenic block has been produced, butadiene is added to the vessel. The polymerization of butadiene is also very exothermic, and as butadiene is a gas with a boiling point of −4.4 °C, care must be taken to not let the reaction become hot enough to reach dangerous conditions. In non-polar conditions, 1,4-polybutadiene is predominately produced, in which each repeat unit contains four carbons with an allylic double bond in the chain backbone. The addition of a polar solvent such as diethyl ether or tetrahydrofuran results in the microstructure changing to a mixture of 1,4 and 1,2-isomers; the 1,2 isomer has a vinyl pendant group which is far more reactive towards crosslinking reactions. If a triblock copolymer is desired, it is less common for styrene to be added again because of the long time required to initiate styrene from the butadienyl anion. Thus, it is more common for coupling agents to be added, a type of molecule that has a difunctional group present and thus couples two S–B blocks together to give the final SBS polymer.

Pure polyisoprene is also produced by this polymerization method and is commonly used to make a wide variety of products such as medical devices, adhesives, sponges, and many more. This and the above-mentioned polymerizations are exceedingly efficient, reaching well above 95% conversion within minutes. The molecular weight can be precisely controlled and the dispersities of the polymers are quite low, often being less than 1.1. While there are a few other dienes and styrenics that are used, such as divinylbenzene, these materials tend to follow the same procedure. When different structures are desired, such as star shaped rather than linear, the most common method of doing this is with a special initiator. This can be an initiator that starts with multiple functional groups, as depicted in Figure 2.1. The other option is to use a standard initiator and then a few equivalents of divinylbenzene in order to control the number of star arms desired, as is shown in Scheme 2.11. While this does add a more complicated initial step, it does not change much else, and once this initiator has been synthesized, the polymerization occurs as detailed before.[18]

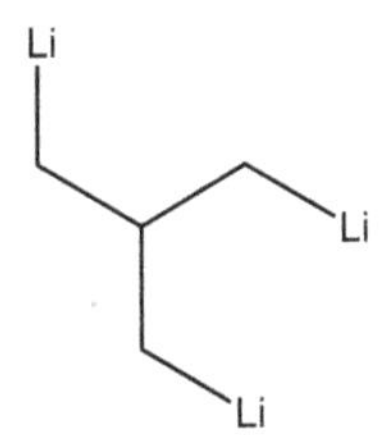

Figure 2.1 Example of a multifunctional initiator.

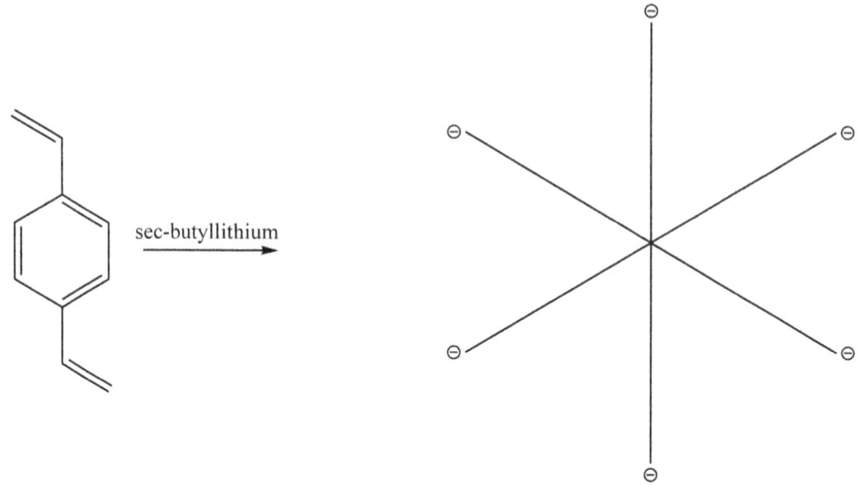

Scheme 2.11 Scheme showing an example of a divinylbenzene initiator.

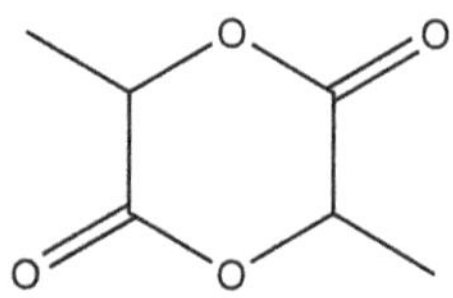

lactide

Figure 2.2 Figure showing an example of an anionically polymerizable lactide ring.

2.4.4.2 *Anionic Ring Opening*

Another common type of anionic polymerization is the ring opening anionic polymerization. This is completed with monomers such as lactide (a bio-based monomer), see Figure 2.2.

In the work done by Bhaw-Luximon *et al.*, they reviewed the anionic ring opening polymerization of lactide using initiators such as lithium diisopropylamide, potassium *t*-butoxide, or butyllithium. These polymerizations are completed in dioxane, tetrahydrofuran, or toluene, and what is found is that polymers of molecular weights lower than 30 kDa can be produced in less than 4 hours. However, although the dispersities were higher than is traditional for an anionic polymer, such as the above example of styrene or butadiene, they were lower than polymers that would be synthesized by condensation polymerization.[19]

In another work, ethylene oxide is polymerized to poly(ethylene glycol) by use of an organolithium in THF first at cryotemperatures and then at near room temperature for 24 hours.[20] This affords a polymer that has a low dispersity. In another work, ε-caprolactam is polymerized by anionic polymerization at 69 °C. As it is polymerized, it is blended with carbon fibers in a mold to provide a carbon reinforced polyamide-6.

2.4.4.3 Methacrylates

In the cases of more complex vinyl monomers, such as acrylates, cryogenic temperatures are required. Typically, in these polymerizations, an alpha methyl styrene or diphenylethylene anion is formed using a lithium compound in THF at −78 °C. The monomer is then added and allowed to react for a short time – usually less than 10 minutes – to achieve very high conversions. If it is desired to control tacticity, lithium salts can be added and the solvent can be changed from pure THF to a blend of THF and toluene.[21] However, this slows the polymerization so that it takes hours instead of minutes.

In work done by Li *et al.*, a styrene–isoprene–methyl methacrylate block copolymer was produced using a small amount of tetrahydrofuran at 0 °C.[20] This polymer was demonstrated to have clear microphase separation as it had three distinct glass transition points, an indication that the polymerization did occur as expected with a well-defined methyl methacrylate block.

2.4.5 Common Petroleum Cationic Polymerization Products and Methods

It is far less common for cationic polymerization to be used for most polymer synthesis. While it is capable of polymerizing many of the same monomers as anionic and free radicals, it is a method that does not usually allow accurate control of the molecular weight. There are usually high dispersity values as a result of the high degree of chain transfer reactions. While there are many publications making use of this method, it is not a very widely utilized method for the industrial production of polymers.

There are a handful of exceptions to this, however. One such example is the polymerization of isobutylene which is usually initiated with a Lewis acid such as boron trifluoride, aluminium chloride, or titanium chloride. In the work by Vasilenko *et al.*, isobutylene was polymerized to molecular weights less than 4 kDa using an $AlCl_3$ ether complex as the initiator. This produced polymers with reasonably low dispersities, typically less than 2, and decent conversions, typically greater than 50%.[22] The general understanding, however, is that most materials can either be made more efficiently and with a much narrower dispersity using anionic polymerization or more affordably through radical polymerization. These factors make cationic polymerization a fairly rare method; however, this is less true in regards to biopolymers.

2.4.6 Anionic Polymerization of Biopolymers

Unsurprisingly, lactone monomers that are produced from biological pathways are some of the most relevant anionic monomers. In work by Jedliński *et al.*, they take (*S*)-β-butyrolactone and polymerize it anionically using a crown ether complexed sodium or potassium initiator. This allows for production of polymers above 10 kDa with dispersities of less than 1.1.[23] In another work, anionic polymerization of a menthone-derived lactam using sodium hydride/potassium *tert*-butoxide with a bulk monomer at 200 °C provided polymers with molecular weights under 2 kDa and dispersities between 2 and 3.[24]

Beyond the production of polyesters or polyamides from lactones/lactams, the literature is remarkably thin in biopolymers that are produced anionically. This is likely because of a number of factors. First, organic anions are exceedingly reactive and tend to react with almost any acidic or electrophilic group. Most biorenewable molecules contain ethers, alcohols, unsaturates, ketones, aldehydes, and a wide variety of acidic protons. These issues make polymerizing biorenewables *via* anionic polymerization challenging. Additionally, many biorenewable monomers can either be directly polymerized or modified slightly and then polymerized using other methods such as cationic or free radical polymerization. Any monomer that is not polymerized by the ring opening polymerization method, and even some that are, would require extremely cold temperatures in order to prevent unwanted side reactions. This leads to an extra layer of difficulty and cost which makes it unappealing as a method of producing biopolymers.

2.4.7 Cationic Polymerization of Biopolymers

The amount of work done on cationic polymerization of biopolymers is significantly more substantial than the work done on anionic polymerization. There are several reasons for this; however, most significant is the fact that cationic polymerization is less sensitive to common functional groups that are present in bio-derived molecules. The first example that can be considered is a polymer produced from anethole, which is a biorenewable plant oil. Using CF_3SO_3H as an initiator and dichloromethane as a solvent at −5 °C for 24 h, Fang *et al.* produced a polymer with a molecular weight of approximately 2 kDa and with a dispersity of 1.5.[25] It was shown that it was possible to crosslink the product to produce a fluorinated polymer, as shown in Scheme 2.12. The polymer was then cured, first at 150 °C, then 200 °C, and finally 220 °C. This produced a polymer that had a low uptake of water and a low dielectric constant.

In another example, castor oil was reacted with hydrogen peroxide to produce epoxidized castor oil. This then provided an oil that had roughly 2.3 epoxide groups per triglyceride. With polymerization using a photocatalyst under UV light the material produced a crosslinked polymer.[26] This is an example of why it can be challenging to obtain thermoplastics from biorenewable sources. While the anethole oil is an exception, most biorenewable sources include more than one functional group which allow unpredictable reactions. Additionally, these feedstocks are rarely one well-defined molecule but a mixture of various molecules. For example, soybean oil can vary in chemical composition depending on the location in which it is grown, the method of obtaining the oil, and any processing that is done on the oil. Cationic polymerization has not been shown to be capable of producing thermoplastics of any substantial molecular weight from multifunctional biorenewable molecules. This means that either the biochemical

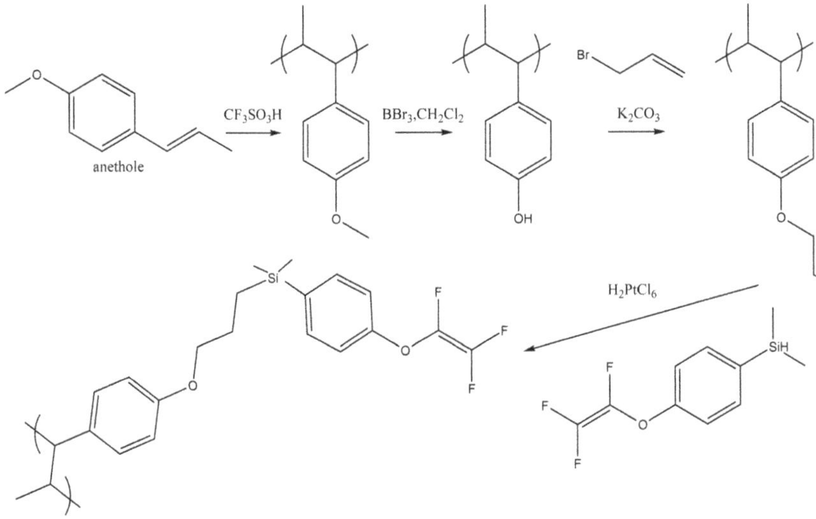

Scheme 2.12 Scheme showing the conversion and polymerization of anethole to a fluoropolymer *via* cationic polymerization process.

Table 2.4 Table showing the glass transition temperature of soybean oil-*co*-dicyclopentadiene polymers as a function of their composition. Reproduced from ref. 26 with permission from American Chemical Society, Copyright 2006.

Copolymer	T_g (°C)
Soy50-DCP42	39.9
Soy55-DCP37	16.5
Soy60-DCP32	2.9
Soy65-DCP27	−6.9
Soy70-DCP22	−13.0
Soy75-DCP17	−17.7

must be monofunctional or the intended application must be that of a thermoset.

In the work by Andjelkovic and Larock, the cationic copolymerization of dicyclopentadiene and soybean oil was shown to produce polymers that varied in properties from soft rubbers to tough and ductile plastics. Table 2.4 shows the effect that composition has on glass transition temperature. With increasing soybean oil content comes a decrease in glass transition and the material behaves more like a soft rubber. As the content of cyclopentadiene increases, the glass transition goes up and the material becomes more tough.[27]

2.4.8 Ionic Polymerization of Biopolymers: Conclusions

Comparing anionic and cationic polymerization of biopolymers, cationic tends to be more applicable while anionic tends to have many limitations.

This is because of the fact that organo-anions are typically so reactive that polymerization of most bio-derived chemicals is impossible. Cationic polymerization, while not the most efficient method of polymerizing most monomers because of its tendency to lead to highly disperse polymers, does allow for polymerization of many bioadvantaged materials. The methods used for this polymerization vary widely depending on the type of monomer, type of functional groups, desired properties, and a wide variety of other parameters. While there are many examples in the literature using these polymerization methods, it is far more common that step-growth polymerization or radical polymerization is used in order to produce bioadvantaged polymers because of the limitations with ionic polymerization.

2.5 Metathesis Polymerization

A metathesis reaction is the interchange of atoms between different chemicals through the assistance of a catalyst. Through the improvement of catalyst selectivity in past decades, currently metathesis is opened to monomers with various functional groups instead of just olefins. Metathesis is used widely in polymer science, both in the preparation of monomers, by metathesis or cross-metathesis, and in the production of polymers, by ring opening metathesis polymerization (ROMP), acyclic diene metathesis (ADMET) polymerization, as well as corresponding acyclic triene metathesis (ATMET) polymerization.

2.5.1 Mechanism

The interchange of carbon atoms in metathesis involves the formation of an intermediate with a metal carbene complex as the catalyst. Because the interchange of alkyl groups in metathesis is reversible, as shown in Scheme 2.13,[28] the structure of reagents and the choice of catalyst are crucial to the feasibility of metathesis reactions. The metal carbene complex is not only the catalyst but also the initiator in olefin metathesis polymerization. The degree of polymerization in metathesis is expressed as

$$N_n = \frac{[M]}{[Cat]} \tag{2.12}$$

Categorized by monomer structures, acyclic alkenes undergo acyclic diene/triene metathesis polymerization while cyclic alkenes polymerize through

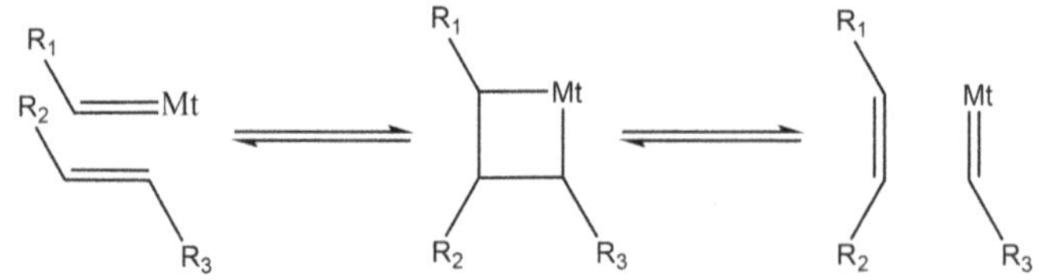

Scheme 2.13 Olefin metathesis mechanism.

ring opening metathesis. The thermodynamics of metathesis reactions are strongly dependent on the monomer architecture; for example, factors such as ring strain strongly dominate the behavior of ring-opening reactions.[29]

Given that olefin metathesis is an equilibrium reaction, multiple metatheses occur amongst reacting species, often resulting in unwanted byproducts and/or oligomers. Common examples of metatheses are shown in Scheme 2.14 detailing the intermolecular chain transfer and intramolecular chain transfer reactions.[30] While the intermolecular chain transfer reaction broadens the molecular weight distribution of the final product, the intramolecular chain transfer reaction decreases the molecular weight. Because the occurrence of these side reactions depends on monomers, the suppression of these side reactions is possible by precise choice of catalyst, reaction conditions, and solvent.

2.5.2 Catalyst Choice

The catalysts that are typically used for metathesis are generally constituted of transition metal alkylidenes, of which molybdenum (Mo), tungsten (W), and ruthenium (Ru) are the most effective types. The popular catalysts commonly used in polymerizing biobased monomers are shown in Figure 2.3, which are Grubbs and Grubbs-related catalysts. These types of catalysts are Ru-based metal carbenes with bulky ligands, which show better tolerance towards different functional groups and are less sensitive to air.[31] The dosage amount of catalyst employed is generally 1–10 mol% to the substrate. It is noted that the choice of catalyst has an impact on the occurrence of side reactions such as olefin isomerization and ring closing metathesis.

2.5.3 Acyclic Alkene Metathesis Polymerization

Depending on the number of vinyl groups, linear polymers can be produced by acyclic diene metathesis polymerization (ADMET). In addition, branched polymers can be produced from acyclic triene metathesis (ATMET) polymerization. Acyclic alkene metathesis polymerization is also a type of step-growth polymerization. The mechanism of ADMET of α,ω-dienes is detailed in Scheme 2.15.[33] The polymerization is usually accompanied by the elimination of ethylene or other heavier hydrocarbons.

To drive the reaction towards polymerization, ADMET is usually conducted under reduced pressure or under continuous inert flow along with elevated temperature to remove the hydrocarbons. If heavier hydrocarbons other than ethylene are eliminated, even higher temperatures are needed, which is more challenging to catalysts because of thermal stability issues. This is not an uncommon challenge with the direct polymerization of unmodified plant oils whose vinyl groups are not at the α,ω-positions. For example, Butilkov *et al.* used ADMET to polymerize jojoba oil, which is a 42-C hydrocarbon with two unsaturated bonds.[34] This molecule eliminates an 18-C hydrocarbon, and thus, the reaction was conducted at 200 °C, significantly harsher than that of most ADMET polymerizations. The Hoveyda–Grubbs II (HG-II) catalyst was

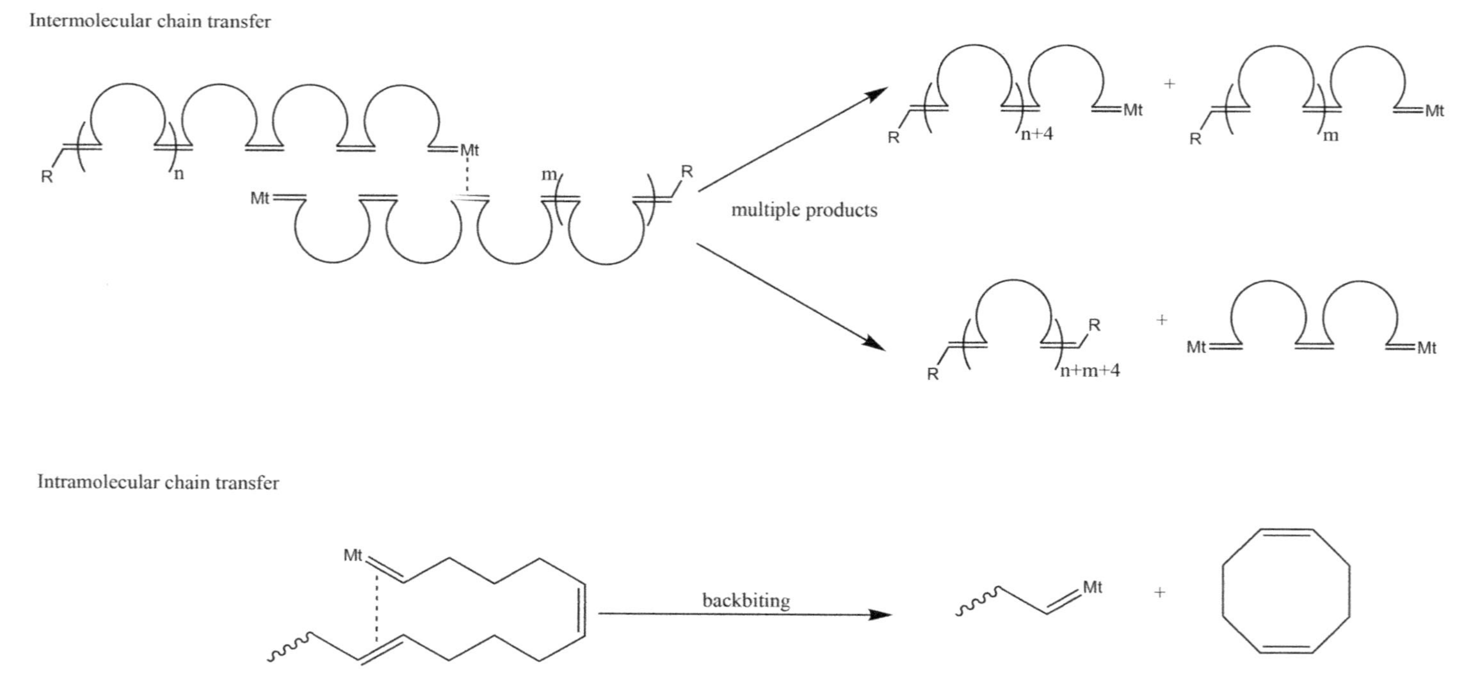

Scheme 2.14 Other pathways of metathesis. In intermolecular chain transfer reaction, the random interchange of carbon atoms between growing polymer chains causes the increase or decrease of the length of polymer chains. In intermolecular chain transfer reaction, a cyclic olefin is released accompanied by the decrease of chain length.

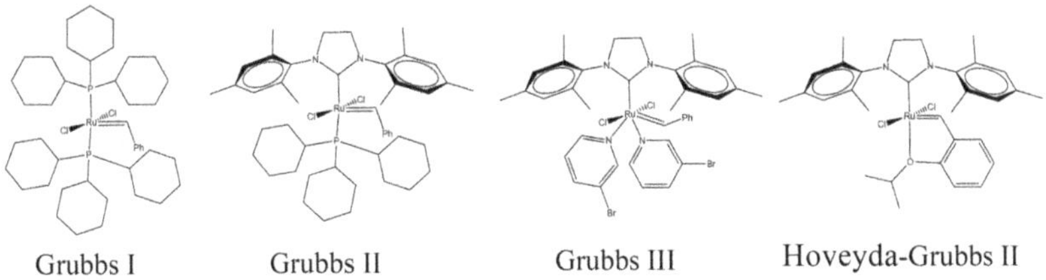

Figure 2.3 Common catalysts used in biobased monomer polymerization.

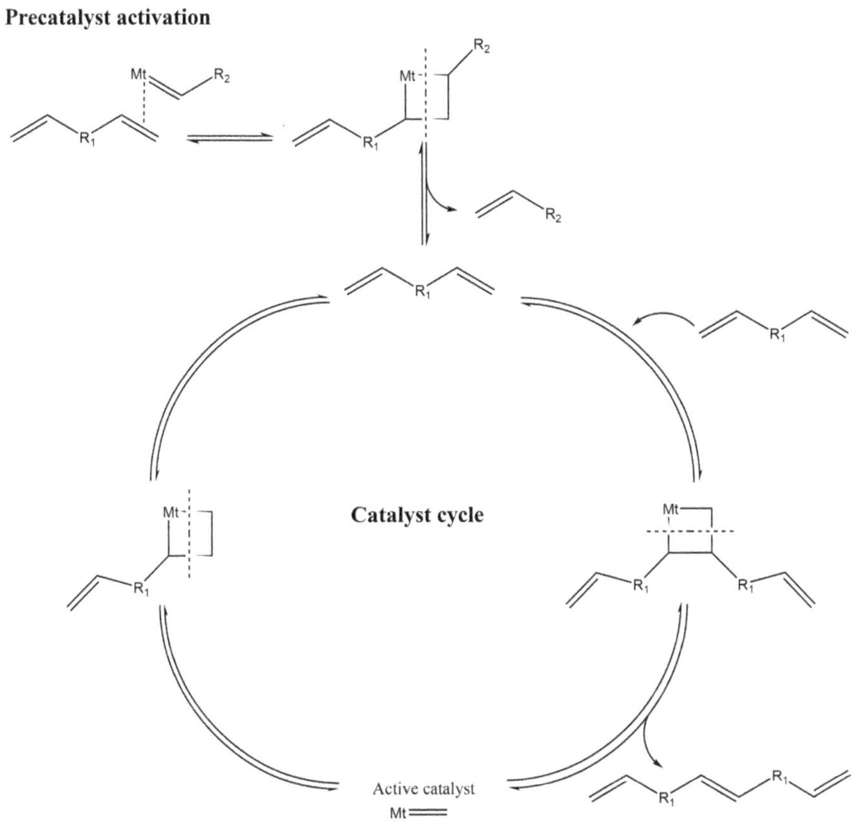

Scheme 2.15 ADMET polymerization mechanism.

shown to have the best stability amongst the catalysts screened, yielding jojoba oil oligomers with 9 repeating units.

Besides plant oils, which are ready candidates of ADMET polymerization because of their aliphatic unsaturated fatty acids, other chemical modifications are often employed for biobased chemicals into ADMET-ready monomers with desired properties. Castor oil derivatives are one of the actively utilized compounds in ADMET polymerization. For example, Ronda *et al.* incorporated phosphorus groups into a castor oil-derived 10-undecenoic acid to produce highly thermally stable polymers by ADMET.[35,36] The functionalized product has potential as a flame retardant material. Meier's group exemplified the possibility of producing polyamides *via* ADMET polymerization through castor oil derivatives.[37] The final products give a variety of desirable properties such as increased strength, toughness, and chemical resistance.

ADMET polymerization is usually conducted in bulk when monomers are liquid, and the reaction is usually performed under reduced vacuum or with continuous inert flow. However, in some cases, solvent is required if monomers are solids or co-monomers are not miscible. Meier and coworkers reported the use of dimethyl carbonate in cross metathesis polymerization of allylated lignin and plant oil.[38] The solvent choice can have a large impact on the polymerization; for example, *N*,*N*-dimethylformamide (DMF) is not an ideal solvent for ADMET. The stability or reactivity of metal carbenes in DMF is low because of its coordinative nature which gives undesirable polymerization efficiency.[39,40] Solvents such as dichloromethane and dimethyl carbonate are much more effective, with a preference for dimethyl carbonate because of it being renewable and less hazardous.

The degree of polymerization of ADMET can be controlled with the aid of a "chain stopper", a mono-functional molecule that yields a dormant chain end. "Chain-stoppers" are generally mono-unsaturated hydrocarbons. Therefore, the ratio of monomers to chain stopper results in the adjustment of the degree of polymerization. In polymerization of biobased monomers, the residual unsaturated fatty acid and its derivatives can act as a chain stopper. Akintayo *et al.* show the growth of polymer chains terminated by the residual oleic acid in direct polymerization of *Plukenetia conophora* oil.[41] The olefin metathesis polymerization of acyclic trienes (ATMET) gives branched polymers. Meier and coworkers polymerized triglycerides by ATMET with methyl acrylate as the chain stopper.[32,42] The introduction of the chain stopper indicates the controllability of branching in ATMET polymerization. Moreover, the concept of incorporating a chain stopper into ADMET polymerization can be extended to the preparation of block copolymers by introducing a macro-chain terminator.[43] A telechelic ABA triblock copolymer is obtained through one-pot ADMET polymerization of castor oil derivative and oligomeric poly(ethylene glycol) methyl ether acrylate ($M_n = 454$ Da) as the chain stopper.

As a plethora of reports demonstrates the feasibility of ADMET polymerization of biorenewable monomers, researchers are still facing obstacles as biobased polymers from ADMET generally have low molecular weight and sometimes only oligomers are obtained often owing to catalyst efficiency issues. Efforts are needed for ameliorating the performance of catalysts so that polymers with a higher degree of polymerization and different functional groups can be obtained.

2.5.4 Non-olefin-based Acyclic Metathesis Polymerization

Miller and coworkers recently broadened the concept of metathesis polymerization from olefins toward different types of functional groups such as oxalates, carbonates, and acetals.[44–46] In their studies, modified aliphatic diols derived from castor oil are functionalized as monomers. Mirrored ADMET polymerization, non-olefin-based acyclic metathesis polymerization undergoes a similar mechanism, but the exchange center is either a carbonyl or ether moiety catalyzed by a Lewis acid or base. As an example, the proposed mechanism of carbonate metathesis polymerization is shown in Scheme 2.16,[46] in which the carbonyl group is the center of metathesis where the interchange of oxygen atoms condenses dimethyl carbonate. A one-pot synthesis is demonstrated starting from diol feedstock. The product molecular weight is up to 50 kDa with more than 70% yield.

2.5.5 Ring-opening Metathesis Polymerization (ROMP)

As mentioned previously, the structure of the olefin plays a critical role in metathesis polymerization. The polymerizability of ROMP monomers is distinctly dependent on their chemical bond ring strains. If the monomer contains a highly constrained ring, it will favor polymerization to reduce the overall potential energy. However, olefins with 5, 6, and 7-membered rings and rings with congested substitutions generally are harder to polymerize because of their low bond strain energy.[47] ROMP is preferably conducted at the lowest possible temperature with the highest monomer concentration to suppress the formation of cyclic species which, according to the Jacobson–Stockmeyer theory, is inevitable in producing high molecular weight polymers.[30,48]

Examples of natural abundant substrates potentially undergoing ROMP are derivatives of α-pinene, as shown in Scheme 2.17.[49] Despite its low ring strain inhibiting ROMP directly, apopinene, the derivative with the removal of the vinylic methyl substitute, is able to undergo ROMP by reducing the steric hindrance of α-pinene. Strick and coworkers reported poly(apopinene) of 15.6 kDa with a dispersity of 1.6 using a Grubbs III catalyst.

In addition to directly polymerizing biobased derivatives by ROMP, another common ROMP-based strategy for producing biopolymers is to graft biobased substrates, such as plant oils, to classical ROMP monomers including norbornene,[50–56] cyclooctene,[57,58] and dicyclopentadiene.[59–62] This method typically has a better control and reactivity toward polymerization. However, one should keep in mind that these ROMP-functional-group-decorated (ROMP-FGD) monomers are not considered as fully biorenewable materials because of the decreased biomass content in product materials. Precursors such as bicyclo[2.2.1]hept-2-ene-5-carbonyl chloride,[56] 5-norbornene-2,3-dicarboxylate anhydride,[50] and 5-norbornene-2-carboxylic acid[52,54] have been reported in functionalization of plant oils and their alcohol derivatives through Fisher esterification and ring opening of epoxy groups. Other functional groups such as cyclopentadiene are modified onto plant oils *via* Diels–Alder reaction.[60]

ROMP is utilized in the synthesis of both thermosets and thermoplastics of ROMP-FGD materials. ROMP-FGD plant oils and alcohols such as linseed oil, palm oil, castor oil, and soybean alcohol are reported in the synthesis of

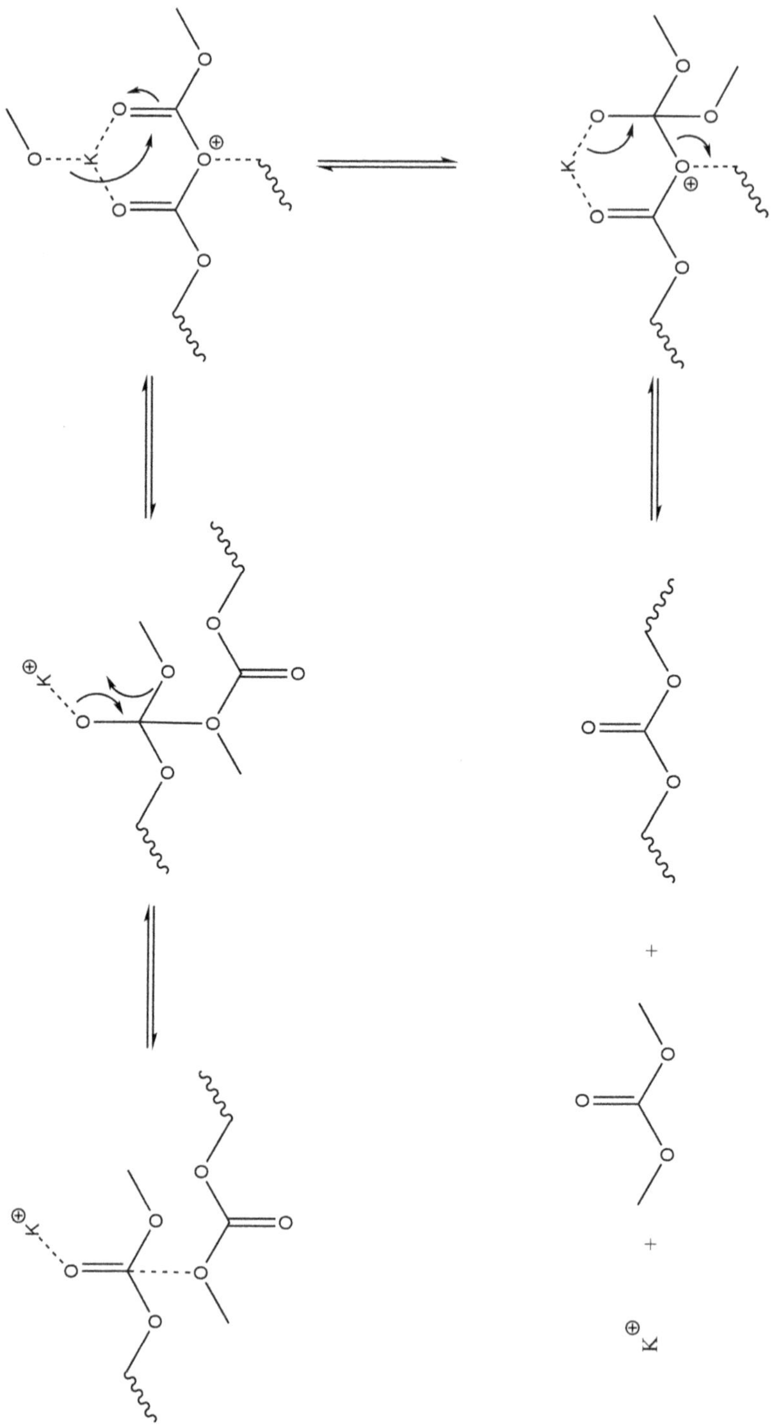

Scheme 2.16 Carbonate metathesis proposed mechanism.

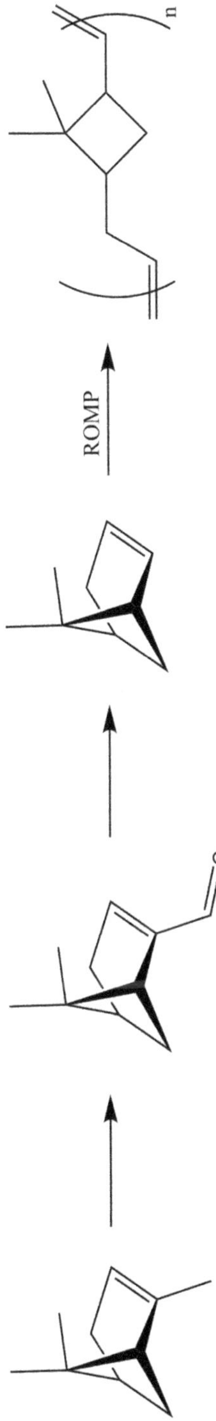

(-)-α-pinene (-)-myrtenal (-)-apopinene

Scheme 2.17 Synthetic route from α-pinene to polymer.

thermosets by ROMP.[28,59,63] Typically Grubbs catalysts are used in polymerization of these materials. In order to increase the reactivity of the catalyst, a freeze-drying process is employed to recrystallize the catalyst. The dosage of catalysts is often within 0.5 wt% in thermosets. ROMP-FGD monomers are often copolymerized with other monomers because of the slow kinetics of polymerization with a single monomer. Fernandes *et al.* gave an example of the low reactivity of norbornene functionalized palm oil.[54] They suspect that the large steric hindrance given by the aliphatic side chain tails hinders the polymerization considering the spatially-oriented nature of the metal carbene. Thus the incorporation of norbornenes as spacer groups in main chains is necessary to enhance the kinetics. However, ROMP-FGD monomers also act as the chain transfer agent to control the degree of crosslinking because of their relatively low reactivity in ROMP. For example, monoterpenes such as β-pinene, carvone, D-limonene, and limonene oxide are investigated as the chain transfer agent in thermosets of dicyclopentadiene polymerized by ROMP.[61] This allows for control of mechanical properties by controlling the crosslink density.

In thermoplastics, the dosage of catalyst in reagents is set by the molar ratio of monomer to catalyst assuming each catalyst initiates one polymer chain. ROMP can be considered a "controlled" or "living" polymerization if an appropriate catalyst is chosen.[30] In a "living" ROMP, chain initiation is rapid compared to chain propagation, side-reactions such as ADMET are suppressed, and the ROMP catalyst remains active such that further propagation will occur with the replenishment of fresh monomer feed. Block copolymers have been synthesized with ROMP-FGD monomers by living ROMP. Instead of adding ethyl vinyl ether to quench the reaction after consuming all monomers, multi-block copolymers are obtained by sequentially adding different monomers into the reaction mixture. Tang and Robertson's group reported triblock and pentablock copolymers from norbornene-functionalized rosin and soybean oil monomers by using Grubbs III or Hoveya–Grubbs II catalysts.[52,54]

2.6 Radical Polymerization

Radical polymerization propagates polymer chains by sequential addition of vinyl groups by radicals and therefore is a type of chain growth polymerization. This type of polymerization has been used for decades and is a highly mature technique in the polymerization of petroleum-based monomers such as vinyl aromatics, (meth)acrylates, vinyl amides, and so on. Propagated by free radicals, radical polymerization is generally conducted under an oxygen-free environment. Depending on whether the radical polymerization is controlled or not, radical polymerization can be sorted into free radical polymerization (FRP) and controlled radical polymerization (CRP). CRP utilizes the addition of one or more controlling agents to limit termination.

2.6.1 Free Radical Polymerization

The basic mechanism of FRP is shown in Scheme 2.18. Standard FRP contains three steps: initiation, propagation, and termination. At the initiation step, an initiator is utilized to generate radicals. There are numerous thermal initiators such as azobisisobutyronitrile (AIBN), benzoyl peroxide, or potassium persulfate, and many photoinitiators such as benzophenone and ferrocene. When activated, the initiator decomposes and gives free radicals. These free radicals are then transferred onto vinyl groups of the monomer. During the propagation step, the growing polymer radicals react with monomers to propagate the chain. Assuming the total radical concentration is stable throughout the reaction, the propagation rate can be expressed as

$$R_{\text{p}} = k_{\text{app}}[\text{M}][\text{I}]^{\frac{1}{2}} \tag{2.13}$$

in which k_{app} is the apparent rate constant of the propagation, [M] is the monomer concentration and [I] is the initiator concentration. The propagation repeats and polymer chains grow until termination occurs. There are two termination mechanisms in FRP among polymer chains: recombination and disproportionation. Recombination occurs when two active polymer chains bridge together, form a covalent bond, and thus lose their activity. Disproportionation occurs when an active radical disproportionates onto the other active polymer chain, a π bond is formed on one of the chains, and at least one chain loses its activity. In many cases, both chains become inactive; however, in some cases, this can lead to branching. Termination also occurs

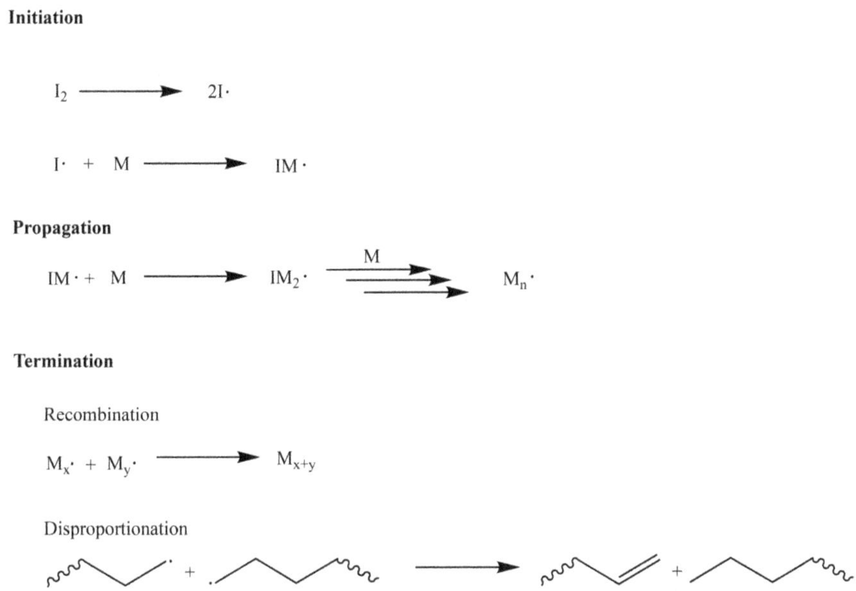

Scheme 2.18 Free radical polymerization mechanism.

when radicals transfer to species other than monomers and polymers and lose their activity.

Many biobased chemicals have unsaturated double bonds which are considered as natural candidates to go through FRP. Plant oils with multiple unsaturated fatty acids can be simply polymerized by FRP into thermosets.[64] The further modification of plant oil, such as conjugating by isomerization, gives more reactive monomers. Robertson *et al.* demonstrated that conjugated poly(soybean oil) shows faster kinetics and reaches gel point earlier than conventional poly(soybean oil) under the same reaction conditions.[64] Other reactive monomers and crosslinkers including styrene, acrylonitrile, divinylbenzene, and dicyclopentadiene are also copolymerized with plant oils into thermosets by FRP.[65]

To polymerize other biobased chemicals without unsaturated double bonds through FRP, chemical modification is necessary to functionalize those with reactive vinyl groups. Biomass-derived chemicals such as ferulic acid,[66] lignin,[67] terpenes,[68] and plant oils[69,70] are polymerizable after chemical modification. The dominating functional groups chosen for chemical modification are (meth)acrylates through pathways of (trans)esterifications. Though plant oils can be polymerized by FRP directly, their low reactivities prompt researchers to investigate other more effective ways to polymerize them. Scheme 2.19 exemplifies a modification that converts the inactive vinyl groups into active acrylates. La Scala and Wool reported the increase of viscosity of monomers with respect to the extent of epoxidation and acrylation resulting from their increased polarity.[71] The viscosity of the reaction mixture is highly related to reaction kinetics during polymerization and gelation because of the mass transfer limit and results in a heterogeneous product.

2.6.2 Controlled Radical Polymerization

Controlled radical polymerization, a polymerization category that has the "living" characteristic, has become important in recent decades because of its ability to produce well-defined and precise products with a variety of structures. The termination step is suppressed in a controlled radical polymerization by drastically reducing the concentration of free radicals. This not only allows for lower dispersities but also, because these polymerization techniques are pseudoliving, allows for the growth of new monomers; thus, well-defined block copolymers can be synthesized.[72,73] Two types of controlled radical polymerizations dominate the field: radical addition–fragmentation chain transfer (RAFT) polymerization and atom transfer radical polymerization (ATRP). Each technique uses a unique way to control the polymerization.

2.6.2.1 Reversible Addition–Fragmentation Chain Transfer (RAFT) Polymerization

RAFT polymerization introduces a chain transfer agent (CTA) into its system. The structures of CTAs are shown in Figure 2.4. The key component

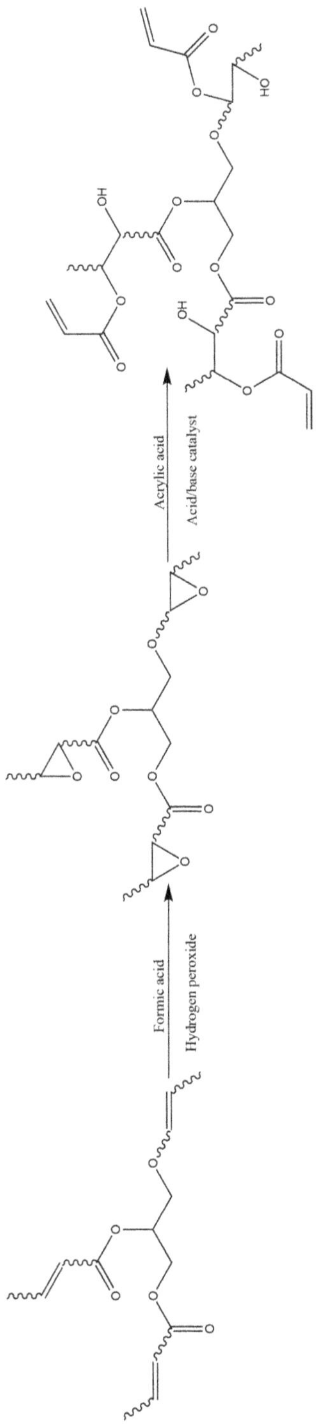

Scheme 2.19 An example pathway of plant oil modification.

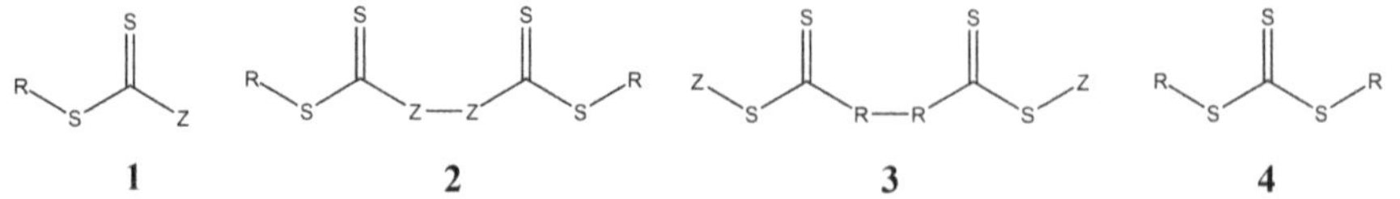

Figure 2.4 Basic skeletons of different types of CTA: classic structure of CTA, **1**; difunctional CTA coupling through *Z* groups, **2**, and coupling through *R* groups, **3**; and telechelic CTA, **4**.

in CTA is the thiocarbonylthio group along with the *R*- and *Z*-groups. The *R*-group is chosen to be a good "leaving" group, while the primary role of the *Z*-group is to stabilize the radical intermediate that serves as a dormant state. The equilibrium mechanism of CTA in polymerization is illustrated in Scheme 2.20. By establishing a fast equilibrium with active polymer chains, the CTA controls the propagation step by forcing most chains to be dormant at all times, while initiation and propagation steps follow the conventional FRP. Because the equilibrium is skewed towards dormancy, the active radical concentration in the system is low. This reduces the propagation rate considerably but essentially eliminates the termination step because its rate scales as the square of the polymer radical concentration. Detailed information about the RAFT process is organized by Moad and coworkers.[74–76] Ideally, when all of the monomers are consumed, polymer chains are dormant and therefore avoid the presence of termination. Because of this nature, any linear polymer chain produces one CTA. The theoretical degree of polymerization of a RAFT polymerization is simply the molar ratio between the monomer and CTA multiplied by the conversion. The governing equations in RAFT polymerization are expressed as

$$\frac{[\text{Initiator}]}{[\text{CTA}]} = C \tag{2.14}$$

$$\frac{[\text{Monomer}]}{[\text{CTA}]} p = N_{\text{n}} \tag{2.15}$$

where p is the extent of reaction, and C commonly lies in between 0.1 and 0.5. The choice of C plays an important role in controlling the polymerization kinetics and the molecular weight distribution of products, while eqn (2.15) gives the theoretical degree of polymerization when reaction is completed. By increasing C, the reaction kinetics will be faster; however, the tradeoff is the potentially higher dispersity product.

The *R* and the *Z* group can dramatically affect several factors in RAFT systems. One of the most important factors is the controllability of chain transfer agents toward various classes of monomers. Common CTAs include dithioesters, xanthates, trithiocarbonates, and dithiocarbamates. Desirable choices of *R* and *Z* groups toward different classes of monomers can be found in reviews by Moad *et al.*[75–77]

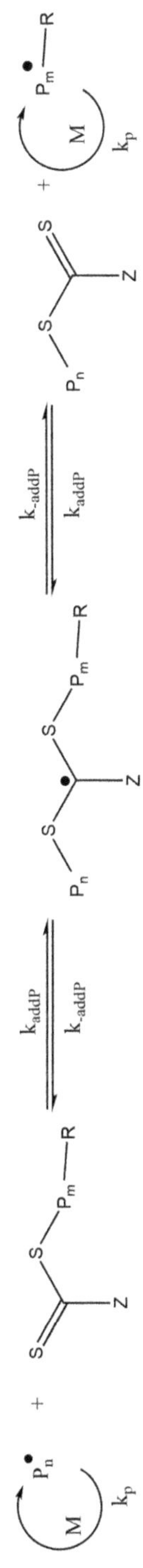

Scheme 2.20 Radical equilibrium between CTA and polymer chains during propagation.

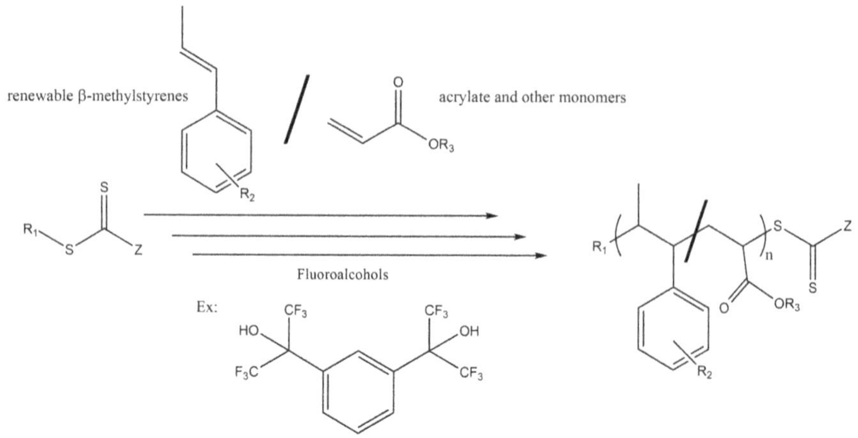

Scheme 2.21 RAFT copolymerization of renewable β-methylstyrenes with acrylate and other monomers.

Monomers that undergo FRP are candidates for controlled radical polymerization providing that vinyl groups in the monomers have adequate reactivity. Some monomers with naturally abundant vinyl groups, without active vinyl group decoration, are polymerizable by RAFT. Examples include β-methylstyrene,[78] ferulic acid derived bio-styrene,[79] and itaconic acid derivatives.[80]

Solvent choice is critical in low reactivity monomer systems. In Scheme 2.21, Nonoyama *et al.* exemplified RAFT copolymerization of β-methylstyrenes, which are typically non-polymerizable because of steric hindrance, with acrylates.[78] In cases of significant reactivity difference between β-methylstyrene and acrylates, fluoroalcohols enhance radical propagation between β-methylstyrene monomers by reducing the electron affinity of the acrylate vinyl group through hydrogen bonding interaction between the carbonyl group of the acrylates and protic fluoroalcohols.[81] The composition of β-methylstyrenes are thus increased in copolymers if protic solvents are used. Chemical modification is also involved in promoting the reactivity of bio-based styrenes. Recently, Takeshima and coworkers reported ferulic acid derivatives polymerized by RAFT.[79] Derived from ferulic acid by decarboxylation, 4-vinylguaiacol is ready for radical polymerization. Nevertheless, the free phenolic groups slightly inhibited the polymerization and led merely to oligomers. Different protections thus have been completed to cap hydroxyl groups by acetyl and silyl groups. Products with molecular weights up to 20 kDa are harvested after protection.

The strategy of incorporating active vinyl groups into biobased derivatives to facilitate RAFT polymerization has been widely studied because it gives products with various functionality and structure. Examples include isosorbide,[82] fatty acids,[83,84] lignins,[85] plant oils,[86,87] and more.[88] Recently, the Cochran group has reported RAFT polymerization of acrylated epoxidized soybean oil (AESO).[86,87,89] To apply AESO-based thermoplastics, the gel point suppression and the control of RAFT polymerization have been thoroughly investigated. Monomer concentration is critical in gel suppression regarding the competition between intermolecular and intramolecular crosslinking. Appropriate monomer concentration leads to PAESO molecular weights up to 100 kDa without macrogelation.

RAFT polymerization features CTA remaining dormant in the polymer chain, which benefits subsequent polymerizations with other monomers to provide block copolymers with different structures. Block copolymers can be formed by the sequential addition of monomers or by reinitiation of macro-CTA with monomers in a step-wise fashion. Either fully bio-based or partially bio-based and partially petroleum-based bio-based block copolymers obtained by RAFT polymerization have been widely reported.[79,80,82–84,86] Prepared by two sequential additions of monomers by RAFT with ethyl 2-(ethoxycarbonothioylthio)-2-methylpropanoate as CTA, the Cochran group presented a thermoplastic elastomer (TPE) of poly(styrene-*block*-AESO-*block*-styrene) (PS-PAESO-PS) which shows microphase separation in TEM.[86] Gallagher and coworkers reported telechelic triblock copolymers of isosorbide derivative and acrylates which showed their potential in pressure sensitive adhesives.[82] Satoh *et al.* demonstrated that after chemical modification, glassy and rubbery polymers are harvested from itaconic esters and itaconimide, respectively.[80] In Scheme 2.22, the further copolymerization of these itaconic acid derivatives gave a triblock copolymer possessing typical thermoplastic elastomer properties and showing microphased separation under AFM.

2.6.2.2 *Atom Transfer Radical Polymerization (ATRP)*

While RAFT polymerization utilizes a CTA to effectively reduce the active radical concentration in the system, ATRP shares the same concept but uses a metal catalyst/ligand complex to control polymerization. The conventional ATRP mechanism for radical activation/deactivation is shown in Scheme 2.23. Initiators used in ATRP are commonly alkyl halides, (RX, X = Cl or Br).

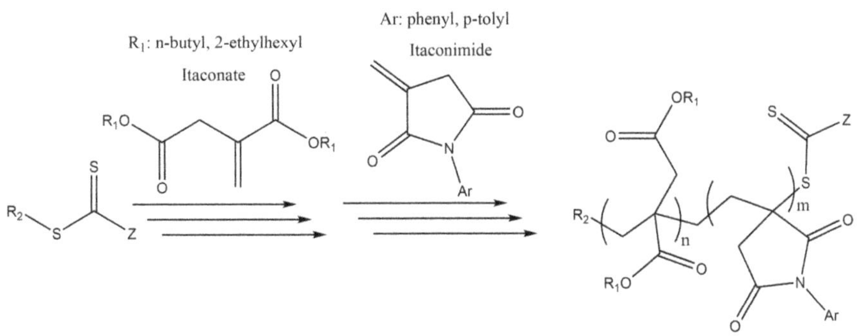

Scheme 2.22 Block copolymer of itaconic acid derivatives. The soft and hard blocks are itaconate and itaconimide, respectively.

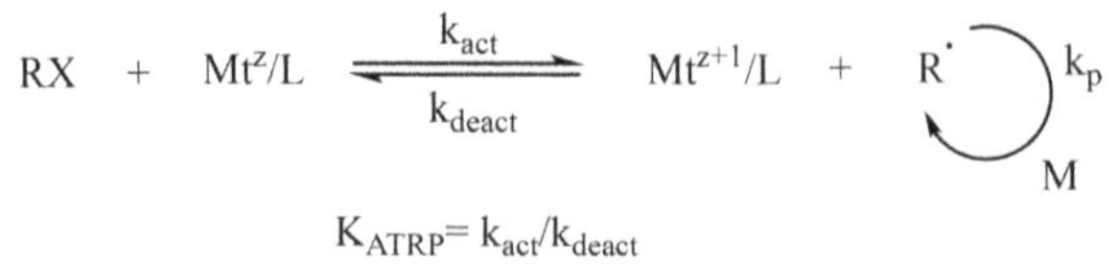

Scheme 2.23 ATRP radical equilibrium mechanism.

Catalysts are transition metals with two oxidation states such as CuBr and CuCl. This section will focus on copper chemistry because of the massive amount of study that has been completed on this system. Ligands are typically nitrogen-based species. In ATRP radical equilibrium, the initiator is activated by oxidizing Cu(I) to the Cu(II) state. Radicals then have the opportunity to propagate with monomers on the right side of the equilibrium reaction. On the left side, the polymer chain is dormant owing to its halide terminus. Through the equilibrium, the radical concentration in the system is kept low and the polymerization is controlled. The living characteristic in ATRP is carried by halides, the dormant species, which remain as an end-group and can be reinitiated. The theoretical degree of polymerization in ATRP is thus equal to the ratio of [monomer] : [initiator].

The equilibrium constant, K_{ATRP}, is highly related to the choice of ligand. A wide range of possible ligands can be found in the study by Matyjaszewski's group.[90] Ligands with a high K_{ATRP} are appropriate for applications because of the low required dosage. In addition, a suitable value of k_{deact} keeps most of the polymer chains in the dormant state and thus the polymerization is controllable. In general, the most effective ligands are those with high K_{ATRP} and high k_{deact}.

Because of the sensitivity and unstable nature of Cu(I), the deoxygenation level of conventional ATRP is more selective than that of FRP and RAFT polymerization though the newer generation of ATRP is reportedly more oxygen-tolerant. Equimolar amounts of initiator, catalyst, and ligand are commonly used. The polymerization is generally conducted between room temperature and 80 °C depending on the polymerization kinetics. ATRP can be run in bulk but polymerization in solvent is preferred in non-polar monomer systems. The solubility of the catalyst has a key impact on the control of polymerization. Resulting from the polar nature of the catalyst, the assistance of the more polar solvents such as *N*,*N*-dimethylformamide, ethyl acetate, and anisole in polymerization sometimes is necessary to keep the system homogeneous and additionally improves the kinetics of the reaction.[91]

To engage biomass derivatives to undergo ATRP, moderate chemical modifications are often used. Because the active functional groups of ATRP are generally (meth)acrylates, bio-based monomers are prepared by (trans)-esterification with some exceptions such as itaconic acid derivatives[92,93] and tulipalin A[94] which naturally contain the requisite chemical functionality. For example, Wang *et al.* reported ATRP of rosin-acid-derived (meth)acrylates with a lignin-decorated macro-initiator.[95] The product from the methacrylate ester of rosin acid shows a higher conversion and molecular weight than those from acrylate esters though the molecular weight distribution of all is high because of the use of a polydisperse macro-initiator.

The flexibility of the structure and functionality of alkyl halide initiators opens wide classes of polymer architectures and applications. The further chain extension by ATRP from macro-initiators provides opportunities for the construction of block copolymers. Block copolymers containing bio-based monomers are obtained either by one-pot sequential addition of monomers[96] or reinitiating the halide terminated macro-initiators prepared by ATRP.[91,97–100]

Block copolymers are also obtained by using macro-initiators prepared by other polymerization techniques. Several reports have presented bio-based block copolymers through ATRP subsequent to ring opening polymerization (ROP) or polycondensation. Liu and coworkers exemplified the synthesis of lysine- and itaconic acid-derived block copolymers by sequential polymerization of ROP and ATRP, illustrated in Scheme 2.24.[92] Lysine *O*-carboxy-anhydride, the derivative of the animal abundant amino acid, was polymerized by ROP. The terminal hydroxyl groups of polyesters were then esterified with 2-bromoisobutyryl bromide to prepare the macro-initiator for ATRP. Polymer chains were successfully extended with *N*-phenyl itaconimide by ATRP. Wilbon *et al.* also demonstrated the ATRP of rosin acid-derivatives followed by ROP of ε-caprolactone.[101] In addition, one-pot simultaneous and sequential polymerizations of both techniques were presented and yielded products with good control of the MW distribution and molecular structure by using 2-hydroxyethyl 2-bromoisobutyrate as the initiator which provided two initiating sites. If sequential polymerization is performed, the order in which ROP and ATRP take place has an impact on the polydispersity of the final products. The product taken the route of ROP and then ATRP shows a better dispersity of 1.38, indicating a better control of polymerization.

The chain extension of polymers depends on the number of initiating sites. Therefore, by utilizing multi-functional alkyl halides, different novel polymer architectures can be designed. For example, Whittaker *et al.* studied the ATRP of glycerol derivatives with methyl acrylate to present 4-arm star amphiphilic block copolymers by using pentaerythritol tetrakis(2-bromopropionate) as the initiator.[98] The influence of different initiator architectures toward K_{ATRP} and k_{act} was investigated by Tang *et al.*[90] They found that both structure and electrophilicity affect the stability of alkyl groups, and the reactivity of initiators thus varies.

Utilizing the multi-functional feature of the ATRP initiator, surface initiated-ATRP (SI-ATRP) is possible which is an application that FRP and RAFT cannot reach. Naturally abundant macromolecules such as cellulose,[102,103] wood,[104] lignin,[95,105] and chitosan[106] can be incorporated with ATRP surface modification utilizing the feature of the ATRP initiator. Figure 2.5 illustrates how SI-ATRP typically occurs. Starting from the hydroxyl groups on the surface, ATRP initiators with at least difunctional halides are bound onto the surface by esterification. The modified surface then becomes the

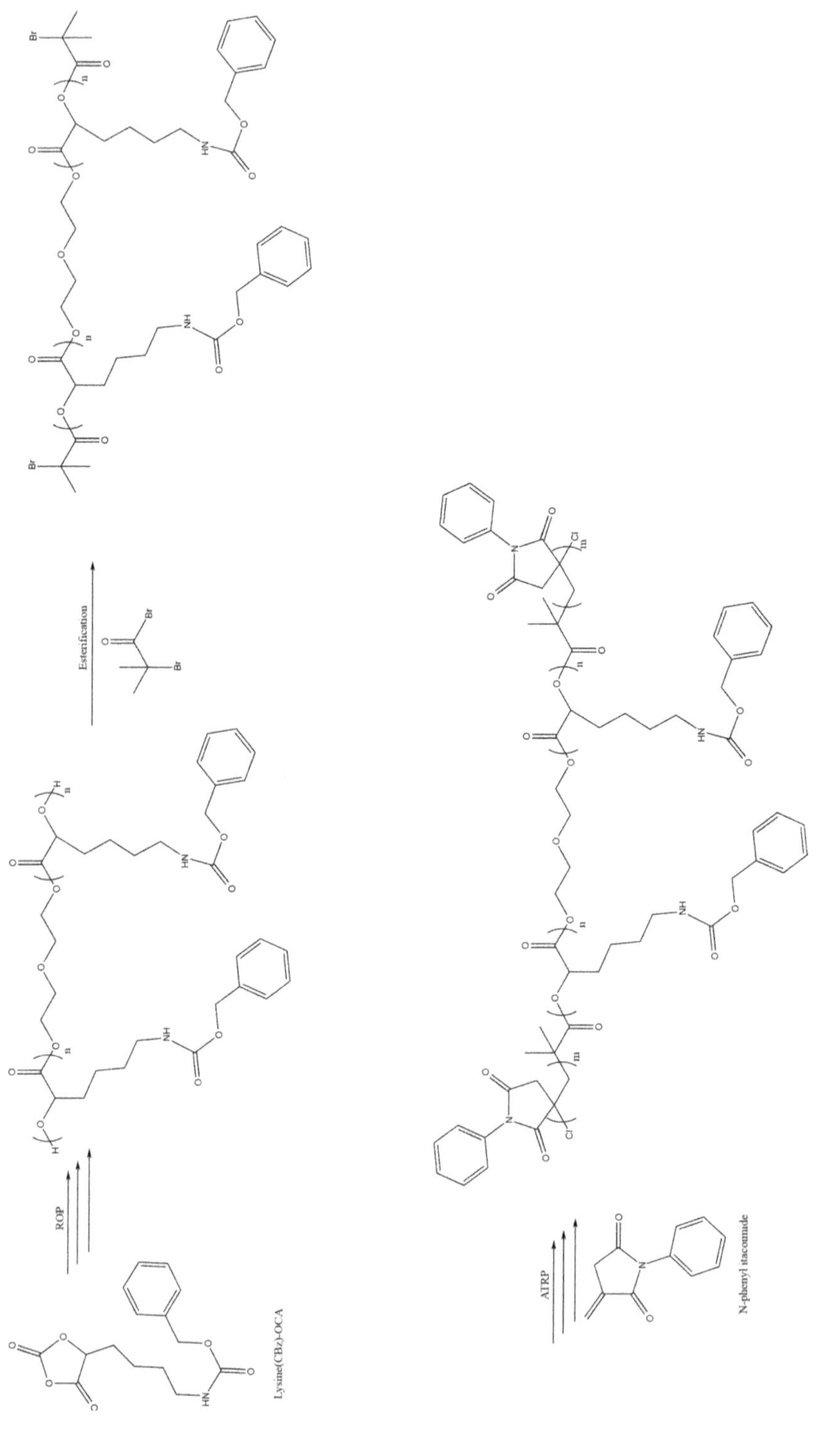

Scheme 2.24 Block copolymer of lysine and itaconic acid derivatives prepared by ATRP followed by ROP.

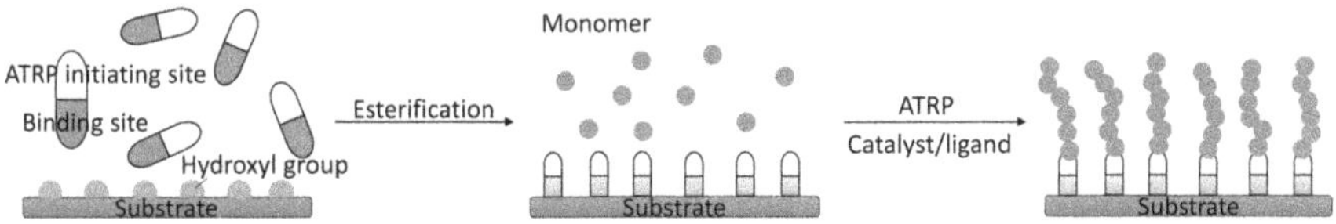

Figure 2.5 Illustration of surface-initiated ATRP.

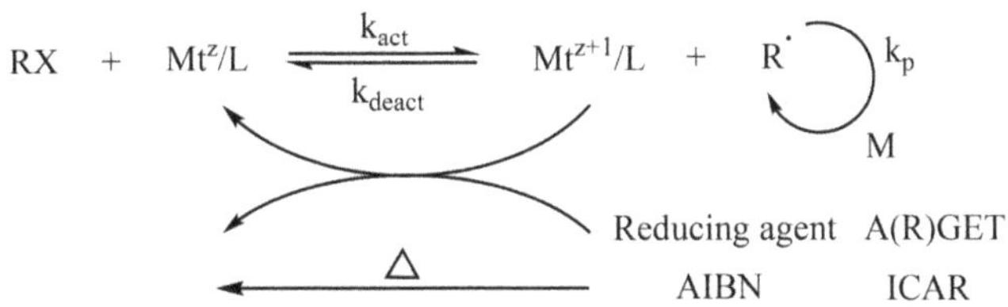

Scheme 2.25 ARGET scheme for "green" ATRP polymerization that uses reducing agents to regenerate active catalyst from deactivated catalyst, requiring only parts per million levels of catalyst overall.

macro-initiator of ATRP which allows grafting of either petroleum-based or bio-based monomers. The hydrophobicity and other properties of substrates are adjusted after surface modifications.[102,107] For instance, Ganewatta and coworkers reported polymerizing resin acid-derivatives by SI-ATRP to create antibacterial and antibiofilm coatings.[107] Once hydrophilic monomers such as 2-dimethylaminoethyl methacrylate (DMAEMA) are grafted onto the hydrophobic lignin surface, the modified block copolymers are applied in gene delivery.[107]

Even though it possesses versatile and unique properties, the drawback of ATRP is the requirement for relatively high amounts of metal which limits its applications in the fields of pharmacy, biology, packaging, and more. To eliminate these catalysts after ATRP, the typical procedure is to pass the reaction mixture through an alumina column which is a time-consuming but necessary process for applications such as drug release[108] and gene delivery.[105] The effort of handling the unstable Cu(I) catalyst also reduces the appeal of ATRP because Cu(I) loses its reactivity by oxidizing to Cu(II) in the presence of low levels of oxygen. To ameliorate these disadvantages, researchers have devoted effort to developing new types of ATRP which include AGET, ARGET, ICAR, and reverse ATRP. These new types of ATRP have two features: a reduced dosage of catalyst to the extent that the need for column purification is eliminated with most effective ligands such as Me6TREN and TPMA; and the *in situ* generation of the Cu(I) catalyst by reduction of Cu(II) at the onset of the polymerization, thus easing the handling of oxygen-sensitive Cu(I). Both approaches result in more ecofriendly and cost-effective types of ATRP, see Scheme 2.25.

Detailed mechanisms of the new types of ATRP can be found elsewhere by Matyjaszewski's group.[109,110] Utilization of new types of ATRP is profound in petroleum-based monomers but not so much in bio-based materials at the present stage. Examples include the polymerization of an itaconic acid derivative by ICAR ATRP[93] and grafting hydrophobic polymer chains onto cellulose surfaces by ARGET ATRP.[103,111] Recently, Tang and coworkers also reported bio-based methacrylates polymerized by metal free ATRP.[112]

2.7 Conclusion

In this chapter, various methods of producing biopolymers utilizing either bioreplacement, the replacing of common petroleum chemicals with those from biological sources, or by utilizing bioadvantaged materials, the substitution of common petroleum chemicals with biologically derived chemicals that have superior properties to their petroleum counterparts, have been covered. With the current state of technology, bioadvantaged molecules are by far more common and scalable, and materials made from these bioadvantaged molecules are already popping up in industry and marketplaces all over the world. These materials can offer several advantages: more biodegradable, tunable properties, low toxicity, low price, and lower price volatility. There are many shortcomings that bioadvantaged biopolymers must overcome however: short lifetime, high residual reactivity, source-dependent properties, or colored impurities. Given the nature of bioadvantaged materials, one individual molecule will find it hard to be as universally useful as some of the common petroleum materials, such as SBS. These are things that must be addressed; however, there is an intense motivation for materials that are cheaper, less toxic, better for the environment, and derived from biological sources. The methods and materials outlined above show a small sampling of the work that is currently being done towards achieving those goals, and further development of these technologies may lead to these pursuits being accomplished.

References

1. N. Hernandez, R. Christopher Williams and E. W. Cochran, *Org. Biomol. Chem.*, 2014, **12**, 2834–2849.
2. P. C. Hiemenz and T. P. Lodge, *Polymer chemistry*, CRC Press, 2007.
3. F. L. Bayer, *Food Addit. Contam.*, 2002, **19**, 111–134.
4. L. N. Ji, *Appl. Mech. Mater.*, 2013, **312**, 406–410.
5. F. E. Wiria, C. L. Tham, A. S. Subramanian, J. N. Tey, X. Qi, C. K. Cheng and B. Salam, *J. Solid State Electrochem.*, 2016, **20**, 1895–1902.
6. T. Baghaee Moghaddam, M. Soltani and M. R. Karim, *Mater. Des.*, 2014, **53**, 317–324.
7. R. J. Palmer, in *Encyclopedia of Polymer Science and Technology*, John Wiley & Sons, Inc., 2002.
8. J. J. Pacheco and M. E. Davis, *Proc. Natl. Acad. Sci. U. S. A.*, 2014, **111**, 8363–8367.
9. M. Colonna, C. Berti, M. Fiorini, E. Binassi, M. Mazzacurati, M. Vannini and S. Karanam, *Green Chem.*, 2011, **13**, 2543–2548.

10. C.-H. (Clayton) Zhou, J. N. Beltramini, Y.-X. Fan and G. Q. (Max) Lu, *Chem. Soc. Rev.*, 2008, **37**(3), 527–549.
11. H. Liu, K. R. M. Ramos, K. N. G. Valdehuesa, G. M. Nisola, W.-K. Lee and W.-J. Chung, *Appl. Microbiol. Biotechnol.*, 2013, **97**, 3409–3417.
12. T. Polen, M. Spelberg and M. Bott, *J. Biotechnol.*, 2013, **167**, 75–84.
13. T. Beardslee and S. Picataggio, *Lipid Technol.*, 2012, **24**(10), 223–225.
14. M. Suastegui, J. E. Matthiesen, J. M. Carraher, N. Hernandez, N. Rodriguez-Quiroz, A. Okerlund, E. W. Cochran, Z. Shao and J.-P. Tessonnier, *Angew. Chem., Int. Ed.*, 2016, **55**, 2368–2373.
15. P. Qi, H.-L. Chen, H. T. H. Nguyen, C.-C. Lin and S. A. Miller, *Green Chem.*, 2016, **18**, 4170–4175.
16. L. Mialon, A. G. Pemba and S. A. Miller, *Green Chem.*, 2010, **12**, 1704–1706.
17. D. Sun, F. Ye, F. Shi and W. Lu, *Pet. Sci. Technol.*, 2006, **24**, 1067–1077.
18. R. F. Storey, K. A. Shoemake, J. W. Mays and S. Harville, *J. Polym. Sci., Part A: Polym. Chem.*, 1997, **35**, 3767–3778.
19. A. Bhaw-Luximon, D. Jhurry, N. Spassky, S. Pensec and J. Belleney, *Polymer*, 2001, **42**, 9651–9656.
20. W. Li, G. Zhang, W. Sheng, Z. Liu and X. Jia, *J. Nanosci. Nanotechnol.*, 2016, **16**, 576–580.
21. S. K. Varshney, J. P. Hautekeer, R. Fayt, R. Jérôme and P. Teyssié, *Macromolecules*, 1990, **23**, 2618–2622.
22. I. V. Vasilenko, A. N. Frolov and S. V. Kostjuk, *Macromolecules*, 2010, **43**, 5503–5507.
23. Z. Jedliński, P. Kurcok and R. W. Lenz, *Macromolecules*, 1998, **31**, 6718–6720.
24. M. Winnacker, M. Neumeier, X. Zhang, C. M. Papadakis and B. Rieger, *Macromol. Rapid Commun.*, 2016, **37**, 851–857.
25. F. He, Y. Gao, K. Jin, J. Wang, J. Sun and Q. Fang, *ACS Sustainable Chem. Eng.*, 2016, **4**, 4451–4456.
26. S. Chakrapani and J. V. Crivello, *J. Macromol. Sci., Part A: Pure Appl. Chem.*, 1998, **35**, 691–710.
27. D. D. Andjelkovic and R. C. Larock, *Biomacromolecules*, 2006, **7**, 927–936.
28. Y. Xia and R. C. Larock, *Green Chem.*, 2010, **12**, 1893–1909.
29. K. J. Ivin, J. C. Mol, K. J. Ivin and J. C. Mol, in *Olefin Metathesis and Metathesis Polymerization*, 1997, pp. 1–11.
30. C. W. Bielawski and R. H. Grubbs, *Prog. Polym. Sci.*, 2007, **32**, 1–29.
31. P. R. Bachler and K. B. Wagener, *Monatsh. Chem. – Chem. Mon.*, 2015, **146**, 1053–1061.
32. U. Biermann, J. O. Metzger and M. A. R. Meier, *Macromol. Chem. Phys.*, 2010, **211**, 854–862.
33. M. D. Schulz and K. B. Wagener, *Macromol. Chem. Phys.*, 2014, **215**, 1936–1945.
34. D. Butilkov and N. G. Lemcoff, *Green Chem.*, 2014, **16**, 4728–4733.

35. L. M. de Espinosa, M. A. R. Meier, J. C. Ronda, M. Galià and V. Cádiz, *J. Polym. Sci., Part A: Polym. Chem.*, 2010, **48**, 1649–1660.
36. J. C. Ronda, G. Lligadas, M. Galià and V. Cádiz, *Eur. J. Lipid Sci. Technol.*, 2011, **113**, 46–58.
37. H. Mutlu and M. A. R. Meier, *Macromol. Chem. Phys.*, 2009, **210**, 1019–1025.
38. L. C. Over, M. Hergert and M. A. R. Meier, *Macromol. Chem. Phys.*, 2017, **218**, 1700177.
39. H. Mutlu and M. A. R. Meier, *Macromol. Chem. Phys.*, 2009, **210**, 1019–1025.
40. L. M. de Espinosa and M. A. R. Meier, *Eur. Polym. J.*, 2011, **47**, 837–852.
41. C. O. Akintayo, H. Mutlu, M. Kempf, M. Wilhelm and M. A. R. Meier, *Macromol. Chem. Phys.*, 2012, **213**, 87–96.
42. P. A. Fokou and M. A. R. Meier, *Macromol. Rapid Commun.*, 2008, **29**, 1620–1625.
43. A. Rybak and M. A. R. Meier, *ChemSusChem*, 2008, **1**, 542–547.
44. J. J. Garcia and S. A. Miller, *Polym. Chem.*, 2014, **5**, 955–961.
45. A. G. Pemba, J. A. Flores and S. A. Miller, *Green Chem.*, 2013, **15**, 325.
46. R. Vanderhenst and S. A. Miller, *Green Mater.*, 2013, **1**, 64–78.
47. K. J. Ivin, J. C. Mol, K. J. Ivin and J. C. Mol, in *Olefin Metathesis and Metathesis Polymerization*, 1997, pp. 224–259.
48. H. Jacobson and W. H. Stockmayer, *J. Chem. Phys.*, 1950, **18**, 1600–1606.
49. B. F. Strick, M. Delferro, F. M. Geiger and R. J. Thomson, *ACS Sustainable Chem. Eng.*, 2015, **3**, 1278–1281.
50. Y. Xia, Y. Lu and R. C. Larock, *Polymer*, 2010, **51**, 53–61.
51. K. Haman, P. Badrinarayanan and M. R. Kessler, *Polym. Int.*, 2009, **58**, 738–744.
52. M. S. Ganewatta, W. Ding, M. A. Rahman, L. Yuan, Z. Wang, N. Hamidi, M. L. Robertson and C. Tang, *Macromolecules*, 2016, **49**, 7155–7164.
53. M. A. Rahman, H. N. Lokupitiya, M. S. Ganewatta, L. Yuan, M. Stefik and C. Tang, *Macromolecules*, 2017, **50**, 2069–2077.
54. H. Fernandes, R. M. Souza Filho, J. L. Silva Sá, B. S. Lima-Neto, C. A. Lenz, A. G. Ferreira and M. R. Kessler, *RSC Adv.*, 2016, **6**, 75104–75110.
55. T. C. Mauldin, K. Haman, X. Sheng, P. Henna, R. C. Larock and M. R. Kessler, *J. Polym. Sci., Part A: Polym. Chem.*, 2008, **46**, 6851–6860.
56. Y. Xia and R. C. Larock, *Polymer*, 2010, **51**, 2508–2514.
57. P. H. Henna and R. C. Larock, *Macromol. Mater. Eng.*, 2007, **292**, 1201–1209.
58. Y. Feng, S. Jie and B.-G. Li, *J. Polym. Sci., Part A: Polym. Chem.*, 2017, **55**, 2211–2220.
59. P. H. Henna, M. R. Kessler and R. C. Larock, *Macromol. Mater. Eng.*, 2008, **293**, 979–990.
60. P. Henna and R. C. Larock, *J. Appl. Polym. Sci.*, 2009, **112**, 1788–1797.
61. J. M. Delancey, M. D. Cavazza, M. G. Rendos, C. J. Ulisse, S. G. Palumbo and R. T. Mathers, *J. Polym. Sci., Part A: Polym. Chem.*, 2011, **49**, 3719–3727.
62. H. Cui and M. R. Kessler, *Compos. Sci. Technol.*, 2012, **72**, 1264–1272.

63. L. M. de Espinosa and M. A. R. Meier, *Eur. Polym. J.*, 2011, **47**, 837–852.
64. M. L. Robertson, K. Chang, W. M. Gramlich and M. A. Hillmyer, *Macromolecules*, 2010, **43**, 1807–1814.
65. M. Valverde, D. Andjelkovic, P. P. Kundu and R. C. Larock, *J. Appl. Polym. Sci.*, 2008, **107**, 423–430.
66. H. Sun, Y. D. Young, S. Kanehashi, K. Tsuchiya, K. Ogino and J. Sim, *J. Fiber Sci. Technol.*, 2016, **72**, 74–79.
67. B. M. Upton and A. M. Kasko, *Chem. Rev.*, 2016, **116**, 2275–2306.
68. M. F. Sainz, J. A. Souto, D. Regentova, M. K. G. Johansson, S. T. Timhagen, D. J. Irvine, P. Buijsen, C. E. Koning, R. A. Stockman and S. M. Howdle, *Polym. Chem.*, 2016, **7**, 2882–2887.
69. T. Eren and S. H. Kusefoglu, *J. Appl. Polym. Sci.*, 2005, **97**, 2264–2272.
70. P. Saithai, V. Tanrattanakul, W. Chinpa, K. Kaewtathip and E. Dubreucq, *Mater. Sci. Forum*, 2011, **695**, 320–323.
71. J. L. Scala and R. P. Wool, *J. Appl. Polym. Sci.*, 2005, **95**, 774–783.
72. M. Szwarc, *Nature*, 1956, **178**, 1168–1169.
73. T. R. Darling, T. P. Davis, M. Fryd, A. A. Gridnev, D. M. Haddleton, S. D. Ittel, R. R. Matheson, G. Moad and E. Rizzardo, *J. Polym. Sci., Part A: Polym. Chem.*, 2000, **38**, 1706–1708.
74. G. Moad, E. Rizzardo and S. H. Thang, *Aust. J. Chem.*, 2009, **62**, 1402.
75. G. Moad, E. Rizzardo and S. H. Thang, *Aust. J. Chem.*, 2005, **58**, 379–410.
76. G. Moad, E. Rizzardo and S. H. Thang, *Aust. J. Chem.*, 2012, **65**, 985–1076.
77. G. Moad, E. Rizzardo and S. H. Thang, *Aust. J. Chem.*, 2009, **62**, 1402–1472.
78. Y. Nonoyama, K. Satoh and M. Kamigaito, *Polym. Chem.*, 2014, **5**, 3182–3189.
79. H. Takeshima, K. Satoh and M. Kamigaito, *Macromolecules*, 2017, **50**, 4206–4216.
80. K. Satoh, D.-H. Lee, K. Nagai and M. Kamigaito, *Macromol. Rapid Commun.*, 2014, **35**, 161–167.
81. K. Koumura, K. Satoh and M. Kamigaito, *Macromolecules*, 2009, **42**, 2497–2504.
82. J. J. Gallagher, M. A. Hillmyer and T. M. Reineke, *ACS Sustainable Chem. Eng.*, 2016, **4**, 3379–3387.
83. B. Maiti and P. De, *RSC Adv.*, 2013, **3**, 24983.
84. S. Wang, S. Vajjala Kesava, E. D. Gomez and M. L. Robertson, *Macromolecules*, 2013, **46**, 7202–7212.
85. Y. Xu, L. Yuan, Z. Wang, P. A. Wilbon, C. Wang, F. Chu and C. Tang, *Green Chem.*, 2016, **18**, 4974–4981.
86. N. Hernandez, M. Yan, R. C. Williams and E. Cochran, *ACS Symp. Ser.*, 2015, 183–199.
87. M. Yan, Y. Huang, M. Lu, F.-Y. Lin, N. B. Hernández and E. W. Cochran, *Biomacromolecules*, 2016, **17**, 2701–2709.
88. S. L. Kristufek, K. T. Wacker, Y.-Y. T. Tsao, L. Su, K. L. Wooley, K. L. Wooley, K. L. Wooley, C. Boyer, T. P. Davis, M. Keller, P. Langan,

A. K. Naskar, J. N. Saddler, T. J. Tschaplinski, G. A. Tuskan and C. E. Wyman, *Nat. Prod. Rep.*, 2017, **34**, 433–459.
89. M. Yan, F. Y. Lin and E. Cochran, *Polymer*, 2017, **125**, 117–125.
90. W. Tang, Y. Kwak, W. Braunecker, N. V. Tsarevsky, M. L. Coote and K. Matyjaszewski, *J. Am. Chem. Soc.*, 2008, **130**, 10702–10713.
91. Y. Zhen, S. Wan, Y. Liu, H. Yan, R. Shi and C. Wang, *Macromol. Chem. Phys.*, 2005, **206**, 607–612.
92. S. Liu, X. Zhang, M. Li, X. Ren and Y. Tao, *J. Polym. Sci., Part A: Polym. Chem.*, 2017, **55**, 349–355.
93. S. Okada and K. Matyjaszewski, *J. Polym. Sci., Part A: Polym. Chem.*, 2015, **53**, 822–827.
94. J. Mosnáček and K. Matyjaszewski, *Macromolecules*, 2008, **41**, 5509–5511.
95. J. Wang, K. Yao, A. L. Korich, S. Li, S. Ma, H. J. Ploehn, P. M. Iovine, C. Wang, F. Chu and C. Tang, *J. Polym. Sci., Part A: Polym. Chem.*, 2011, **49**, 3728–3738.
96. V. Raghunadh, D. Baskaran and S. Sivaram, *Polymer*, 2004, **45**, 3149–3155.
97. S. Wan, Y. Zheng, Y. Liu, H. Yan, K. Liu, Y. Zhang, T. Tsakalakos and M. Muhammed, *J. Mater. Chem.*, 2005, **15**, 3424.
98. M. R. Whittaker, C. N. Urbani and M. J. Monteiro, *J. Polym. Sci., Part A: Polym. Chem.*, 2008, **46**, 6346–6357.
99. D. P. Chatterjee and B. M. Mandal, *Polymer*, 2006, **47**, 1812–1819.
100. V. Anand, S. Agarwal, A. Greiner and V. Choudhary, *Polym. Int.*, 2005, **54**, 823–828.
101. P. A. Wilbon, Y. Zheng, K. Yao and C. Tang, *Macromolecules*, 2010, **43**, 8747–8754.
102. G. Morandi, L. Heath and W. Thielemans, *Langmuir*, 2009, **25**, 8280–8286.
103. G. Li, H. P. Yu and Y. X. Liu, *Adv. Mater. Res.*, 2011, **221**, 90–94.
104. Y. Fu, G. Li, H. Yu and Y. Liu, *Appl. Surf. Sci.*, 2012, **258**, 2529–2533.
105. X. Liu, H. Yin, Z. Zhang, B. Diao and J. Li, *Colloids Surf., B*, 2015, **125**, 230–237.
106. V. K. Thakur and M. K. Thakur, *ACS Sustainable Chem. Eng.*, 2014, **2**, 2637–2652.
107. M. S. Ganewatta, K. P. Miller, S. P. Singleton, P. Mehrpouya-Bahrami, Y. P. Chen, Y. Yan, M. Nagarkatti, P. Nagarkatti, A. W. Decho and C. Tang, *Biomacromolecules*, 2015, **16**, 3336–3344.
108. X. Tang and C.-Y. Pan, *J. Biomed. Mater. Res., Part A*, 2008, **86A**, 428–438.
109. N. V. Tsarevsky and K. Matyjaszewski, *Chem. Rev.*, 2007, **107**, 2270–2299.
110. K. Matyjaszewski, *Isr. J. Chem.*, 2012, **52**, 206–220.
111. S. Hansson, E. Östmark, A. Carlmark and E. Malmström, *ACS Appl. Mater. Interfaces*, 2009, **1**, 2651–2659.
112. J. Wang, L. Yuan, Z. Wang, M. A. Rahman, Y. Huang, T. Zhu, R. Wang, J. Cheng, C. Wang, F. Chu and C. Tang, *Macromolecules*, 2016, **49**, 7709–7717.

CHAPTER 3

Overview of Bioplastics

D. GREWELL

College of Engineering and Engineering Technology, Northern Illinois University, 590 Garden Rd, Dekalb, IL 60115, USA
Email: Dgrewell@niu.edu

3.1 Bioplastics

As the name implies, bioplastics are large molecules consisting of repeating (poly) units (mers) in which at least a portion, if not all, of the repeating units are derived from a sustainable or renewable feedstock. These feedstocks can vary from corn, sugar cane, vegetable oils, and algae as well as others, such as woody biomass.[1] In general, a sustainable feedstock is one that is a biological material. This is in contrast with the way traditional plastics are produced from petrochemical feedstocks, including oil and natural gas.

As previously noted, in most cases the repeating units are hydrocarbon based and the source of this hydrogen and carbon is from petrochemical plastics which are carbon sources that are fossilized carbon. While this fossilized carbon was initially bio-based (plants and/or animals) from dead and decayed biological material, the life cycle of this carbon includes being stored in the ground for millions of years. Thus, it may be confusing to distinguish between renewable and petrochemical feedstocks in that they both originate from biological materials; however, the major difference is the time scale. In the case of renewable feedstocks, the carbon life cycle is typically less than a decade. In this case, the life cycle of carbon is one in which the carbon is cycled between the atmosphere, land, and water.[2] For example, in the case of the carbon cycle for renewable materials, shown in Figure 3.1, only a few years are required, while in contrast, the time required

Green Chemistry Series No. 79
Bioplastics and Biocomposites: A Practical Introduction
Edited by David Grewell

Published by the Royal Society of Chemistry, www.rsc.org

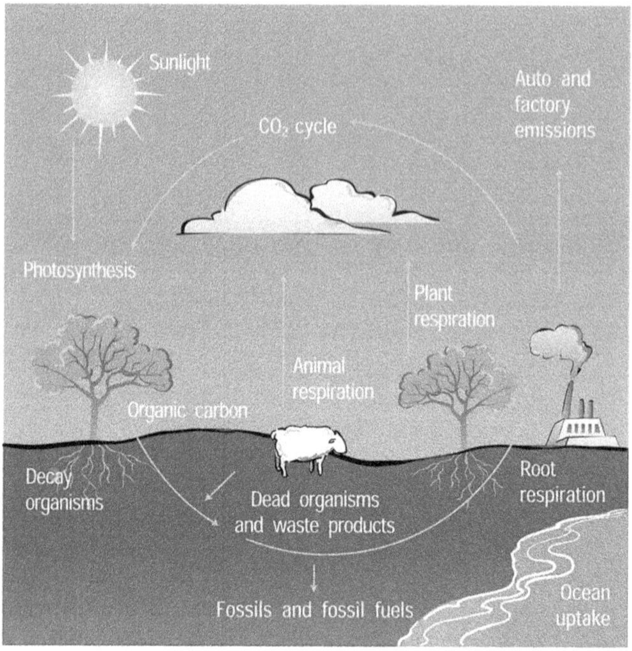

Figure 3.1 Illustration of carbon life cycle.

to produce fossil fuels, also shown in the lower portion of the figure, is millions of years.[3]

Thus, the major difference between a plastic and bioplastic is the source of carbon. For the most part, there is no physical difference in this renewable carbon and petrochemical carbon; however, renewable carbon-based plastics have a small amount of C14 (less than one ppt), which is a radioactive isotope of carbon that decays over time. In the case of petrochemical plastics, the time scale of fossilization results in a great reduction in the amount of C14 because of radiative decay. The measurement and characterization of this carbon is detailed in a following section.

While many assume that bioplastics are degradable in the environment, this is not always true. In fact, one of the most widely produced bioplastics, bio-based PET (polyethylene terephthalate), which is partially derived from sugar (corn and sugarcane), has exactly the same chemical structure as petrochemical PET and, thus, is not degradable under traditional environmental conditions, and in particular, not by microbial activities. In contrast, PLA (polylactic acid), derived from sugar (corn and sugarcane) is compostable. In addition, there are petrochemical plastics that are degradable. An illustration of the "bioplastic region" is illustrated in Figure 3.2.

There are many bioplastics and they have been studied for many years. One could easily argue that humans have used natural polymers such as leather and bones since cave men started using tools. However, in more

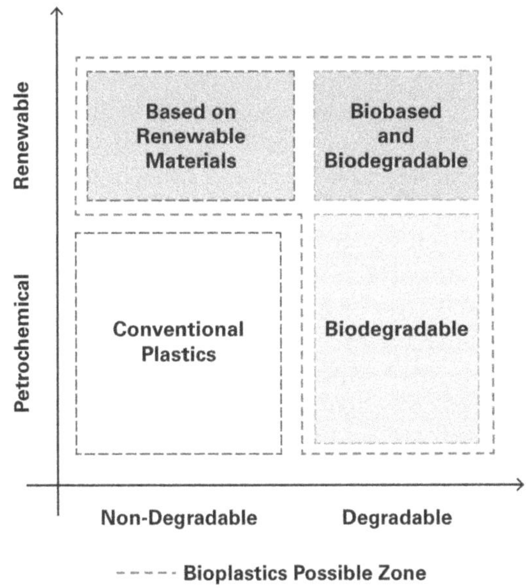

Figure 3.2 Illustration of bioplastic and plastic based on feedstock and degradation.

recent times, we have used sustainable feedstock to produce plastics from a range of materials including cellulose, sugars, and proteins. In some cases, society has modified natural polymers, such as cellulose, by modifying its chemistry to produce materials with desired thermal mechanical properties.[4] It is also possible to use natural polymers without chemical modifications, such as proteins, by simply compounding them with additives, such as plasticizers, to produce desirable mechanical properties. Other bioplastics use feedstock as a food source for microbes that produce monomers as well as polymers for the use of bioplastics. It is also possible to use bio-based chemicals, such as glycine, from the biodiesel industry to produce polymers. These plastics are detailed in later chapters; however, this section will generally review bioplastics.

The use of natural resources to produce plastics dates back to 1865 when Paul Schützenberger produced cellulose acetate for coatings, films, and paints from cellulose rich biomass.[5] Later in 1869, John Wesley Hyatt produced nitrocellulose. This material was used to produce billiard balls.[6] These materials use cellulose which is a polymer of sugars (polysaccharides) as backbone for their chemical structure. In the case of cellulose acetate, pure cellulose is extracted from biomass then chemically treated with acetic acid, sulfuric acid, and acetic anhydride. This reaction replaces many of the OH (a hydroxide) groups with sulfate and acetate groups. In order to improve the mechanical properties, the product is partially hydrolyzed (addition of water) to replace some of these side groups with OH groups. These will biodegrade under the proper conditions.

A common, fully bio-based bioplastic is PLA.[7] This plastic is produced by a fermentation process in which sugar is fed to microbes that produce lactic acid. The sugar is typically produced from corn starch that is hydrolyzed into sugar; however, it is possible to extract sugar directly from feedstocks, such as sugar cane. Once the lactic acid is extracted and purified from the microbes, it is polymerized through a condensation (removal of water) process. PLA is mass produced by companies such as NatureWorks, Total Corbion, and Myriant and is considered compostable. In addition, it is a polyester and water sensitive (needs to be dried prior to processing). It is also possible to directly produce other bio-based polyesters, such as polyhydroxyalkanoates (PHA) and polyhydroxybutyrates (PHB), through microbial activity. In more detail, select species of microbes produce polymers such as PHAs and PHBs directly as a food storage molecule. This has the advantage that polymerization is not required as a secondary process because the microbes complete this process.

There are a number of natural polymers (biopolymers) that exist in nature. For example, deoxyribonucleic acid (DNA) is polymerized nucleotides, starch is polymerized sugar, and proteins are polymerized amino acids. A common bioplastic on the market is plasticized starch commonly referred to as thermoplastic-like starch (TPS). These materials are often plasticized with water and glycerol. In addition, it is possible to plasticize corn protein (zein) or soy protein with various solvents, such as alcohols, glycerin, or water, to produce thermoplastic materials.

Another common natural plastic is natural rubber.[8] Natural rubber is extracted as a latex from rubber trees, dandelions, gutta percha trees, guayule, as well as other plants. Latex is a stable emulsion of a polymer in water. The polymer in the emulsion is polymerized isoprene ($CH_2{=}C(CH_3){-}CH{=}CH_2$), which has a structure as seen in Figure 3.3, where n is the degree of polymerization, or the number of repeating units.

The polymer is coagulated out of the emulsion by the addition of acids, and the water is removed through rolling and heating. The resulting "green" rubber is then compounded with a vulcanizing agent, sulfur, and through a heating process these structures are converted into a cross linked structure. While the structure is a thermoset, the relatively flexible sulfur bonds give the final product a high degree of flexibility and elasticity. While synthetic rubber is more commonly used in most products, the natural fire retardancy of natural rubber lends itself to often be used in fire prone

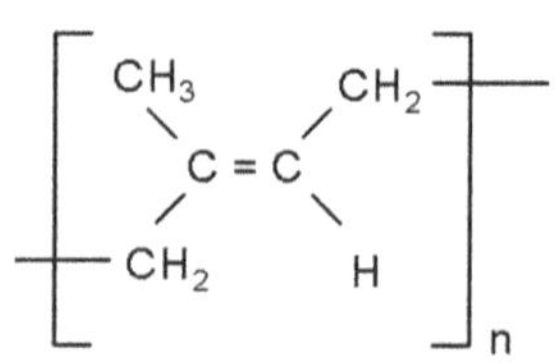

Figure 3.3 Chemical structure of polyisoprene.

Table 3.1 Common bioplastics and bio-based content.

Bio-based plastics	Average bio-based content of the polymers/%
Cellulose acetate (CA)	50
Polyamide (PA)	Rising to 60
Polybutylene adipate terephthalate (PBAT)	Rising to 50
Polybutylene succinate (PBS)	Rising to 80
Polyethylene (PE)	100
Polyethylene terephthalate (PET)	Up to 35
Polyhydroxyalkanoates (PHAs)	100
Polylactic acid (PLA)	100
Polypropylene (PP)	100
Polyvinyl chloride (PVC)	43
Polyurethane (PUS)	30
Starch blends (in plastic compounds)	40

applications, such as tires for aircraft on carriers where aggressive lands can occur.

Other bioplastics include the conversion of biomass to produce traditional petrochemical plastics. For example, Braskem produces polyethylene from sugar cane and PET can be partially synthesized from biomass.[9]

As noted, the bio-based content for various bioplastics can vary depending on the type of plastic as well as the source/manufacturers. Table 3.1 summarizes common bioplastics and their typical bio-based content.[10]

3.2 Evaluating Bio Content

As previously noted, the carbon source defines the renewable content of bioplastics and in particular the existence of C14.[5,11] Because C14 decays to non-radiative carbon over time, it is possible to use this as a measure of the amount of bio content in a plastic. In more detail, by assuming that the time scale for the life cycle of the product is short (less than 50 years), it is possible to determine the amount of renewable carbon by measuring the amount of C14, which has a half-life of 5730 years. That is to say, that after 5730 years, half of the original amount of C14 decays to a non-radiative carbon (C12). The source of the radiative C14 is natural and the result of cosmic radiation found high in the atmosphere. As illustrated in Figure 3.4 and detailed in eqn (3.1), nitrogen, the predominant gas in the atmosphere, is converted to C14.

$$n + {}^{14}_{7}N \rightarrow {}^{14}_{6}C + p \tag{3.1}$$

where n is a neutron and p is a proton. In summary, in the upper atmosphere where radiation levels are relatively high compared to the radiation at the earth's surface, nitrogen adsorbs a neutron which has no electrical charge. This new and unstable atom then releases a proton and produces a

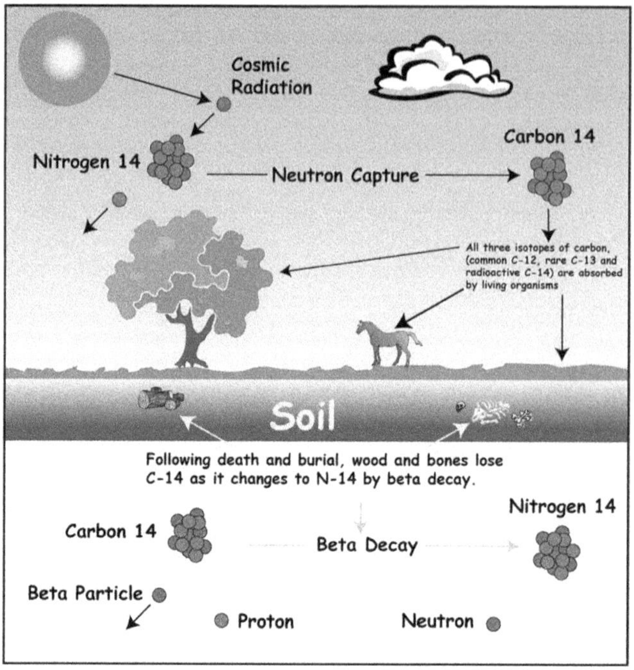

Figure 3.4 Illustration of C14 life cycle.

carbon atom with an atomic mass of 14 ($g\,mol^{-1}$) as compared to the most common isotope of carbon which has an atomic mass of 12 ($g\,mol^{-1}$).

The exponential decay of carbon can be expressed as seen in eqn (3.2).

$$N(t) = N_0 e^{-\lambda t} \tag{3.2}$$

In this case, $N(t)$ is the amount of C14 as a function of time (t), and N_0 is the original amount of C14 at time zero and is a decay constant. It is commonplace to define carbon dating based on the half-life which correlates to the time required for half of the original C14 to decay. The half-life of C14 is 5730 years, thus by rearranging eqn (3.2), it is possible to determine the decay constant for C14 as 0.00012 per year as seen in eqn (3.3).

$$N(t) = N_0 e^{-\lambda t} \rightarrow \frac{N(t)}{N_0} = e^{-\lambda t} \rightarrow \ln\left(\frac{N(t)}{N_0}\right) = -\lambda t \therefore \frac{\ln\left(\frac{50\%}{100\%}\right)}{t} \tag{3.3}$$
$$= -\lambda = \frac{-0.693}{5730} = -\lambda \therefore \lambda = 0.00012\ \text{per year}$$

To measure the amount of C14 in an unknown sample, it is common to measure the radioactivity of the sample compared to a known sample. Because the level of radioactivity is low, this test usually requires a measurement time in the range of days. The standard method for testing bio-based content is detailed by the American Society for Testing and Materials D6866 standard.

3.3 Regulations

One of the most influential policies in the United States is the U.S. Department of Agriculture's (USDA) BioPreferred® Program which was enacted in 2002.[12] Driven to enhance the economy of rural America, it helps drive the use of bio-based materials including plastics. While the act covers a very broad range of products, bioplastics is included as one of the products. In more detail, the act simply states that federal government agencies, including the military, park services, and postal services, must purchase a bio-based product as long as it has the same or better performance compared to a non-bio-based product and costs the same. In order to get a product certified there is a web-based application process and includes a verification process of the bio-based content. Once certified, the product can be labeled with the USDA Certified Biobased Product label, as seen in Figure 3.5.

3.4 Composites

A composite is a system of two or more dissimilar materials that retain a corresponding number of phases once mixed.[5] Often composites are used when materials for lightweight, high strength applications are required. A common example is the use of graphite fibers in the aerospace industry. A less critical application is the use of glass fibers in the production of skis. One of the most common types of composites is fiber-based composites. In general, strength and stiffness are proportional to the length of the fibers and filler level of the composite. For example, in the case of the aerospace industry, often the graphite fibers, which are some of the strongest industry fibers, are continuous and thus very long. While the final product is relatively strong and stiff, the manufacturing of components produced with continuous fibers is labor-intensive as the fibers have to be laid in place, typically on a mold, either manually or using semi-automated processes. Because most components are comprised of multiple layers of fibers, this process is repeated a number of times. Once all layers are deposited, the component is typically cured into a consolidated product with a vacuum oven process. To facilitate the process, the fibers are typically

Figure 3.5 Example of a USDA Certified Biobased Product label.

received as a "pre-preg" which is a tape of fibers with a thermosetting resin coating on the surface to serve as a binding agent.

In contrast, short fiber composites, in which fibers are 2–3 mm long, are typically not as strong as continuous fiber composites; however, they are relatively easy to process. The base resin, or matrix, can be thermoplastic and thus produced with common thermoplastic processes, such as extrusion and injection molding. These processes are further detailed in Chapter 4. The orientation of the fibers is random, and thus the strength of the components is relatively isotropic (not depending on load orientation).

Another common type of fiber-based composite is manufactured using long (from a few mm to +100 mm), randomly orientated fibers. Unlike continuous fibers, where the individual layers of fiber have a distinctive direction, random long fibers exhibit random orientation, and they are placed into a mold in the form of mats. Again, the resins (or matrix) used for these composites are typically thermosets.

Other composites are manufactured using particles or plates (sheets). A common example of a particle-filled composite is talc-filled plastics. Often talc, a solid phase, is added to plastics to reduce the thermal expansion as well as to increase the toughness of the material. Similar to short fiber composites, particle-based composites can be processed with standard polymer processing equipment and thus manufacturing with these materials is relatively easy.

Abbreviations

PET	Polyethylene terephthalate
PLA	Polylactic acid
PHA	Polyhydroxyalkanoates
PHB	Polyhydroxybutyrates
DNA	Deoxyribonucleic acid
TPS	Thermoplastic-like starch
USDA	United States Department of Agriculture

References

1. H. Karan, C. Funk, M. Grabert, M. Oey and B. Hankamer, Green bioplastics as part of a circular bioeconomy, *Trends Plant Sci.*, 2019, **24**(3), 237–249.
2. R. Narayan, Carbon footprint of bioplastics using biocarbon content analysis and life-cycle assessment, *MRS bull.*, 2011, **36**(9), 716–721.
3. G. Ross, S. Ross and B. J. Tighe, Bioplastics: new routes, new products, *Brydson's Plast. Mater.*, 2017, 631–652.
4. A. Biswas, G. Selling, M. Appell, K. Woods, J. Willett and C. Buchanan, Iodine catalyzed esterification of cellulose using reduced levels of solvent, *Carbohydr. Polym.*, 2007, **68**(3), 555–560.

5. J. W. S. Hearle, and C. Woodings, Fibers related to cellulose, *Regenerated Cellulose Fibers*, Woodhead Publishing, Sawston, Cambridge, UK, 2001, pp. 156–173.
6. J. W. Hyatt, Address of acceptance, *Ind. Eng. Chem.*, 1914, **6**(2), 158–161.
7. M. Murariu and P. Dubois, PLA composites: From production to properties, *Adv. Drug Delivery Rev.*, 2016, **107**, 17–46.
8. S. Kohjiya, *Natural rubber*, Smithers Rapra, 2015.
9. A. J. Eerhart, A. P. Faaij and M. K. Patel, Replacing fossil based PET with biobased PEF; process analysis, energy and GHG balance, *Energy Environ. Sci.*, 2012, **5**(4), 6407–6422.
10. L. Dammer, M. Carus, A. Raschka, and L. Scholz, *Market Developments of and Opportunities for biobased products and chemicals*, Nova-Institute for Ecology and Innovation, Hürth, Germany, 2013.
11. S. Thakur, J. Chaudhary, B. Sharma, A. Verma, S. Tamulevicius and V. K. Thakur, Sustainability of bioplastics: Opportunities and challenges, *Curr. Opin. Green Sustainable Chem.*, 2018, **13**, 68–75.
12. R. Buckhalt, USDA BioPreferred program update, *Ind. Biotechnol.*, 2011, 7(4), 248.

CHAPTER 4

Polymer Processing Principles

ELMAR MORITZER,[a] VOLKER SCHÖPPNER,*[a] MICHAEL KRÖKER,[a] CLAUDINE REISDÖRFER,[a] JOHANNA MÜLLER,[a] MAIKE ENGELEN[a] AND DAVID GREWELL[b]

[a] Paderborn University, Germany; [b] Northern Illinois University, Dekalb, IL
*Email: volker.schoeppner@ktp.uni-paderborn.de

4.1 Introduction

4.1.1 Heating Thermoplastics

Plastics tend to be relatively good insulators for both heat and electricity. These insulation properties render them well suited for such things as electrical insulation of wires and as thermal barriers. However, during manufacturing good thermal insulators can limit production rates because of their relatively low thermal conductivity. The heat flux (q), measured in $\mathrm{W\,m^{-2}}$, is defined in eqn (4.1) as the product of thermal conductivity (λ) and temperature (θ) gradient through a distance (x).

$$q = -\lambda \frac{\mathrm{d}\theta}{\mathrm{d}x} \tag{4.1}$$

The thermal conductivity of a typical solid plastic ranges from 0.2 to 1.0 $\mathrm{W\,m^{-1}\,K^{-1}}$. In contrast, metals, such as aluminium, have a thermal conductivity of over 200 $\mathrm{W\,m^{-1}\,K^{-1}}$, which is more than 200 times that of plastics. The difference in the thermal gradient between a typical plastic and a typical metal is illustrated in Figure 4.1. It shows that for a given heat flux of 1000 $\mathrm{W\,m^{-2}}$, which is a typical heat flux from sunlight at the equator at

Green Chemistry Series No. 79
Bioplastics and Biocomposites: A Practical Introduction
Edited by David Grewell

Published by the Royal Society of Chemistry, www.rsc.org

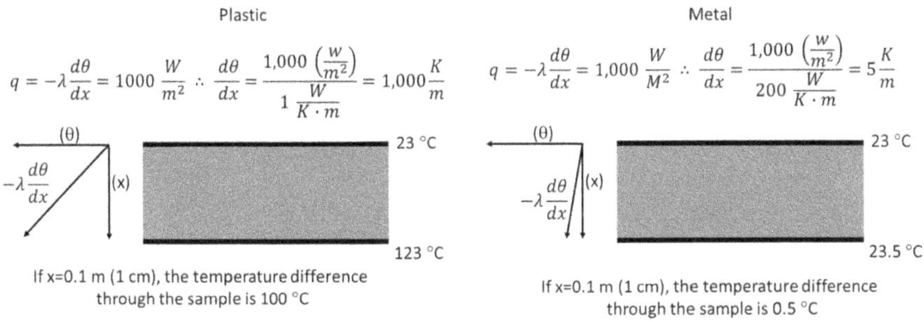

Figure 4.1 Illustration of thermal gradient difference between a typical plastic and a typical metal.

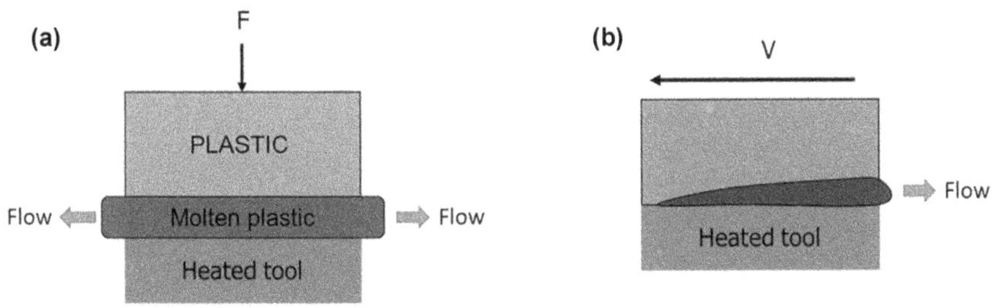

Figure 4.2 Illustration of (a) force-induced melt removal and (b) drag-induced melt removal.

high noon, the temperature difference in a piece of plastic is 100 K while in a piece of metal it is 0.5 K. In other words, one side of a plastic sample can be relatively hot at 123 °C while the other side is relatively cool at 23 °C. Thus, only one side is heated and softened when exposed to a heat source. This makes the use of a traditional heating/softening process, such as a heated plate, a time-intensive process.

In order to ensure that a plastic material is heated and softened effectively and uniformly throughout, industry has adopted the fundamental processes of pressure- and drag-induced melt removal.[1] A good analogy illustrating these two processes is the melting of a slab of butter in a frying pan. If the butter is allowed to melt on its own, the time required to melt the slab is limited by the thermal conductivity of the butter. It is possible to apply a force (F) to the butter so that the molten butter flows outward, exposing unheated butter to the heated surface, see Figure 4.2(a). To further improve the heating process, it is possible to add a dragging motion at a velocity (V) to the butter so that the melted butter is removed by both a squeeze flow (force-induced) as well as by a drag-induced mechanism.

The polymer processing industry standard for melt removal is the use of extrusion screws. As a pellet-filled screw rotates, the melt removal process is promoted at the heated barrel surface. In addition, as the screw rotates, the pellets are sheared, and this results in internal heating of the plastic material, which further increases the heating efficiency of the process. In fact, it is possible for shear heating to be the dominant heating mechanism in an extrusion process. While extruders have electrically heated band heaters on the barrel to preheat the system during startup, it is common to turn these heaters off once the process is operating in a continuous extrusion process. The shear heating process can produce sufficient heat to plasticize the plastic material without additional heating from the band heaters.

While it is common to use the word "melting" or "melt" when describing polymer processing, it is more accurate to use the word "plasticizing." When plastics are heated sufficiently to promote flow, they behave similarly to highly viscoelastic fluids. That is to say, they exhibit viscoelastic properties similar to Silly-Putty®; thus, they do not flow like a fluid and do not "melt" but rather are plasticized.

While plastic pellets are typically heated through screw extrusion processes, plastic sheets are commonly plasticized using infrared (IR) heaters. IR heating offers the advantage of also providing bulk heating. In more detail, wavelengths in the near visible range (800–1200 nm) can penetrate the surface and heat the interior of the sheet. Although the penetration depth is typically limited to a few hundred microns, this heating mode is effective for heating and softening sheets of plastics. These wavelengths can be generated using lasers, quartz halogen lamps, or ceramic heaters.

4.2 Supply Chain

Plastic products are typically manufactured through multiple steps during the value chain process with each step increasing the value of the product. In general, there are three steps in producing plastic products: polymerization, compounding, and molding or extrusion. Initially, the polymer material is synthesized from monomers from either renewable or petrochemical raw materials. These monomers, which are generally liquid at room temperature, are subsequently polymerized in a material-specific reaction (polymerization, see Chapter 2) to create a base polymer by significantly raising the molecular weight. This polymer is generally in powder or pellet form. The chemical reactors typically have throughputs of several tons per hour and are an integral part of the chemical industry. Polymers are formed with a molecular weight distribution (see Chapter 1) that is defined by product specification and customer needs. For example, extrusion grade plastics typically have a higher molecular weight (promoting higher viscosities and melt strength) compared to injection molding grades.

These plastic materials are modified through the addition of a wide range of additives which ensures that the finished material meets customer specifications. Examples of additives include stabilizers to increase service temperature, UV stabilizers, processing aids, fillers, and reinforcing additives such as glass, carbon, or natural fibers (see Chapter 5). Often, these additives are added to base pellet resins in a secondary extrusion process.

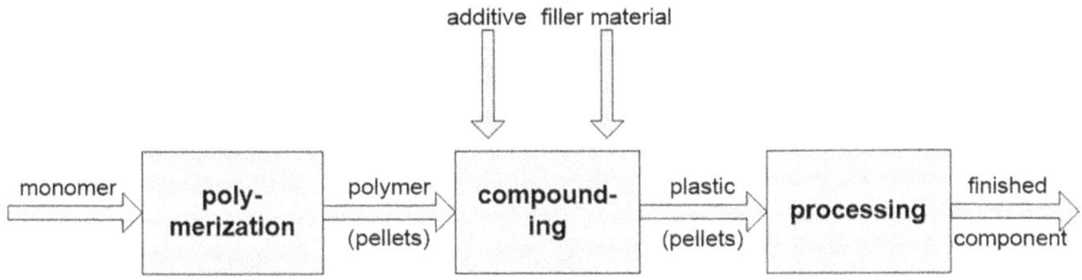

Figure 4.3 Polymerization–compounding–processing.

The third step in the value chain is the actual manufacturing of the product as defined by customer specifications. The core competency of the plastics processor or converter is the development and manufacture of a plastics product that meets the customer's expectations and requirements.

Figure 4.3 shows this value chain. A disadvantage of this common strategy is the frequent melting (plasticizing) and freezing of the material as well as the numerous customer–supplier interfaces between these steps. However, the complexity of each step requires that industry experts focus on their respective core competencies. It is important to note that present developments show a trend towards process integration, *i.e.*, compounding being integrated into the plastics processing step.

This chapter will focus on plastics processing, which does not distinguish between traditional petrochemical plastics and biopolymers/bioplastics. To reduce the scope, this chapter will only review processes used for large-scale production. While there are other processes, such as additive manufacturing, rotational molding, and compression molding, the scale of production is relatively small compared to injection molding and extrusion.

4.3 Material Properties

Plastics are highly viscous, have poor thermal conductivity, and are thermally sensitive. The high viscosity of plastics is the result of the entanglement of the large macromolecules that renders them resistant to mutual slip. The liquid (plasticized) state is characterized by the fact that the molecules can slide past one another. It is important to note again that while "liquid state" or "molten" plastic are terms that are often used for convenience, a better term is "plasticized" because thermoplastics do not melt when heated but rather undergo a softening process. The larger the molecule, the larger the probability that the chains are entangled with one another, reducing the opportunity for sliding past one another. At their processing temperature, plastics have viscosities of approximately 1000 Pa s, while water at room temperature has a viscosity of approximately 0.0018 Pa s. This high viscosity has three consequences when processing plastic materials:

- The flows are laminar and creep at low Reynolds numbers.
- Flow can only be facilitated by using pumps; a plastic melt does not flow under gravity at rates that enable production technology.

- During flow, the plastic melt generates heat by shear heating, which facilitates melting and flow but may also negatively impact final part quality. It makes specific temperature control of the processes more difficult, especially with large machines with a low surface-to-volume ratio.

Figure 4.4 shows a typical viscosity curve of a thermoplastic. The viscosity is highly dependent on temperature and shear velocity. This non-linear relationship between viscosity and shear rate makes the flow characteristics of plastics complex.

The thermal conductivity of plastics is low compared with other materials, see Figure 4.5. Thus, the application of external heat sources to melt plastic materials is not efficient.

While for many plastic products this low thermal conductivity is an advantage, during processing it usually is a limitation. Heating and cooling processes are relatively slow as a result because of the low thermal diffusivity, thus limiting the production speed during both injection molding and extrusion.

Processing above the melt temperature must be carefully controlled. Depending on the plastic material, irreversible thermal damage to the molecular chains (chemical structure) will occur at temperatures between approximately 330 and 400 °C. Therefore, correct temperature control during plastics processing is critical to achieve proper flow properties while reducing thermal degradation.

In summary, while polymer processing is a very effective industrial technology, the techniques are challenged by plastic's high viscosity, poor thermal conductivity, and relatively narrow processing temperature ranges.

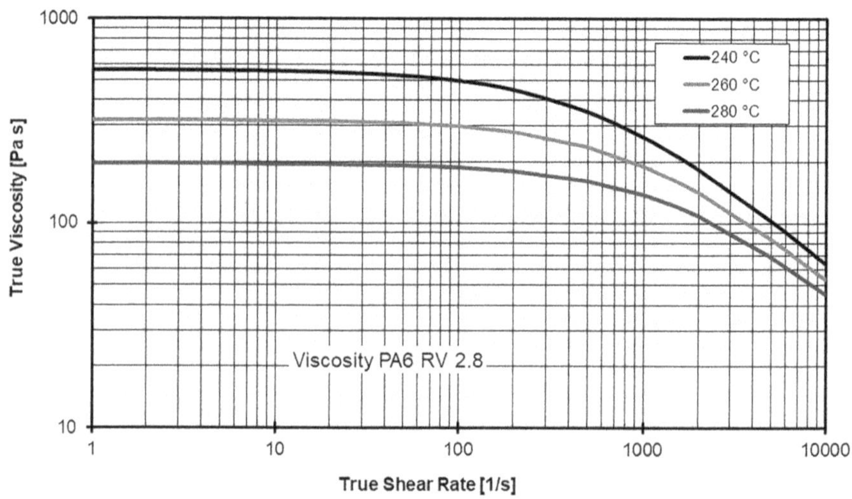

Figure 4.4 Viscosity curve.

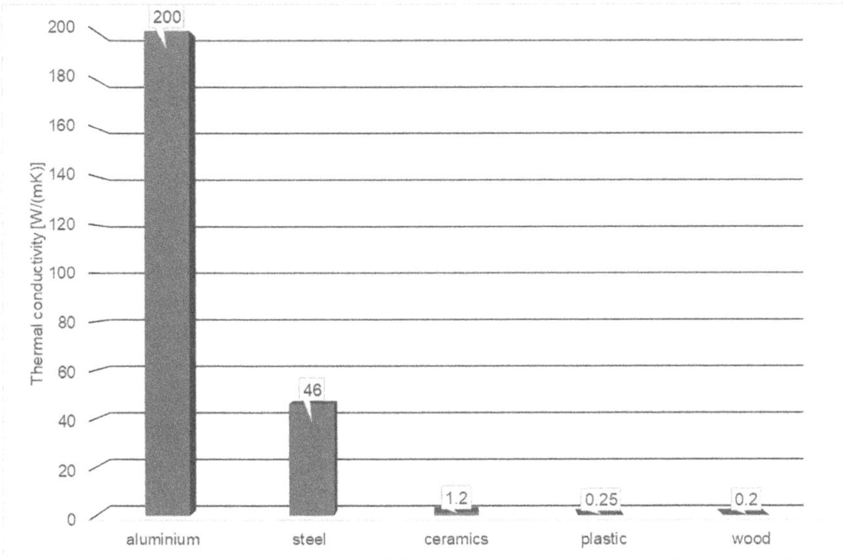

Figure 4.5 Thermal conductivity of typical materials.

4.4 Polymer Processing Machines

The sequence of process steps is usually very similar for all plastics processing machines, as seen in Figure 4.6. A mix of pellets is prepared as a free-flowing bulk product, then melted, and subsequently mixed. The melt is forced into a mold in which it solidifies or freezes. Subsequently, the finished product is cooled so that the material is relatively stiff for subsequent process steps, for example, separation into individual components, welding, or painting. Unlike typical material removal methods, plastic processing produces little scrap as it is a net shape technique. It is economically essential that as much as possible of the feedstock is converted into the final product. Conversion rates, material in *versus* final product material, of above 95% are typical.

A critical component in polymer processing is the screw (see in the next section), which produces a continuous, thoroughly mixed melt flow from a pellet mix. The screw is often called the "heart" of polymer processing.

An extrusion screw can be used to produce continuous products with a constant cross section as well as produce pellets (compounding). Typical products are films, pipes, profiles, filaments, and sheets. The machines operate continuously with a uniform screw speed and extrusion rate. In terms of volume, extrusion processes dominate the plastics processing industry.

Individual plastics parts/components are produced by injection molding. The process works cyclically, *i.e.*, the parts are produced in as short a cycle time as possible in a step-type process. The key component of injection molding machines is a complex and expensive mold. In terms of value, injection

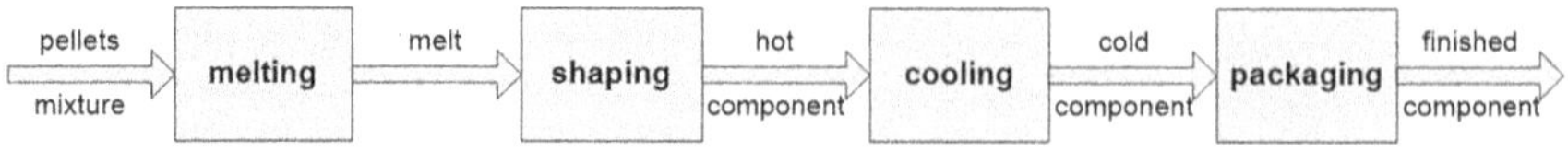

Figure 4.6 Flow diagram.

molding dominates the plastics processing industry. The machines are relatively small and standardized and can be found in any industrial sector.

The manufacture of containers, in particular bottles, requires a special process because it is not possible to produce thin-walled hollow articles using injection molding. During stretch blow molding, preforms are inflated to produce bottles. The process starts either with pellets (extrusion blow molding) or with injection-molded parisons that are heated with infrared lamps and then stretched through pressured air "blown" in a mold/cavity.

Parts with relatively small wall thicknesses and relatively large areas are also difficult to produce by injection molding. For these applications, thermoforming is a possible solution in which extruded film or sheets are heated and then drawn into a forming die with punches and pressure. This process is similar to the deep drawing of metal sheets.

4.4.1 Screw Machines

During manufacturing, most plastics are melted (plasticized) with a screw, which makes all molding processes such as injection molding, extrusion, compounding, calendering, and blow molding efficient and fast.

In a screw extruder, the plastic material is fed into a feeder in the form of free-flowing pellets or powder and then processed by one or more screws. Frictional heating and conduction heating (heater bands) plasticize the material. The melt can be used to manufacture either injection-molded parts or extruded semi-finished products such as pipes, films, threads, profiles, *etc.*

During extrusion processes, the screw melts, mixes, and pressurizes (pumps) the material. The pressure must be higher than the flow resistance of the extrusion die which is typically relatively high. Depending on the viscosity of the melt and the design of the die, pressures between 50 and 500 bar are common.

Figure 4.7 describes characteristics of common screw compounders. In general:

- In the most common case, *i.e.*, a planetary roller screw compounder, the machine has up to 12 screws; however, single-screw machines are most common.
- The raw material can consist of several components and is fed into the machine from a hopper as free-flowing powders or pellets.
- The power (P) describes the drive power of the machine which is applied to rotate the screw. This drive power is consumed by conveying and pumping; however, the majority is consumed by dissipative/frictional heating.
- The main purpose of barrel heating is to reduce heat loss: $T_{barrel} \approx T_{melt}$. To achieve this temperature at the heater bands, the heating

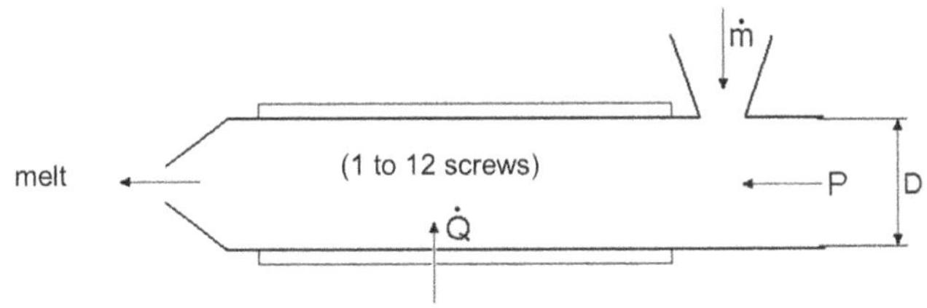

Figure 4.7 Considered sizes of screw compounders.

system must produce a volumetric heat rate $\dot{Q}$ (W) to match the heat loss attributed to convection from the material as well as heat loss from the barrel. Because much of the heat is internally generated, often the heating bands are used only during machine startup. In addition, if excessive heat is generated by the screw and frictional heating, air cooling is often employed.

- The screw compounder produces a melt flow that is homogeneous, hot, and under pressure to overcome the flow resistance of the die or mold.
- For continuous applications, such as extrusion, throughput $\dot{m}$ (kg s^{-1}) is constant and metered by the screw. When accumulation is not possible, the amount of raw material fed into the screw will match the melt output.
- The size of the machine is dependent on the internal barrel diameter D which equals the screw diameter in single-screw extruders. Its length is given as multiples of the diameter. A typical screw length ranges between 15 and 40 times D with recent developments involving longer screws. Screws rotate relatively slowly, usually at speeds between 30 and 1000 rpm, with large machines rotating more slowly.

Various screw designs, see Figure 4.8, are available depending on the respective material and the product to be manufactured. In more detail:

- Screw compounders in extrusion plants are quasi-stationary. Here, the screw or screws revolve at a constant speed, but they do not translate. In contrast, the plasticizing units in cyclical operations, such as in injection molding machines, often translate in the direction of the axis of the screw. Injection molding processes use single screws exclusively.
- The simplest machine is the single-screw extruder. Most single-screw extruders serve as plasticizing extruders in extrusion lines working with pellets or powders. In less common applications, extruders are fed with materials in a molten form. Melt extruders are used to add additives to plastics that take the form of a highly viscous melt after production (*e.g.*, polycondensates PA (polyamide), PET (polyethylene terephthalate)) and which are then processed in-line. In addition, some melt extruders are homogenization and cooling extruders used in

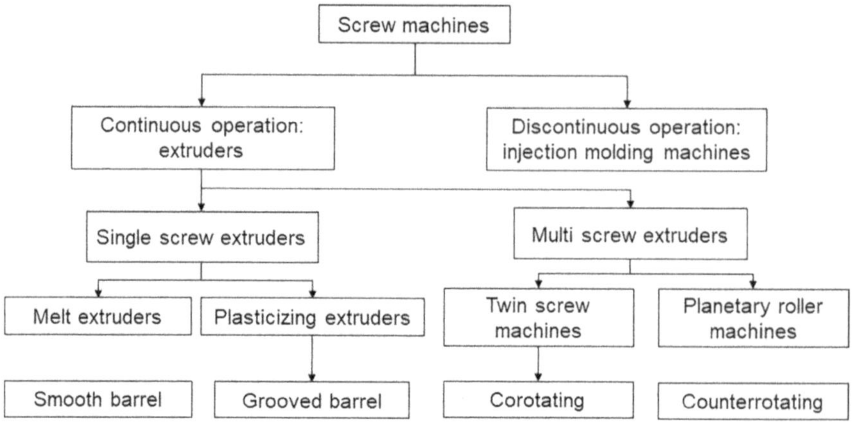

Figure 4.8 Types of screw design.

two-stage plants (cascade plants) which are applied in stretch film and foaming production units.

- Processing polyolefin pellets (PE (polyeythylene), PP (polypropylene)) requires a special design such as the grooved barrel extruder. Here, the feed zone is specifically designed for high rates of conveying. These types of extruders act as pellet pumps, whereas regular screw geometries act more as melt pumps.
- More demanding tasks, such as better mixing or more uniform residence, may require multi-screw extruders.
- Counter-rotating twin-screw extruders are often used with PVC (polyvinyl chloride) processing. Because PVC can chemically decompose when exposed to long residence times at high temperatures, processing requires screws with good self-cleaning properties and without dead spots, such as counter-rotating twin-screw systems. In addition, counter-rotating twin-screw extruders have separate feeding chambers, thus providing for highly homogeneous residence times.
- Co-rotating twin-screw extruders are also used to mix large quantities of additives, *e.g.*, fillers and reinforcing materials, that require a high level of homogeneity quickly and without extensive heating.
- During extrusion, the pressure at the screw tip is a function of the die geometry, the material's viscosity, and the mass flow. When injection molding, the pressure can be selected by the machine's control unit. Figure 4.9 shows a typical pressure profile in a single-screw machine.

In order to ensure that all pellets are fully melted, it is important to apply sufficient energy (heat) to make sure that the product quality is not compromised. Figure 4.10 shows a melting profile where the amount of melt increases from 0–100% from the hopper to the screw tip.

Figure 4.11 details the different zones of a conventional three-section screw.

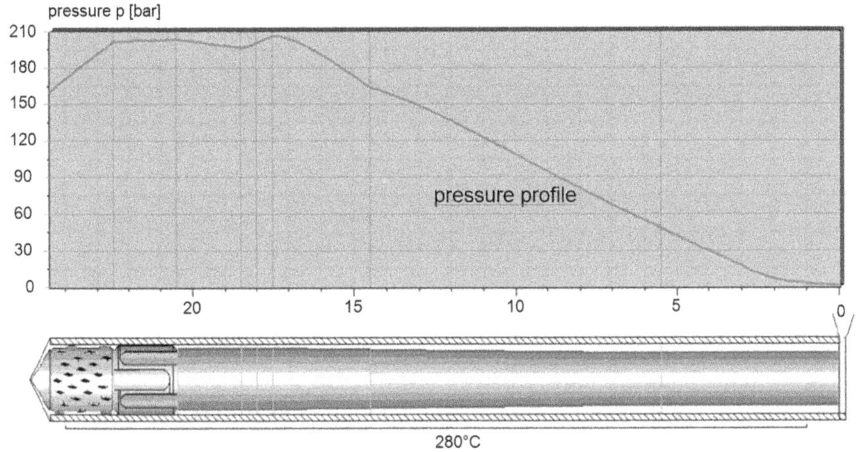

Figure 4.9 Pressure profile.

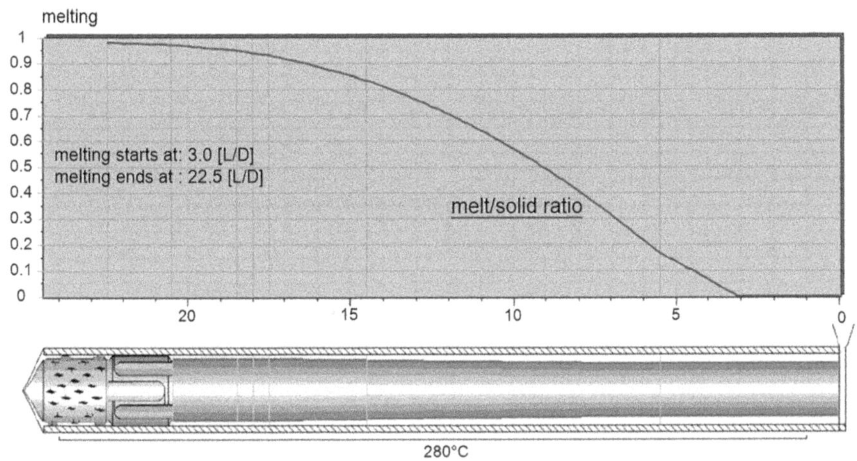

Figure 4.10 Melting profile.

While the functions of each zone are not totally separated from each other, each has a particular primary function. For example, pressure build-up and homogenization start as early as the feed section.

Modeling of the flow processes is based on the geometry of the screw and typically uses computational fluid dynamics. However, simple designs can be modeled analytically. For this purpose, the geometry of the screw channel must be simplified to enable conservation equations to be solved. To achieve this aim, a rectangular section is assumed in the single-screw extruder, which is obtained by developing the screw geometry and neglecting the edge radius of the screw bars between the channels.

The geometry variables of a screw channel are shown in Figure 4.12.

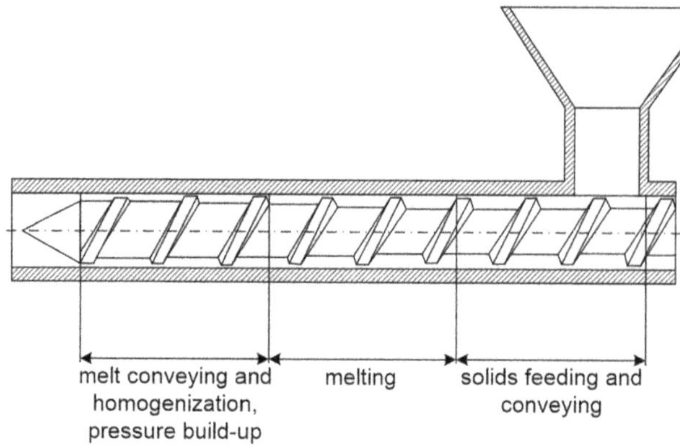

Figure 4.11 The different zones of a conventional three-section screw.

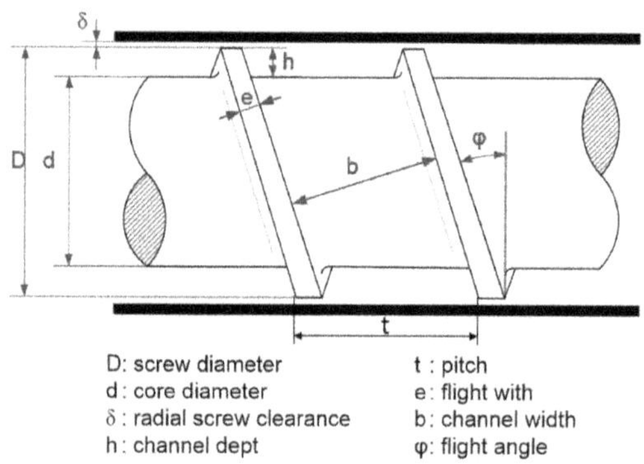

Figure 4.12 Screw channel.

The model uses the following terms shown in eqn (4.2a), (4.2b), (4.3), (4.4), and (4.5).

$$\tan \varphi = \frac{t}{\pi D}, \tag{4.2a}$$

$$\tan \overline{\varphi} = \frac{t}{\pi (D - h)} \tag{4.2b}$$

$$d = D - 2h \tag{4.3}$$

$$b = \frac{t \cos \varphi}{i} - e \tag{4.4}$$

$$Z = \frac{L}{\sin \varphi} \tag{4.5}$$

The helix angle φ is radially variable and is usually adjusted at the barrel wall. The term Z is the unwound screw channel length. For simplification, the screw is imagined to be fixed with the barrel wall around the screw. In addition, the model is assumed to be in a single plane. This assumption is valid because the flight depth is small compared to the screw diameter. This model is called a "channel model." For the components in the z and x coordinate direction, the following is valid, see eqn (4.6).

$$V_0 = \pi ND, \quad V_{0Z} = v_0 \cos \varphi, \quad V_{0X} = V_0 \sin \varphi \tag{4.6}$$

4.5 Cooling

4.5.1 Introduction

Cooling begins when the material exits the extruder and enters the mold and/or cavity. It involves freezing or solidifying the material and subsequent cooling until the end of the production cycle is reached. Because of the low thermal conductivity of plastic materials, cooling is relatively slow. In many processes the cooling time governs the cycle time of the machine. This applies in particular to thick-walled products.

In injection molding, the part is cooled by contact with the mold, which is often water-cooled. Cooling in film extrusion occurs in contact with the die that is also often water-cooled. Thicker extrudates are often pulled through chilled air flow or through a water bath so that convection cooling reduces cycle times.

Accurate calculation of the cooling rate is important for planning the production capacity. This calculation is computationally intensive and typically completed through numerical methods.

4.5.2 Calculation of the Cooling Process

The cooling of plastic materials is calculated using the energy balance equations based on fundamental principles. It states that a time-related change in the internal energy can occur by dissipation, by volume change energy, by heat flow, or by internal heat sources or sinks. Using these principles, the following simplifications can be used:

- Only single unit operations.
- The material properties are temperature independent.

With these simplifications, the energy theorem becomes eqn (4.7).

$$\rho c \frac{\partial T}{\partial t} = \lambda \frac{\partial^2 T}{\partial y^2} Q \tag{4.7}$$

The internal heat source Q in eqn (4.7) can be used to describe the influence of crystallization, comparable to an exothermal chemical reaction, in which heat is generated and the cooling rate is lower. In most cases, however, this is neglected. In this model, T is temperature and t is time.

Eqn (4.8) can be used when dimensionless standardization is considered:

$$\frac{\partial \theta}{\partial \zeta} = \mathrm{Fo}\frac{\partial^2 \theta}{\partial \xi^2} \tag{4.8}$$

where θ is the dimensionless temperature: $\theta = \frac{T - T_U}{T_0 - T_U}$, T_U is the ambient temperature, T_0 is the temperature of the plastic at the beginning of the cooling, Fo is the Fourier number is the $\mathrm{Fo} = \frac{\lambda t_0}{\rho c d^2} = \frac{\lambda L}{\rho c v_0 d^2}$, v_0 is the take-off speed, L is the length of the observed section, d is the product thickness perpendicular to the cooled surface, t_0 is the observed duration, ζ is the dimensionless coordinate $\zeta = \frac{z}{L} = \frac{t}{t_0}$ in time direction, ξ is the dimensionless coordinate in thickness direction $\frac{y}{d}$.

4.5.2.1 *Initial Condition*

To solve the differential equation, an initial temperature distribution $T_0(y)$ or θ_0 (ξ) must be defined. If the cooling is calculated directly after leaving the die, a uniform initial temperature distribution is defined in eqn (4.9) as:

$$\theta_0(\xi) = 1 \quad \text{for} \quad \zeta = 0 \quad \text{and} \quad 0 \le \xi \le 1 \tag{4.9}$$

The screw should be cut to elements for higher precision. Then the temperature of the profile of the previous element (i) of the unit is taken as the temperature profile for the element being determined ($i+1$) where there are I number of elements being considered in the model.

4.5.2.2 *Underlying Conditions*

To calculate the cooling profile, thermal boundary conditions can be defined at surfaces of the extrudate. The simplest boundary condition is defining the surface temperature T_0 (Figure 4.13) as the ambient temperature T_U, as shown in eqn (4.10).

$$T(y = d) = T_0 \quad \text{or} \quad \theta_0(\xi) = 0 \quad \text{for all} \quad \xi \tag{4.10}$$

A boundary condition of a constant heat flux (so-called 2nd kind) can be characterized by a modified Biot number Bi* (Figure 4.14) as defined in eqn (4.11):

$$\lambda \frac{\mathrm{d}T}{\mathrm{d}y}(y = d) = \dot{q}_0 \quad \text{or} \quad \frac{\partial \theta}{\partial \xi}(\xi = 1) = \text{const} = \frac{\dot{q}_0 d}{T_0 - T_U} = \mathrm{Bi}^* \quad \text{for all} \quad \zeta \tag{4.11}$$

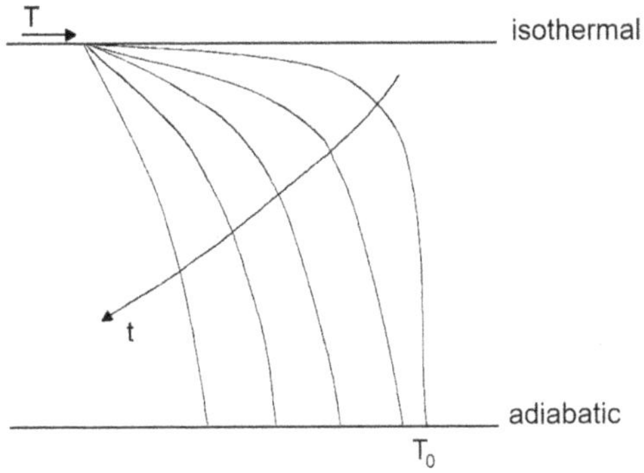

Figure 4.13 Boundary condition of the 1st kind.

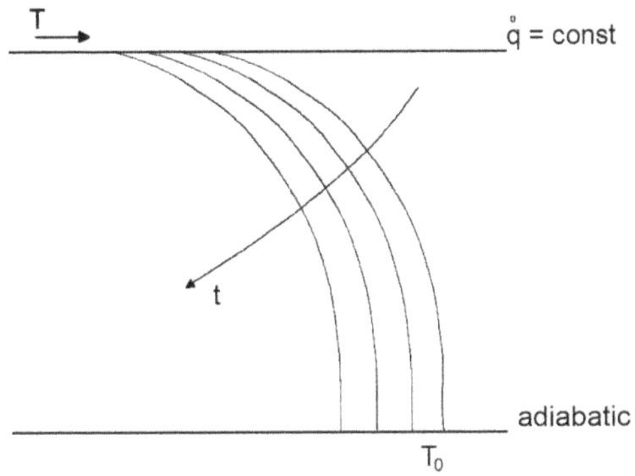

Figure 4.14 Boundary condition of the 2nd kind.

This assumption is useful when using radiation heating of plastics (thermoforming units).

With many cooling processes in extrusion technology, convective heat transfer is achieved with the help of a cooling medium such as air, water, or oil. Then we have a boundary condition of the 3rd kind, see Figure 4.14, in which the heat flow at the interface is proportional to the temperature difference between the surface and the cooling medium, as defined in eqn (4.12):

$$-\lambda \left.\frac{\mathrm{d}T}{\mathrm{d}y}\right|_{y=1} = \alpha(T - T_{\mathrm{U}}) \quad \text{or} \quad -\frac{\partial \theta}{\partial \xi}(\xi = 1) = \mathrm{Bi}\, \theta(\xi = 1) \quad \text{for all} \quad \zeta \tag{4.12}$$

where α is the heat transition coefficient and Bi is the Biot number: $\mathrm{Bi} = \frac{\alpha d}{\lambda}$.

In this model, thermal boundary conditions (loads) are applied only to one extruder surface. In addition, the opposite side is assumed to be an ideal insulator (adiabatic surface) and symmetry is assumed (Figure 4.15), *i.e.*, eqn (4.13).

$$\frac{\partial \theta}{\partial \xi} = 0 \quad \text{for} \quad \xi = 0 \quad \text{and all} \quad \zeta \tag{4.13}$$

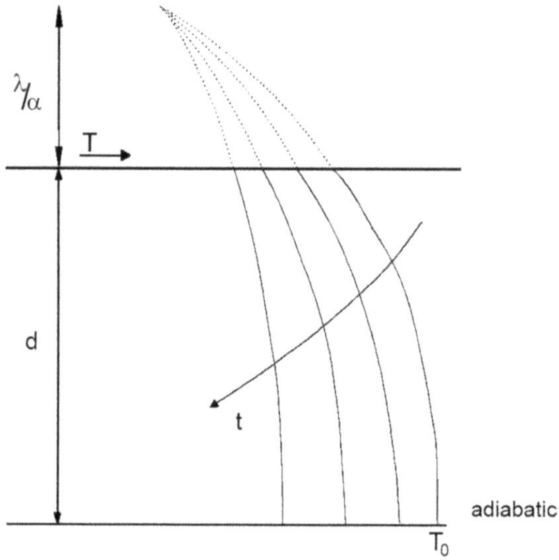

Figure 4.15 Boundary condition of the 3rd kind.

When considering the extrusion process, a distinction can be made between three cases:

1. The assumption of an adiabatic surface is useful for internal surfaces without contact to the tempering agent. This applies to blown films without internal cooling and pipe/hose lines.
2. With the same boundary conditions [see eqn (4.14)–(4.16)] on both surfaces, *e.g.*, two-sided cooling in the air, because of symmetry about the middle of the model, the Biot number and Fourier number must be derived with half of the product thickness (see Figure 4.16).

$$\frac{\partial \theta}{\partial \xi} = (\xi = 0) = -\frac{\partial \theta}{\partial \xi}(\xi = 1) \quad \text{for} \quad 0 \leq \xi \leq 1 \tag{4.14}$$

$$\frac{\partial \theta}{\partial \xi} = 0 \quad \text{for} \quad \xi = 0.5 \tag{4.15}$$

3. The most difficult to treat are cases with different boundary conditions for each side, such as different heat transfer coefficients on each side. In this case, the model must be subdivided into two areas of different thickness, see Figure 4.17.

$$\frac{\partial \theta}{\partial \xi}(\xi = 0) = \mathrm{Bi}_0 \theta \quad \text{and} \quad \frac{\partial \theta}{\partial \xi}(\xi = 1) = \mathrm{Bi}_1 \theta \tag{4.16}$$

$$\frac{\partial \theta}{\partial \xi} = 0 \quad \text{for} \quad \xi = \xi_{\text{limit}} \tag{4.17}$$

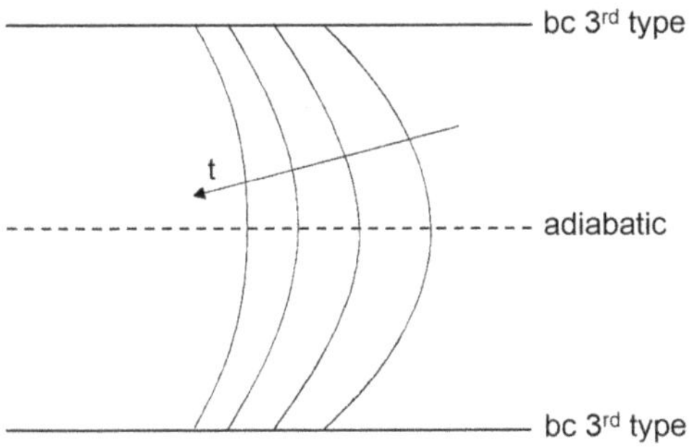

Figure 4.16 Symmetrical profile with two-sided cooling.

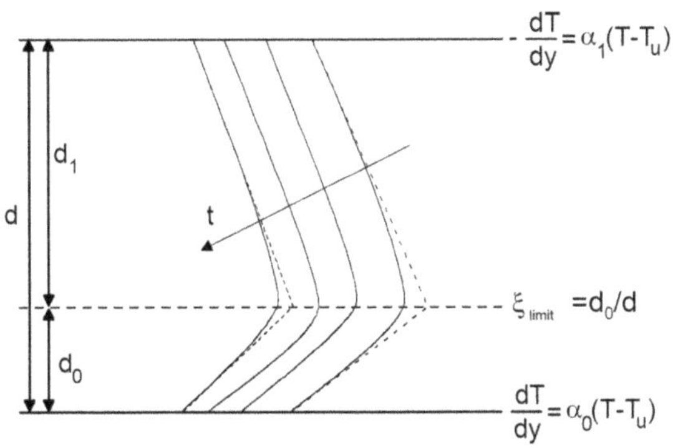

Figure 4.17 Two-sided asymmetrical cooling.

$$\xi_{\text{limit}} = \frac{\text{Bi}_0}{\text{Bi}_0 + \text{Bi}_1} \tag{4.18}$$

It is possible to model this type of cooling through the use of a different Fourier series.

Finding an analytical calculation of the temperature profiles can be difficult. To simplify the model, see eqn (4.19), it is possible to approximate the surface temperature θ_0 to:

$$\theta_0 = e^{\text{Bi}^2 \text{Fo}}\left(1 - \text{erf}\left(\text{Bi}\sqrt{\text{Fo}}\right)\right) \tag{4.19}$$

The temperature difference between the hottest region (core with two-sided cooling or adiabatic wall with one-sided cooling) and the cooled surface amounts to a maximum.

$$\Delta\theta_{\max} \leq \frac{\text{Bi}}{1 + \text{Bi}} \tag{4.20}$$

In eqn (4.19), the so-called error function is used, which is shown in Figure 4.18.

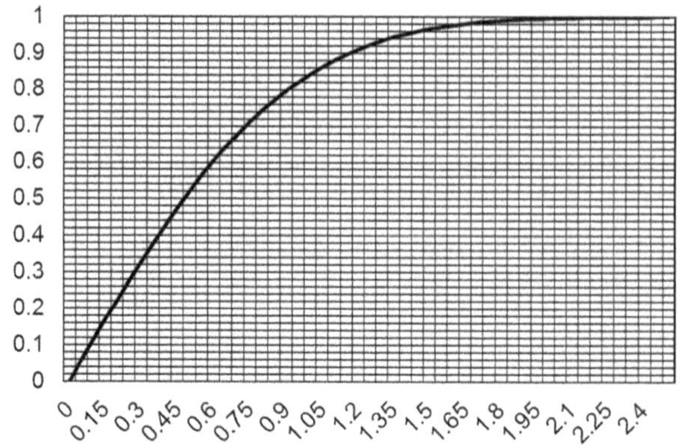

Figure 4.18 Example of error function.

With convective cooling with water or air, the heat transfer coefficient must be calculated. For this, relationships such as the Reynolds number, Prandtl number, or Nusselt number are needed. They can be found in the literature for forced or free convection with parallel or vertical flow.

4.5.3 Cooling Technologies

4.5.3.1 Extrusion

Extrusion processes use three main techniques to cool the products.

1. Film and sheets are cooled by direct contact with water-cooled rolls, also known as chill rolls. When calculating the cooling process, isothermal boundary conditions are assumed, *i.e.*, the surface temperature of the rollers is the same as the temperature of the coolant. While this assumption is not fully accurate, it allows for the models to be simplified and is usually sufficient. Often, the rolls are cooled separately with a single heat exchanger in which the water is conveyed in a closed circuit. Thin films are often cooled on a single large roll, while thick films (sheets) are typically cooled with two rolls for each surface of the film.
2. Profiles, such as pipes, are often passed through a water-filled cooling tank; however, better heat transfer can be achieved by spraying water onto the product through a series of nozzles. Because of partial evaporation, faster cooling rates can be achieved than in a water bath.
3. Blown film and staple fibers are cooled with forced air.

4.5.3.2 Injection Molding

Injection molds for thermoplastics are typically cooled with water. The molds have internal channels in which cooling water flows close to the cavity walls. The water circulates in a closed circuit controlled by heat exchanger(s). The water is typically demineralized to prevent calcification of the cooling channels. The cooling is often segmented, *i.e.*, different areas (zones) of the injection mold halves are cooled independently to allow the temperature of the mold to be varied based on the geometry of the part.

While higher cooling rates can reduce cycle times, an excessively fast cooling rate can reduce crystallization and promote internal stresses. Thus, cooling conditions are often a compromise between part quality and production rate.

4.6 Extrusion Lines

Extrusion lines can achieve relatively high throughput rates (film production with up to 5000 kg h^{-1} and compounding with up to 75 000 kg h^{-1}). The smallest machines are often used in the medical industry, incorporating vitamins or medication into thermoplastic carriers. Some of the largest machines are employed for compounding and extrusion of pipes. However, the common denominator is the product which has a constant cross section and is produced continuously. Typical extrudates include fibers, film, sheets, pipes, profiles, and pellets from round profile extrusion. They can be filled with glass or carbon fibers or foamed. The raw material is fed through the hopper as a pellet mix, and at the end of the extrusion line, the continuous product is typically cut and trimmed into the final product.

Figure 4.19 shows the basic structure of an extruder. A typical pressure and temperature profile for the extrusion of packaging is seen in Figure 4.20.

The hopper is filled with free-flowing plastic material, a mixture of components defined by the formulation. This mix is then conveyed through the extruder and melted (plasticized). The resultant melt is forced through a forming die to produce the geometry. The strand is frozen in the calibrating unit and cooled to ambient temperature in the cooling section. The design of the die and components of the extrusion line are highly dependent on the final product and the material used. The pressure and temperature of the plastic is characterized by the ambient conditions T_a, p_a and the melting temperature of the material.

Figure 4.21 shows a typical curve of the specific enthalpy Δh in a typical extrusion line. The final enthalpy is approximately zero because the bulk material is simply converted into the final product without any chemical reaction.

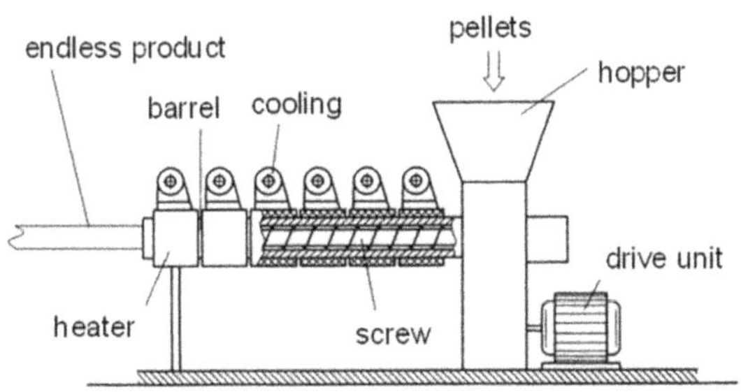

Figure 4.19 Extruder.

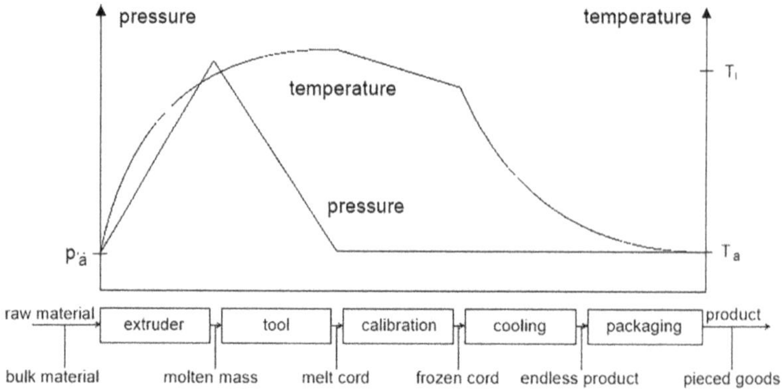

Figure 4.20 Flow diagram of an extrusion line.

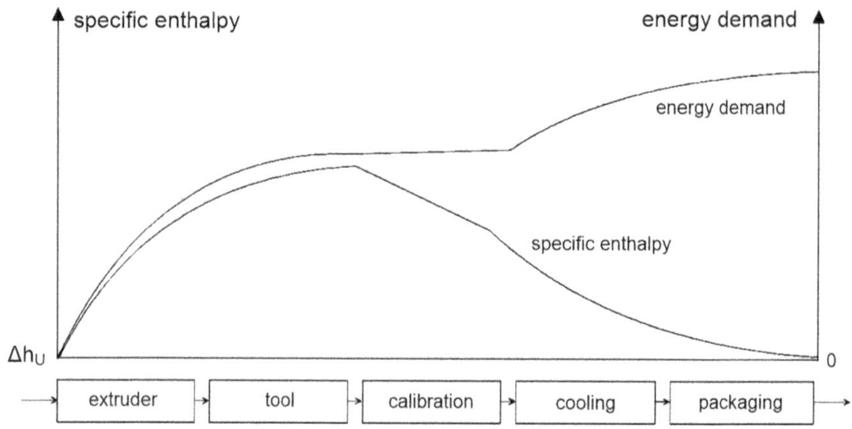

Figure 4.21 Enthalpy curve in an extrusion line.

Following the energy balance, the plastic material is initially melted by introducing energy which is then dissipated through the cooling section after forming. Although it is possible, it is not typical to recycle this energy which could amount to significant energy savings. Without the recycling of energy, an estimated energy density of 300–800 $\mathrm{J\,g^{-1}}$ is required to process most plastics. The energy loss of extrusion lines occurs primarily through the heat radiated to the environment by the heated sections of the machine and the power needed to drive the cooling equipment, take-off unit, and cutting unit.

4.7 Injection Molding

4.7.1 Introduction

Injection molding is one of the most important processes in the plastics industry. Approximately 60% of all plastics processing machines are injection molding machines. Depending on the type of material being processed, injection molding technologies feature several fundamental processing differences.

The reasons for the popularity of injection molding include production rates, part quality, freedom of design, and little waste. For example, the injection molding process allows the production of a wide range of parts varying in weight from just a few milligrams to as much as 30 kg.

Injection-molded parts have become an integral part of present-day society and are now indispensable. The range of products and applications extends from household articles, such as storage containers, toothbrushes, toys, and dustbins, to technological applications, such as switch covers, ventilation flaps, housings, filters, and headlight covers in the automotive industry.

Figure 4.22 shows an injection molding machine and typical injection-molded parts.

4.7.1.1 The Injection Molding Cycle

Figure 4.23 shows the injection molding process schematically. The following section explains the individual process steps, and Figure 4.24 illustrates this with the aid of a pie chart including relative times.

The manufacturing cycle for injection-molded parts begins with the closure of the mold. After the mold is closed, the plasticizing unit is moved to engage the clamping unit. The plasticized melt is injected into the mold through a gating nozzle. The screw advancing over a defined distance allows it to function as a cylinder pump thus injecting the melt. In order to offset the shrinkage of the part caused by cooling of the melt, additional melt is pushed into the mold during a holding pressure phase which compresses the melt in the mold. The part is allowed to cool in the mold while new material is plasticized (melted) for preparation of the next cycle. After a defined cooling time, the mold is opened, the part ejected, and a new production cycle begins.

Figure 4.22 Injection molding machine and everyday injection-molded parts.[2] Reproduced from ref. 2 with permission from Arburg.

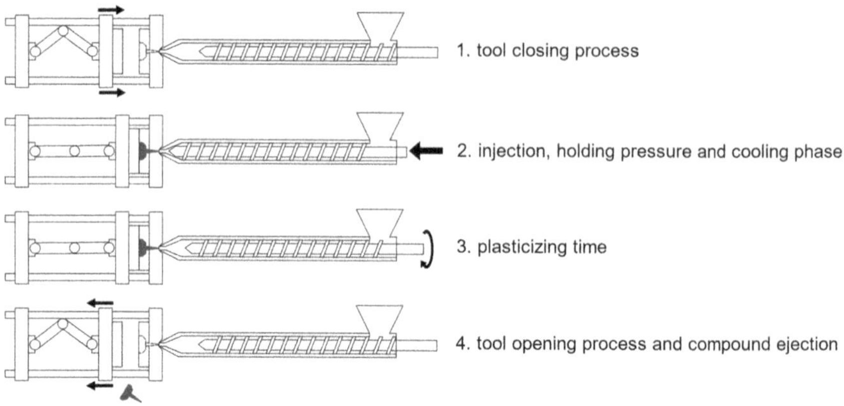

Figure 4.23 Injection molding process.

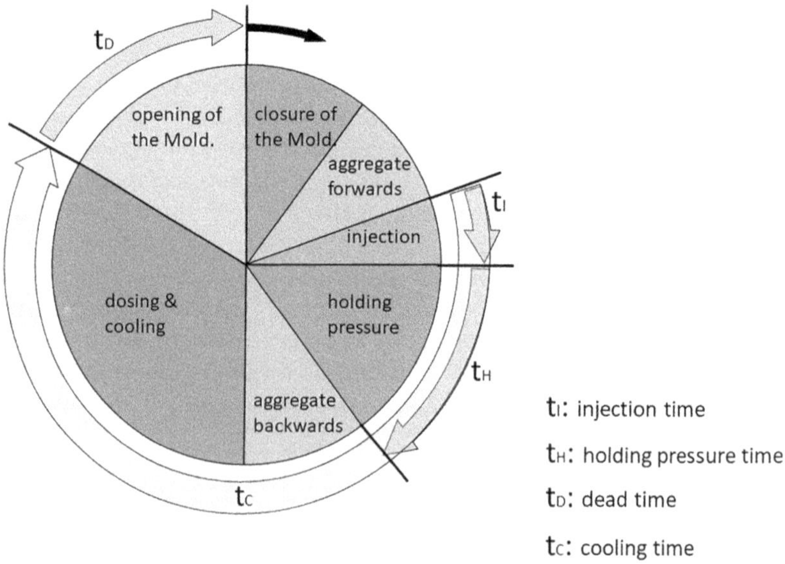

Figure 4.24 The injection molding cycle.

4.7.1.2 Structure of Injection Molding Machines

While many different designs and sizes of injection molding machines are available, the fundamental elements of these machines are the same for all machine designs.

Figure 4.25 shows the basic components of an injection molding machine. It typically consists of a clamping unit with a molding tool, a control unit, and a plasticizing unit. In this standard horizontal configuration design, the clamping and plasticizing units are arranged in the axis on the machine bed.

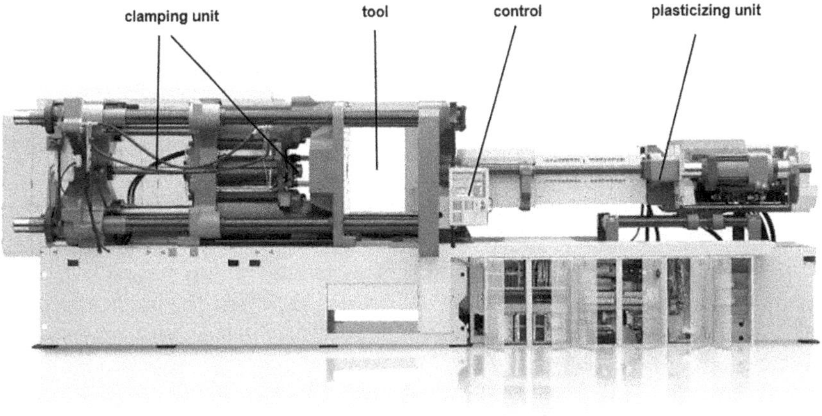

Figure 4.25 Principle structure of an injection molding machine[2]. Reproduced from ref. 2 with permission from Arburg.

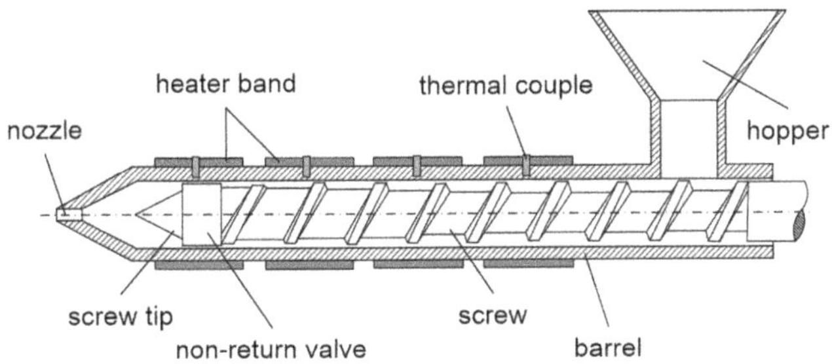

Figure 4.26 Schematic structure of a plasticizing unit.

4.7.1.2.1 The Plasticizing Unit. Figure 4.26 shows a diagram of the fundamental elements of the plasticizing unit.

After the plastic pellets are loaded into the hopper, they are plasticized in the screw. During startup, heating is accomplished with heater bands around the barrel. Once the system is in steady state, most of the heating occurs by shear heating facilitated by the rotating the screw creating friction energy. A homogeneous melt is formed in the screw which is conveyed in the direction of the nozzle through rotation of the screw. The space between the nozzle and the screw tip, the "shot," gradually fills with melt and pushes the screw back. Once the shot is filled with sufficient melt, it is injected into the mold cavity by an axial motion of the screw. To compensate for plastic shrinkage during cooling, holding pressure is maintained by the plasticizing unit. Thus, the tasks of the plasticizing unit are:

- To receive and melt the plastic pellets.
- To convey and homogenize the melt.
- To inject the melt.
- To keep the melt under pressure (apply holding pressure).

Depending on the type and quantity of material to be processed, different screw designs are available. The basic structure of a screw is shown in Figure 4.27 for the example of a three-section screw. This type of screw is very common because of its suitability for a wide range of processable materials.

A non-return valve (check valve) is placed at the nozzle end of the screw to prevent the melt from flowing back from the shot into the area of the screw thread during the injection and holding pressure phase. The structure of a check valve is shown in Figure 4.28.

The check valve shown in Figure 4.28 consists of two rings, the lantern ring and the pressure ring. During metering, *i.e.*, when the melt is conveyed in the direction of the screw tip, the melt pushes the lantern ring forward so that the melt can go through and flow into the screw shot. However, during the injection and holding pressure phases, because of its axial movement, the lantern ring is pushed by the melt in the shot-chamber against the

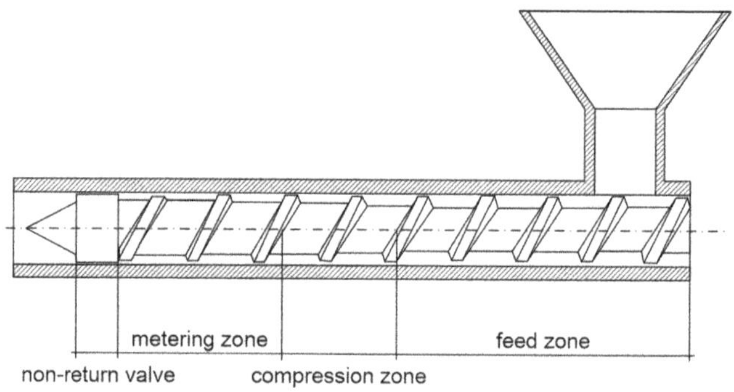

Figure 4.27 Structure of a three-section screw.

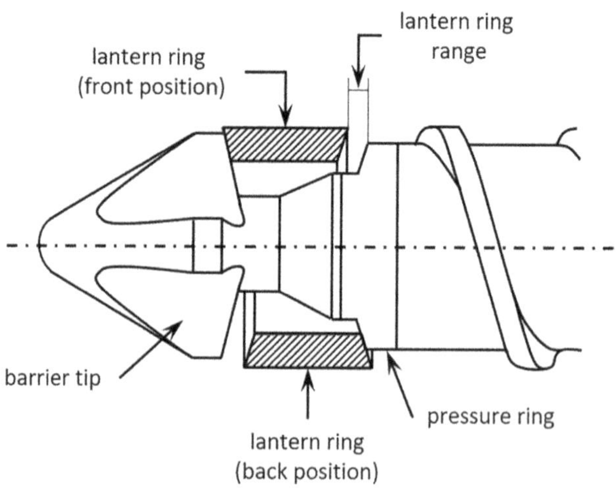

Figure 4.28 Diagram of a check valve (schematic).

lantern ring so that the gap (stroke of the lantern ring) is closed, preventing the melt from flowing back into the screw.

4.7.1.2.2 The Clamping Unit. The clamping unit holds the mold and has the following tasks:

- Opening of the mold.
- Keeping the mold closed during the injection and packing stages.
- Ejecting the molded parts.

There are two clamping systems:

- Mechanical clamping systems or toggle mechanisms which enable positive locking through the toggle.
- Hydraulic clamping systems which use non-positive frictional locking by hydraulic cylinders.

4.7.1.2.3 Mechanical/Toggle Systems. Figure 4.29 shows the design of a toggle system with a single toggle and fixed drive cylinder. Here, the clamping force is applied by hinging of the toggle. An advantage of this mechanism is that its specific force and velocity can be controlled. This allows not only a combination of various speed and force combinations but also for the mold platen to travel at high speed during the initial closing and slowing before the mold is fully closed to prevent mold damage. A disadvantage with this clamping system is its relatively high manufacturing cost.

4.7.1.2.4 Hydraulic Clamping Systems. Figure 4.30 shows the design of a hydraulic clamping unit. Here, a hydraulic cylinder is used to apply the clamping force, thus the travel speed is relatively low compared to a mechanical toggle system. To reduce the opening and closing times, it is possible to add a second smaller cylinder in addition to the large clamping

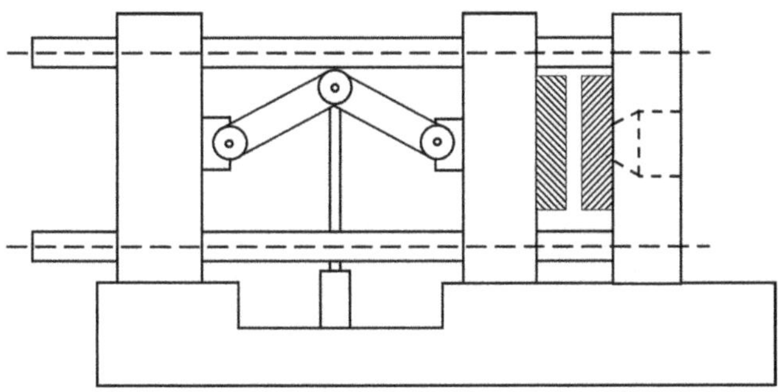

Figure 4.29 Diagram of a toggle clamping unit.

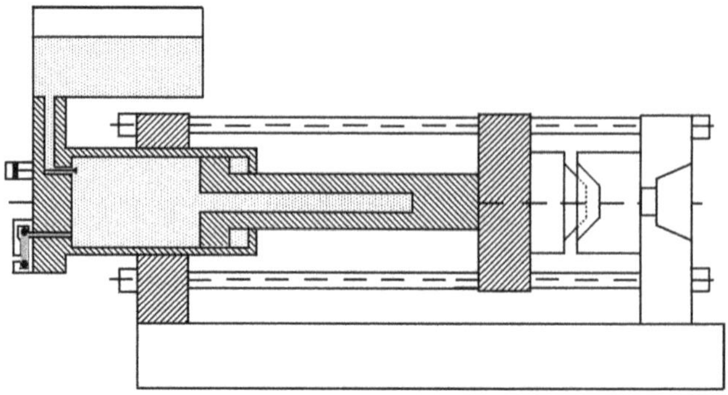

Figure 4.30 Diagram of a hydraulic clamping unit.

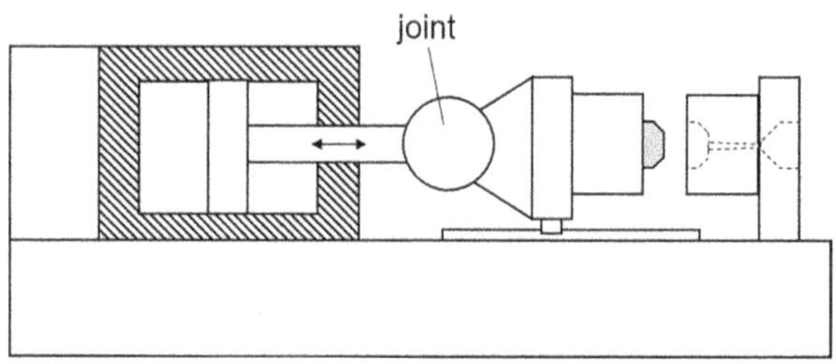

Figure 4.31 Diagram of a clamping unit without tie bars.

cylinder. Hydraulic clamping systems can be closed with less pressure compared to mechanical toggle systems.

In addition to the mechanical and hydraulic clamping systems described above, it is also possible to use mechanical–hydraulic mechanisms as a combination of both systems in which the travel distances are controlled mechanically and the clamping is controlled frictionally or non-positively by hydraulic cylinders.

Clamping units without tie bars have a C-shaped frame where the movable clamping platen is guided on a slide rail. The bending of the C-frame is compensated by means of a joint in order to ensure platen parallelism of the mold halves which requires a high degree of machine bed rigidity. The advantage of this design is the free accessibility of the clamping unit. Figure 4.31 shows the design of a clamping unit without tie bars.

4.7.1.2.5 The Injection Molding Tool. The injection molding tool (the mold) is secured to the clamping unit and determines the shape of the final part. In addition, the mold influences not only the quality of the molded part, such as warpage and surface quality, but also the economics of the process.

An injection molding tool has the following tasks:

- Receiving the melt.
- Distributing the melt within the cavity(ies).

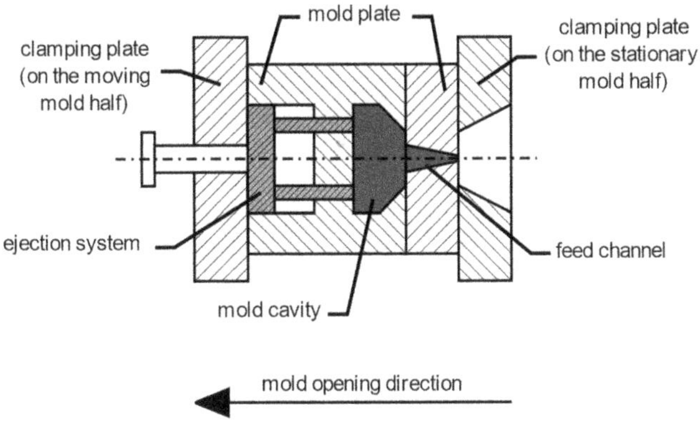

Figure 4.32 Diagram of an injection molding tool.

- Cooling of the part.
- Ejecting the molded part.

The design of a typical injection molding tool is shown in Figure 4.32. The basic elements of an injection mold include:

- Mold plates (platens on the moving and stationary mold halves, mold plates with recessed cavity and feed channel (runner)).
- Demolding system with ejectors.
- Cavity.
- Gating system with runner.

4.7.2 Design of Injection Molding Machines

A wide range of injection molding machines of various designs, levels of control, and sizes is available on the market. To establish the most suitable injection molding machine, three main factors have to be considered: plasticizing unit size (shot size), daylight, and maximum clamping capacity. The shot size is based on the screw diameter and the maximum travel distance of the screw. The daylight is the distance between the tie bars that support the clamping system. It limits the maximum size of mold that can be loaded onto the machine. The clamping capabilities or the locking or clamping force are important to prevent flashing, the flow of plastic out of the cavity between the mold interface, when high pressures are used.

4.7.2.1 Process Variables

The following section summarizes the most important process design considerations and process parameters of injection molding.

4.7.2.2 Number of Cavities

When designing an injection molding process, the number of cavities can vary considerably. For very large, molded parts, the machine capacity can limit the design to a single cavity; however, for smaller parts it is possible to have multiple-cavity molds in order to increase productivity. For example, the production molds for beverage bottles typically have 96 cavities.

4.7.2.3 Clamping Force, Mold Opening Force, and Locking Force

Typical machine clamping forces range from five to +2000 ton. The clamping force of a machine indicates the force that acts on the machine's tie bars after closing the mold. After injecting the melt into the cavity, an internal pressure is created through the injected melt. As a result, a force, the so-called mold opening force, acts against the clamping force. The mold opening force is the product of mold cavity pressure and projected area in the cavity, see eqn (4.21):

$$F_{\text{open}} = p_{\text{i}} \times A_{\text{proj}} \tag{4.21}$$

where F_{open}: mold opening force, p_{i}: mold cavity pressure, A_{proj}: projected area at the interface.

This relationship is shown in Figure 4.33.

The mold opening force acts during the injection process against the clamping force of the mold.

4.7.2.4 Shot Volume and Metering Stroke (Metering Distance)

The required shot volume is the combined volume of a gating system and cavities plus a buffer or "cushion." Because this should correspond approximately to the volume that is fed into the screw shot size, the shot

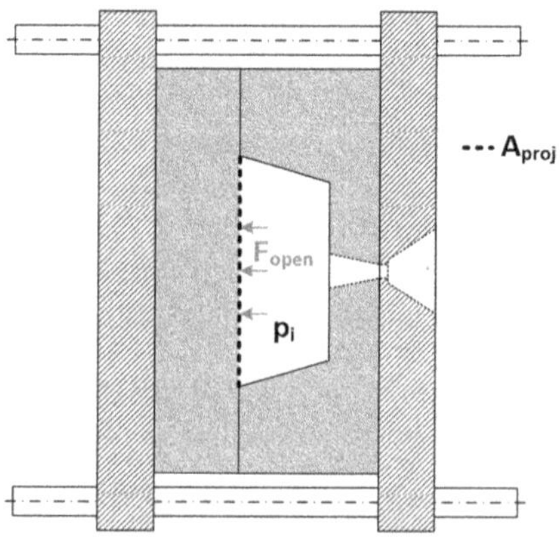

Figure 4.33 Schematic representation of the mold opening force.

volume can also be described as the metering volume and can be approximately calculated as a function of the screw diameter D and the metering stroke s_D as follows in eqn (4.22):

$$V_S = \frac{\pi D^2}{4} s_D \tag{4.22}$$

where V_S: shot volume, s_D: metering length (or shot length), D: screw diameter.

The metering distance s_D is the total stroke of the screw during metering. This should be approximately in the range of one to three screw diameters, see eqn (4.23):

$$1D \leq s_D \leq 3D \tag{4.23}$$

If the metering stroke is too small or too large, surface defects such as streaks, sink marks, blistering, or thermal defects (*e.g.*, inhomogeneity of the melt) can occur. Figure 4.34 shows characteristic elements of the plasticizing unit for calculating the shot volume.

4.7.2.5 Backpressure

Backpressure is the pressure of the plasticized melt in the screw shot-chamber that must be applied during the metering phase to push the screw plunger back in the direction of the hopper to close the check valve.

4.7.2.6 Screw Speed

The screw speed is the number of screw rotations per unit of time, usually per minute. Generally, the rate of plasticizing (melting) can reduce the cycle time; however, it is important to note that the cooling phase is often the limiting factor for cycle times. In addition, an excessive screw rate can promote excessive temperatures and degradation as well as increase the dwell time, the time when melt is stored in the shot between cycle times, which may also promote degradation.

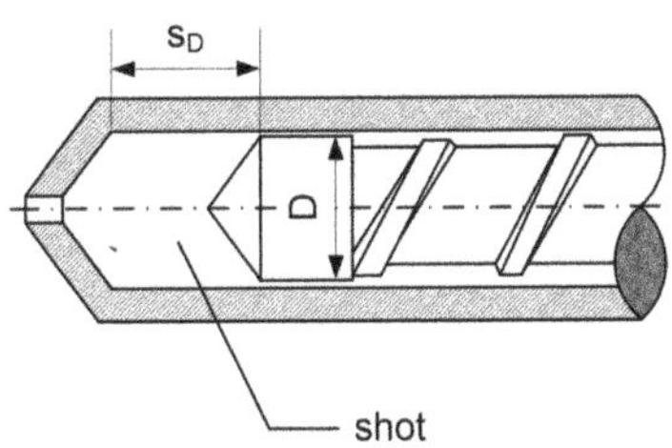

Figure 4.34 Characteristic data for the plasticizing unit.

4.7.2.7 Switchover Point

In order to assure complete filling of the cavities and good part quality with a minimal amount of shrinkage and warpage, a key parameter is the switchover point. This is defined as the time when the pressure is dropped to prevent flashing, plastic flowing out of the cavity, or even a potential opening of the mold. In more detail, because the viscosity of plastic is so high, the injection pressure must also be high to promote flow of the plastic into the mold. However, once the mold is full, this excessively high pressure can result in a force greater than the clamping force, causing the mold to open, sometimes called "blowing the mold." Thus, the pressure is reduced to a lower value, yet still high enough to promote flow so that additional plastic can be "pushed" into the cavity to reduce sink marks and warpage as the plastic cools. This lower pressure portion of the cycle is often called the packing phase. If the switchover is made too early, the cavity may not be fully filled, and a short shot can result (incomplete part). If the switchover point is too late, excessive pressure in the mold cavity can result in opening of the mold and the creation of flash.

4.7.2.8 Cooling Time

The cycle time for an injection-molded part is typically governed by the cooling time. To reduce the cycle time, the cooling time can be reduced by water cooling of the mold. However, excessive cooling can result in warpage of the finished parts.

The cooling time (t_c) depends on the largest wall thickness of the part and behaves proportionally to the square of the thickness (d). To estimate the cooling time, eqn (4.24) can be used.

$$t_c = 50 \text{ s} \quad \text{for} \quad d = 3 \text{ mm} \tag{4.24}$$

where $t_c \sim d^2$.

4.7.2.9 Temperatures

In injection molding, various temperatures can be set on the machine that will influence the process, the cycle time, and part quality. The primary processing temperatures are detailed here:

- Melt temperature: the melt temperature is the temperature of the melt in the plasticizing unit. It is important to note that it is not the temperature at which the plastic melts for semi-crystalline materials but is near to the temperature set point of the barrel temperature control.
- Barrel temperature: the temperature of the barrel in the plasticizing unit is regulated by heater bands and with the aid of thermocouples (see Figure 4.26). The heater bands are primarily used during startup of the machine. Once the machine is at steady-state production, most of

the heat is obtained through rotation of the screw. If needed, air cooling is often employed to prevent overheating.

- Mold wall temperature: the mold wall temperature is the temperature on the surface of the cavities. Raising the mold wall temperature leads to reduced internal stresses and reduced warpage of the part but is also accompanied by an increase in cooling time of approximately 2% $°C^{-1}$.

Abbreviation

IR Infrared

References

1. T. Zehev and C. G. Gogos, *Principles of polymer processing*, John Wiley & Sons, 2006.
2. www.arburg.de, 2022.

CHAPTER 5

Biocomposites

ALPER KIZITAS,*[a] CHAD ULVEN,[b] MITCH MICHEL[c] AND D. GREWELL[d]

[a] Ford Motor Company, Dearborn, Michigan, USA; [b] Department of Mechanical Engineering, North Dakota State University, Fargo, ND 58108, USA; [c] Center for Crops Utilization, Iowa State University, Ames, IA, USA; [d] College of Engineering and Engineering Technology, Northern Illinois University, 590 Garden Rd, Dekalb, IL 60115, USA
*Email: kiziltasalper@gmail.com

5.1 Fiber-based Composites

Fibers are filaments used as reinforcing materials; they can be circular, rectangular, tubular, and hexagonal in cross section. Fiber reinforcement in composites provides great opportunities for the overall enhancement of the composite material's properties as the fibers are often inherently stronger compared to the bulk polymer matrix. The effects of fiber diameter can also contribute to the strength of the composite. As fiber diameter is reduced, the fibers become more flexible and can readily align along the principal axis of flow during composite formation. This alignment in the composite can also help mitigate defects within the composite. Fiber composites are classified into two groups, continuous and discontinuous fiber composites. Continuous fibers have long aspect ratios (length/diameter) and provide continuous reinforcement to enhance mechanical properties. Discontinuous fibers are short fibers with aspect ratios of 100 or less. In both cases the fibers can have either a predetermined or random orientation.[1,2]

Green Chemistry Series No. 79
Bioplastics and Biocomposites: A Practical Introduction
Edited by David Grewell

Published by the Royal Society of Chemistry, www.rsc.org

5.2 Particle-based Composites

Particles are another common filler for composite materials. Particles may vary in size from the microscopic to the macroscopic range (1–20 000 nm). The distribution of particles throughout a matrix polymer is random, and therefore, the composite will tend to have isotropic properties. Dispersed particles within matrices at 15% loading do not distribute stress in the composite because of their short length/size.[1,2] However, they tend to promote strength and hardening of the matrix polymer by restricting polymer chain mobility and thereby strengthening the polymer matrix. As particle sizes increase and loading increases above 25%, the properties can change within the composite. Applied stresses on the composite structure are shared because of the relatively high strength of the particles and proper bonding between the polymer matrix and the filler.[1,2]

5.3 Rule of Mixtures for Density

Composite materials' properties are functions of the base materials. Properties of these composites can be calculated using the rule of mixtures. The rule of mixtures is a calculated weighted average of the base materials. Density is one example of a property calculated by the rule of mixtures. The mass of a composite is the sum of the masses of the matrix polymer and the reinforcing filler, as seen in eqn (5.1).[1,2]

$$m_c = m_m + m_f \tag{5.1}$$

where the mass of the composite (m_c) is equal to the sum of the mass of the polymer matrix (m_m) and the fiber mass (m_f). The volume of composites can also be calculated as the sum of base materials, as seen in eqn (5.2).[1,2]

$$V_c = V_m + V_f + V_v \tag{5.2}$$

where the volume of the composite (V_c) is equal to the sum of the volume of matrix polymer (V_m), the volume of the fiber (V_f), and the volume of voids (V_v). Assuming no voids, the density of composites can also be calculated as the mass divided by volume, as seen in eqn (5.3).[1,2]

$$\rho_c = \frac{m_c}{V_c} = \frac{m_m + m_f}{V_c} \tag{5.3}$$

where the density of the composites (ρ_c) is equal to the mass of the composite divided by the volume of the composite. The masses of the matrix and filler are calculated as their densities multiplied by their volumes, as seen in eqn (5.4) and (5.5).[1,2]

$$m_m = \rho_m V_m \tag{5.4}$$

$$m_f = \rho_f V_f \tag{5.5}$$

Substituting eqn (5.4) and (5.5) into eqn (5.3) gives us the resulting equation:[1,2]

$$\rho_c = f_m \rho_m + f_f \rho_f \tag{5.6}$$

where f_m and f_f are the volume fractions of the polymer matrix and filler respectively.

5.4 Theory of Composites

As previously stated, composites are used to produce materials with higher mechanical properties compared to traditional materials. While ultimate strength is often the primary property that is considered when developing composites, estimating the ultimate strength can be challenging because it is highly dependent on the interfacial strength of the filler and the matrix. However, with relatively long or continuous unidirectional fibers, it is possible to estimate the stiffness of a composite in both the longitudinal and transverse directions.

5.4.1 Stiffness of Composites in the Longitudinal Direction

In the longitudinal direction, the applied force is parallel to the direction of the fibers in a unidirectional continuous fiber composite, as seen in Figure 5.1.

It is possible to remove an imaginary unit cell from the composite that has the same volume fraction of matrix (V_m) and fiber volume fraction (V_f) as the composite, as seen in Figure 5.2.[1,2]

If it is assumed that the element is sufficiently long, the strain (ε) in the fiber (ε_f), matrix (ε_m), and composite (ε_c) are equal, as seen in eqn (5.7).[1,2]

$$\varepsilon_f = \varepsilon_m = \varepsilon_c = \varepsilon \tag{5.7}$$

Under this equilibrium condition, it is possible to derive a relationship for the composite force (F_c) based on the forces in the matrix (F_m) and fiber (F_f), as shown in eqn (5.8).[1,2]

$$F_c = F_m + F_f = A_m \sigma_m + A_f \sigma_f \tag{5.8}$$

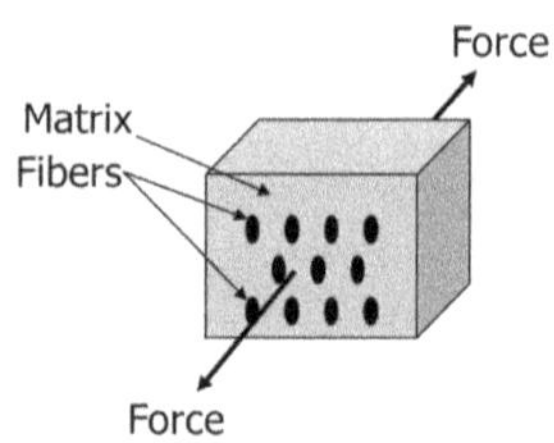

Figure 5.1 Schematic of longitudinal loading of a unidirectional continuous fiber composite.

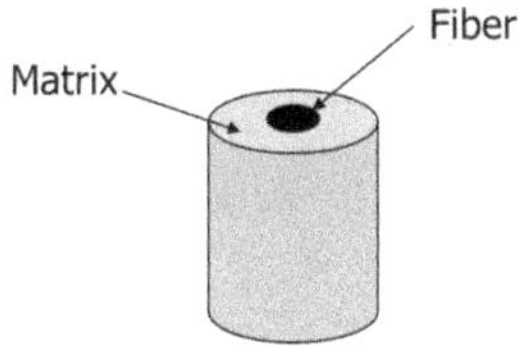

Figure 5.2 Schematic of volume removed from the composite in longitudinal loading.

where A_m and A_f are the cross-sectional areas of the matrix and fiber, respectively, which are directly proportional to the volume fraction of each component as seen in eqn (5.9).[1,2]

$$F_c = A_m\sigma_m + A_f\sigma_f = A_mE_m\varepsilon + A_fE_f\varepsilon = \varepsilon(A_mE_m + A_fE_f) \tag{5.9}$$

In eqn (5.9), E is the modulus of each component. By dividing eqn (5.8) by the total area of the element (A_c) it is possible to derive a relationship for the longitudinal modulus for the composite (E_c) based on the filler level and known modulus of the matrix, as seen in eqn (5.10).[1,2]

$$\begin{aligned} \frac{F_c}{A_c} &= \varepsilon\left(\frac{A_m}{A_c}E_m + \frac{A_f}{A_c}E_f\right) \\ \sigma_c &= \varepsilon(V_mE_m + V_fE_f) \\ \frac{\sigma_c}{\varepsilon} &= \frac{\varepsilon}{\varepsilon}(V_mE_m + V_fE_f) = E_c = V_mE_m + V_fE_f \end{aligned} \tag{5.10}$$

This is another example of the rule of mixture.

5.4.2 Stiffness of Composites in the Transverse Direction

In the transverse loading condition, the loads are perpendicular to the fiber of a unidirectional continuous fiber composite, as seen in Figure 5.3.

Again, it is possible to remove an imaginary unit cell from the composite that has the same volume fraction of matrix (V_m) and fiber volume fraction of fiber (V_f) as the composite, as seen in Figure 5.4.

In this case, the stress for the matrix (σ_m), fiber (σ_f), and composite (σ_c) are all equal. In addition, the summation strain of the fiber (ε_f) and the matrix (ε_m) equal the strain of the composite (ε_c), as seen in eqn (5.11).[1,2]

$$\varepsilon_c = \varepsilon_f + \varepsilon_m \tag{5.11}$$

It is possible to use a transfer function that correlates the stiffness of each component to a spring with stiffness (K), as seen in Figure 5.5, where K_1 and K_2 are the spring stiffness coefficients of each spring. In this model, the

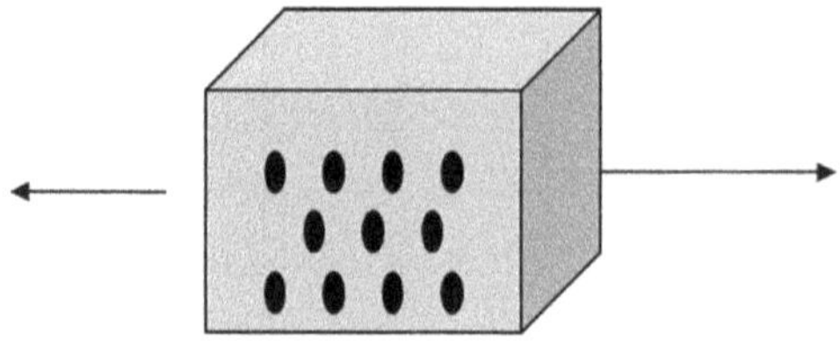

Figure 5.3 Schematic of a unidirectional continuous fiber composite in a transverse loading condition.

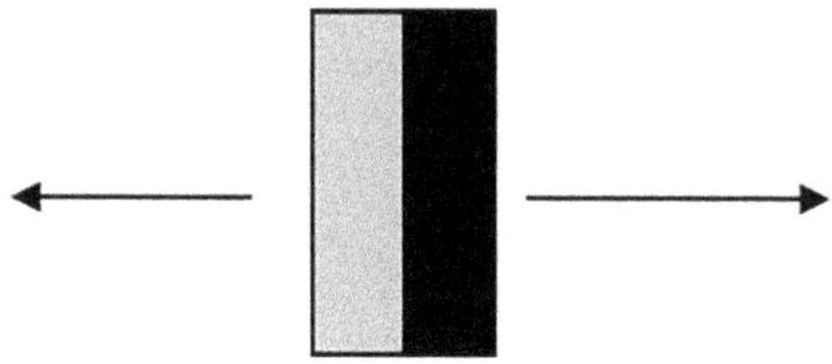

Figure 5.4 Schematic of volume removed from the composite in a transverse loading condition.

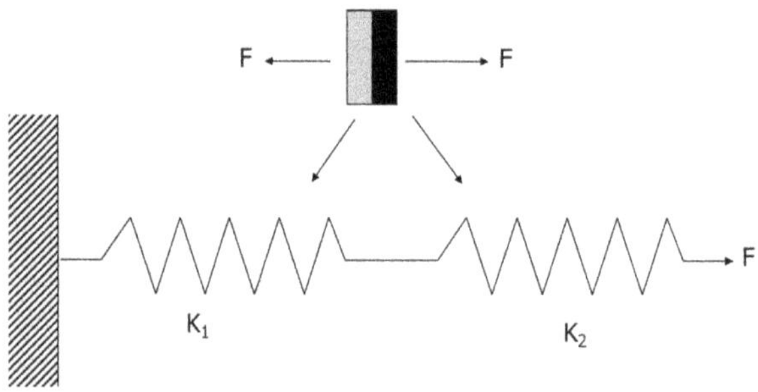

Figure 5.5 Schematic of spring representation of an element from a transverse loaded composite.

stiffness of two springs in series (K), as seen in Figure 5.5, is defined in eqn (5.12).[1,2]

$$\frac{1}{K} = \frac{1}{K_1} + \frac{1}{K_2} \tag{5.12}$$

Knowing the stiffness of each component, K can be modeled as seen in eqn (5.13).[1,2]

$$K = \frac{E_1 A}{L_1} \tag{5.13}$$

It is then possible to determine the overall stiffness of the composites. In this case, the volumetric fraction is directly proportional to the length (L) of the model (width in Figure 5.9). Substituting eqn (5.13) in eqn (5.12), it is possible to derive eqn (5.14).[1,2]

$$\frac{L_c}{E_c A} = \frac{L_f}{E_f A} + \frac{L_m}{E_m A} \tag{5.14}$$

Multiplied by the area (A), eqn (5.14) can be simplified as seen in eqn (5.15).[1,2]

$$\frac{L_c}{E_c} = \frac{L_f}{E_f} + \frac{L_m}{E_m} \tag{5.15}$$

By dividing eqn (5.15) by the entire length of the element (L_c) and assuming the lengths are equal to the volume fraction, it is possible to define the transverse modules of the composite, as seen in eqn (5.16).[1,2]

$$\frac{1}{E_c} = \frac{V_f}{E_f} + \frac{V_m}{E_m} \tag{5.16}$$

5.5 Biocomposites

Biocomposites, as the name implies, are comprised of either bio-based or naturally produced (non-synthetic) materials forming the polymer matrix, the filler materials, and in some cases both. Matrix polymers may be petroleum-derived, such as polypropylene (PP) and polyethylene (PE), or biodegradable, such as polylactic acid (PLA) and polyhydroxyalkanoate (PHA). Filler materials may be synthetic, such as E-glass, or natural fibers or particles, such as agave fibers and lignin. A composite having at least one portion that is bio-based is a biocomposite.

The highest performance composites made from natural fibers to date are mainly based on fibers extracted from bast plants (*i.e.*, flax, hemp, kenaf, jute, *etc.*). Bast fibers perform well as a reinforcement because of their high crystalline cellulose content and relatively low hemicellulose and lignin content as compared to other natural fiber sources (*i.e.*, leaf, seed, grass, wood, *etc.*).[3] However, the organization of the crystalline cellulose, lignin, and hemicellulose in bast fibers is complex, as illustrated in Figure 5.6 below.[4]

The strength and stiffness of bast fibers is mainly controlled by the arrangement of the crystalline cellulose in the secondary wall of this hierarchal structure (Figure 5.6). The angle at which the cellulose grows in relation to the longitudinal axis of the fiber is known as the microfibril angle. For bast fibers, this microfibril angle typically ranges from 5–15° based on numerous factors which are discussed later. These hierarchal structured fibers are sometimes viewed and treated as naturally occurring composites themselves. The lower the microfibril angle, the stronger and stiffer the fiber because when pulled, the crystalline cellulose can directly absorb more of the load. With a greater microfibril angle, the crystalline cellulose needs to

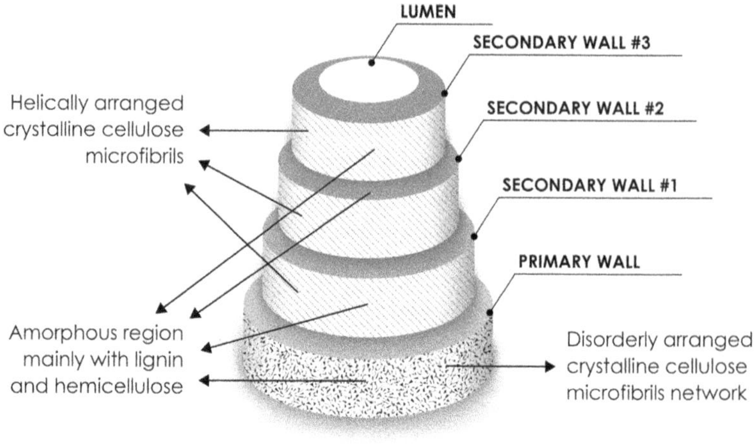

Figure 5.6 Hierarchal bast fiber structure.[4]

first twist and further align in the fiber before absorbing a greater load and therefore appear to have a lower modulus and/or strength.

Growing parameters which influence the quality of bast fiber and their microfibril angle include plant variety, soil type, weather conditions during the growing season, and age of the plant when harvested. Harvesting and processing parameters which influence the quality of the bast fiber include harvesting methods, retting methods, collection methods, and decortication methods employed.[5] Many of these parameters are true for any natural fiber used in biocomposites. Because of the vast number of variables which need to be controlled in order to produce consistent natural fibers for composites, a lot of variability exists in the performance of natural fibers from field to field, region to region, and year to year.[5]

Variation in the performance of natural fibers has led to many challenges for designers of engineered biocomposites. Several researchers over the past couple of decades have attempted to address this deficiency in multiple ways. From a harvesting and processing standpoint, employing methods which handle fibers in a gentler way while refining them for composite use and using controlled retting methods with enzymes rather than traditional water or dew retting have been evaluated.[6] From a fiber quality standpoint, non-destructive methods have been developed to grade fibers after they have been processed as a means to determine which type of composite they would be most appropriate for.[7] Some of these methods include near-infrared spectroscopy and constituent analysis. Finally, from an engineering standpoint, statistical methods have been used to establish different levels of confidence in certain properties such as strength and modulus of natural fibers. The primary approach employed has been Weibull statistics, which is similar to how engineers have used this approach to address the variation of performance seen in ceramics.[8]

Once a natural fiber has been properly processed, the type of architecture for which the composite will be utilized needs to be determined. Natural fibers can be chopped into short fibers, randomly oriented into mat form, or stitched/woven together from spools into textiles as reinforcement for composites. Short fibers can be challenging to handle and consistently feed or distribute evenly when manufacturing short fiber biocomposites. Randomly oriented natural fiber mats can be produced through either wet lay or air lay processes and are relatively easy to handle during composite manufacturing. For most structural composites, a stitched or woven form of engineered fibers is most commonly used; however, with natural fibers this presents some challenges.[9,10]

In order to form an engineered natural fiber textile designed for composite construction, normally natural fibers need to be twisted into yarns so that they frictionally lock together or can be pulled through textile manufacturing processes. The twisting of natural fibers to produce yarns, however, makes them challenging to impregnate with liquid resin during composite processing.[11] A few studies have explored the optimal ratio of natural fiber twist for proper pulling through textile processes *versus* the ability to fully wet with resin during composite processing. The other method of creating textiles from natural fibers is through the use of synthetic carrier fibers (polyester, nylon, *etc.*) as reinforcement for the natural fiber tows while being pulled through textile production processes. Using an additional carrier fiber, however, leads to challenges in polymeric resin bonding to both types of fiber surfaces when manufacturing composites.

The processing of biocomposites using any one of the three fiber architectural types described above as compared to conventional composites has also provided some challenges to increased biocomposite utilization. Trying to properly adhere or bond relatively hydrophilic natural fibers to relatively hydrophobic polymeric resins, achieving high fiber volume fractions with lofty natural fibers, and limiting thermal decomposition of natural fibers while processing composites are some of the major challenges.[3]

Several approaches have been employed to improve the bonding of natural fibers to a variety of polymeric resins in the manufacturing of composites. These range from fiber surface treatments to fiber coatings (*e.g.*, sizing) to modifications of the resins themselves.[12–18] Fiber surface treatments studied to date include acid or base washes to remove certain contaminants, waxes, pectins, *etc.*, typically observed in natural fibers and better expose the hydroxyl groups for improved functionality of the fiber surface. The types of chemical coatings or sizing employed have been similar to those used for conventional composite reinforcement (*i.e.*, silanes, maleic anhydrides, *etc.*). Finally, the blending of additives or polymer types (*e.g.*, acrylic resin in vinyl ester) have shown some promise in improving the bonding of natural fibers to polymeric resins in composite manufacturing.

Most conventional high-performance composite materials utilize fiber volume fractions in the order of 50–70%. Because of the high amount of loft and openness of the natural fiber structure, it is difficult to achieve volume fractions as high without significant modifications to the composite manufacturing process utilized. The use of higher pressures, formation of pre-impregnated layers, and use of yarn-less textiles are a few methods that have been employed to increase fiber volume fraction in biocomposites during manufacturing.[9]

Thermal decomposition in natural fibers initiates around 180 °C as a result of the low degradation temperature of hemicellulose contained within natural fiber structures, followed by significant cellulose and lignin decomposition when temperatures exceed 250 °C as observed in the TGA, see Figure 5.7.[19] Therefore, the selection of an appropriate polymeric resin is critical in trying to form biocomposites. Unfortunately, most engineered thermoplastics such as acrylonitrile butadiene styrene, polycarbonates, and polyamides (nylons) cannot be used because their processing temperatures are higher than the natural fiber decomposition temperatures. However, commodity thermoplastics such as PP, PE, polystyrene, and polyvinyl chloride often have processing temperatures that are similar to if not less than that of the onset of natural fiber degradation. Bio-based thermoplastic resins such as PLA and PHAs also exhibit processing temperatures appropriate for natural fibers. For thermosetting resins such as epoxies, vinyl esters, *etc.*, the typical exothermic heat generated during curing does not exceed the degradation temperature of natural fibers.

5.6 Advantages and Limitations of Biocomposites

Properties that make biocomposites attractive for use in a variety of engineering applications range from societal and environmental to those purely

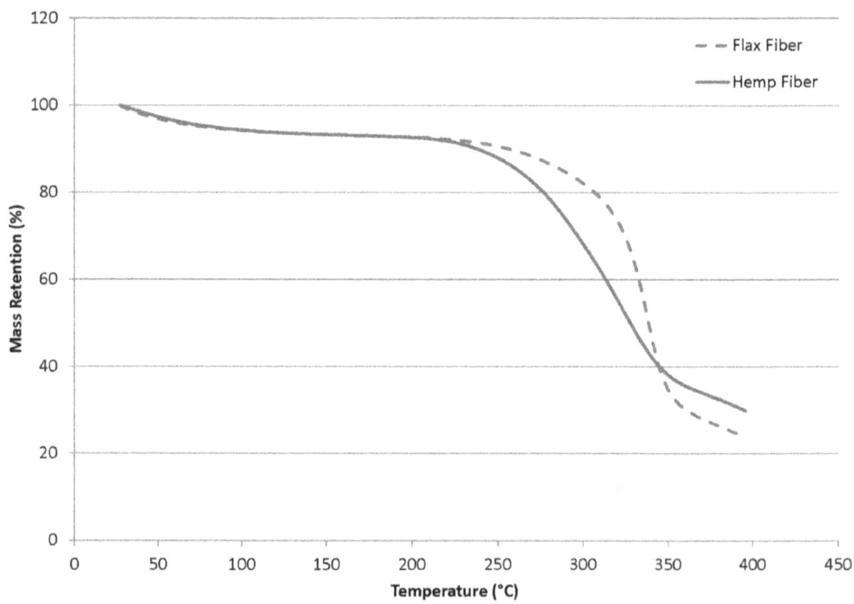

Figure 5.7 TGA curves for two bast natural fibers.[19]

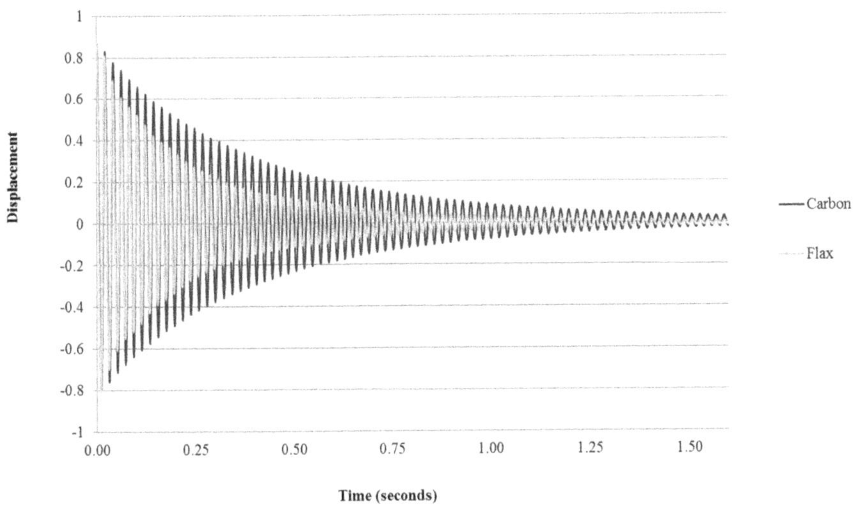

Figure 5.8 Typical unit operations for manufacturing of a biocomposite product.

based on increased mechanical and physical properties. With most natural fiber exhibiting lower densities ($\sim$1.4–1.6 g cm^{-3}) as compared to conventional composite fiber reinforcement (*i.e.*, aramid – 1.5 g cm^{-3}, carbon – 1.7 g cm^{-3}, fiberglass – 2.6 g cm^{-3}), the opportunity to create certain lightweight composite components exists. In addition, the specific strengths and specific modulus that can be achieved are very attractive for sporting goods, automotive, and aerospace applications.[20–22]

In addition to mechanical performance advantages, biocomposites exhibit improved vibration dampening, thermal insulation, and sound absorption owing to their complex hierarchical structure, low density, and loftiness. The following figure depicts the dampening behavior of two composite specimens, one containing epoxy reinforced with flax and one containing the same epoxy but reinforced with carbon fiber (see Figure 5.8). Both specimens were fixed in a cantilever beam configuration and excited with the same impulse. As can be seen, the flax fiber composite dampens the vibrations much more quickly than the carbon fiber composite. Again, it is clear the natural fibers provide an advantage. Finally, many of the natural fiber composites in use for automotive applications provide not only mechanical property advantages but also improved sound absorption for increased cabin comfort.

From a life cycle perspective, biocomposites have recently shown a lot of promise over conventional composite materials. Limited studies have shown bast natural fibers to exhibit lower carbon footprints as compared to fiberglass and other traditional reinforcing fibers. However, the number of growing inputs needed (fertilizer, weed control, and water usage) can prove to make carbon footprint gains negligible when grown in certain parts of the world. In addition, different studies consider the CO_2 absorption of the natural fiber plant during its growth as part of its carbon footprint while others have neglected it and only considered the carbon impact once harvested.

From a societal perspective, consumers are pulling bio-based resins and biocomposites into more and more commercial market applications because on the whole they are striving to become better stewards of the environment and recognize there are limited finite resources available for the products utilized on a daily basis. Evidence of these buying trends can be found in some of the market segment analysis conducted by major retailers around the world. In addition, the natural look provided by most biocomposites is attractive for targeted applications. The ability for biocomposite products to become part of the marketing strategy for those products because they are aligned with the application is a niche concept. For instance, a coffee mug made from coffee or a trash can made from trash are now possible.

Another area where biocomposites have an advantage but one that has not necessarily been proven to date is in cost. Many of the natural fibers available are lower in cost as compared to conventional fibers used in composite construction. The costs currently associated with transporting low bulk density materials long distances between farmland and factory as well as the chemical treatments and alternative techniques needed to process structural biocomposites are high and become prohibitive for certain applications. Any time a brand-new supply chain needs to be established and volumes are low to start, costs of those materials are going to be higher. Ideally, decorticating (*i.e.*, fiber and core refinement) facilities would exist close to where the crops are grown and harvested. Once processed and therefore densified, the raw materials could be more efficiently transported for conversion into biocomposites. Until such facilities are established strategically, the cost of natural fibers will be similar to, if not more expensive than, conventional fibers used as reinforcement in composites.

One area of recent advancement has come in the hybridization of fibers used in composites to enhance mechanical properties while creating a more sustainable, low-cost material. Using carbon fiber or glass fiber in conjunction with natural fibers can improve strength and reduce variability in properties in the composite while maintaining a high renewable content, high stiffness, and low cost. A few studies have shown that as little as 5–10% carbon fiber hybridized with flax fiber can significantly increase average tensile strength and decrease the standard deviation.[23–25] However, one area that still requires attention is deciding the best method of hybridizing the fibers. Some have used a co-mingling approach within the fiber tows themselves, alternating fiber tows within the fabric architectures or using an alternating layer stacking approach.[26] Each approach has its pros and cons, but no definitive approach has emerged as the most favorable.

Another area of improvement for natural fiber use in biocomposites is in the establishment of standards for natural fibers used in composite applications. Bast natural fibers cannot always be tested in the same manner as conventional synthetic fibers used as reinforcement in composites and therefore require new standards to be developed. ASTM Committee D13.17 (Bast Fibers and Plants) has been focused on the development of new standards. Some standard methods have been established to determine color, fineness, cleanliness, widths through image analysis, and density so that decortication facilities can measure and essentially grade bast natural fiber quality prior to a composite manufacturer's procurement.[27] However, additional standards need to be established in order to further improve this grading strategy for such naturally variable materials including strength, modulus, and fracture toughness.

Finally, the last area of improvement needed for biocomposites to be fully embraced by the engineering community is in the understanding of their long-term performance and stability. From a perspective of fatigue, creep, weathering, and hygrothermal performance, designers and engineers need to have additional information and proof of how these materials compare to traditional structural composites. Some studies have evaluated the fatigue performance of flax reinforced epoxy composites and found them to be quite comparable to traditional composite materials.[28,29] There appears to be an initial break in period where stiffness and energy absorbed decrease per cycle until a steady state is reached until failure. When it comes to creep, studies have shown that this property is mainly governed by the polymer matrix whether the reinforcement is synthetic or natural based.[30–32] Finally, accelerated weathering studies have shown similar performance between glass and flax fiber reinforced biocomposites when the same renewable-based polymer matrix was employed.[33,34]

5.7 Applications

5.7.1 Upgrading Biomass for Headlamp Housing

The component is a headlamp housing unit that consists of polypropylene mixed with coffee chaff biocarbon fillers, as seen in Figure 5.9. The coffee chaff is a waste byproduct of McDonald's coffee manufacturing process. Biocarbon made from coffee chaff has a specific gravity of 1.27 $g\,cc^{-1}$ – weighing 55% less than baseline talc filler (magnesium silicate) whose specific gravity is 2.8 $g\,cc^{-1}$. By partnering with Varroc Lighting, Competitive Green Technologies, University of Guelph, and McDonalds, Ford designed a composite that reduced weight by 17% and cost by 5% without changing the manufacturing equipment. There was no need for changes in the mold or processing parameters. Furthermore, only 20% biochar filler was used instead of 40% talc reinforcement. This resulted in composites with a density of 1.02 $g\,cc^{-1}$ as opposed to 1.23 $g\,cc^{-1}$. The main challenges when working with organic fillers are fogging, VOCs, odor, and water absorption. In collaboration with Varroc, Competitive Green Technologies, and McDonalds, a team was able to implement a formulation created at the University of

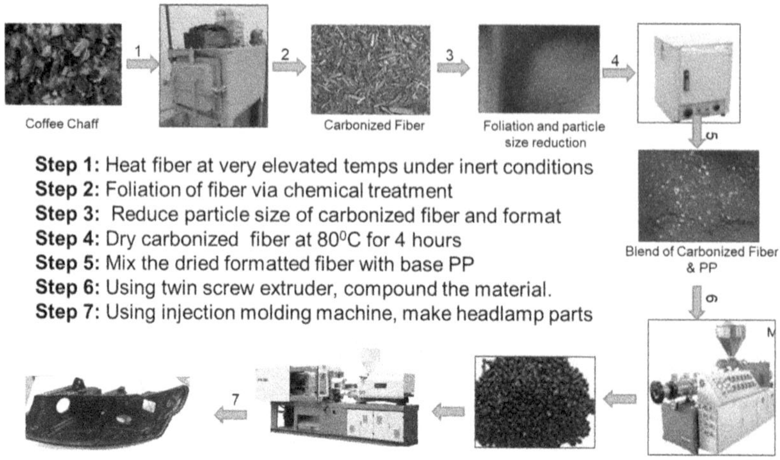

Figure 5.9 Biocarbon-based composite production flow and headlamp housing.

Guelph, licensed to Competitive Green Technologies, using biocarbon processing and biocomposite compounding methods at Competitive Green Technologies that eliminates the above, in particular, odor and fogging. Moreover, carbonization of coffee chaff results in better compatibility with the polypropylene matrix, which coupled with the reduced hydrophilicity of the functional biocarbon filler, results in less moisture absorption by the composite part. The innovative carbonization process, and the accompanying particle treatment and integration with a fossil polymer matrix to create a homogeneous biocomposite that matched the CLTE (coefficient of thermal expansion) and shrinkage of the incumbent material, was a challenge, and in overcoming this lies the material and processing innovation. This solution can be easily transferred to other applications/programs where the automotive part is manufactured using injection molding, compression molding, or extrusion since no new design for tooling or changes in processing parameters are required. In addition to the decrease in density of the composite with coffee chaff fillers, the improvement in material properties allows for more weight reduction because less material can be used to achieve the same mechanical performance, see Table 5.1. As such, the light weight coupled with ease of processing due to increased melt flow index can enable a faster cycle time (8 seconds, 15% reduction) and lower processing temperature resulting in greater manufacturing productivity, therefore resulting in revenue enhancement.[35]

5.7.2 Latest Breakthroughs with Sustainable Hybrid Composites in Lightweight and Semi-structural Applications

Using cellulose as reinforcement in thermoplastic composites has numerous advantages for the environment, such as reducing CO_2 emissions into the atmosphere during their life cycle including CO_2 absorption during plant growth,

Table 5.1 Characterization of biocarbon-filled PP composites.

Material properties	Ford spec requirements WSS-M4D643-B1	Current talc PP grade	CGT developed BioPP HMS3040
Type	40% talc PP	40% talc PP	20% coffee chaff PP
Density (g cc^{-1})	1.18–1.32	1.23	1.02
Mold shrinkage (cm cm^{-1})	—	0.015–0.018	0.014–0.016
Melt flow (g per 10 min)	—	12	16
Tensile strength (MPa)	>24	29	51
Flexural strength (MPa)	—	47	72
Flexural modulus (MPa)	>3200	3800	3500
Notched Izod impact@23 °C (J m^{-1})	>15	24	40
HDT@0.455 MPa (°C)	>62	132	140
Ford odor test at 23 °C, 40 °C, and 65 °C	—	PASS	PASS
Fogging test, fog number	>90	95	99
Resistance of fungi to materials	—	No growth	No growth
Resistance of plastics to bacteria	—	No growth	No growth
Outgassing test per FLTM BZ 157-01	<1	0.4	0.3

emission reduction due to lower energy requirements during production, as well as reduced vehicle emissions due to lower part weight. Additionally, the use of cellulose as reinforcement helps to overcome some of the challenges facing the automotive industries related to end-of-life vehicle laws. However, incorporating cellulose as reinforcement in thermoplastic polymers has also brought its own challenges. These challenges include lower modulus and impact strength, higher shrinkage rate, and inferior moisture resistance as compared to glass fiber reinforced composites for semi-structural applications in the automotive, aerospace, and construction industries. Hybridization of cellulose fibers with inorganic fillers/fibers is a possible solution for improving the mechanical properties and moisture resistance and decreasing the shrinkage of the composites. The advantage of combining two different materials is that the unique properties of one type of fiber could complement what is lacking in the other. In this sense, hybrid composites made with cellulose fibers possess advantages in cost, weight reduction, and sustainability that the auto industry is continuously striving to achieve. It is with these objectives in mind that Ford Motor Company and Celanese and International Paper, together with the Tier 1 molder, Summit Polymers, demonstrated their commitment to the environment by putting forward the industry's first cost effective, lightweight, and sustainable hybrid composites for use in console substrates. Hybridization of cellulose fiber with long glass fibers in polypropylene composites reduced the use of inorganic fibers in console substrates, leading to weight and cost savings, and contributed to sustainability of the composites, as shown in Figure 5.10.

The hybrid composites were prepared by injection molding to achieve the desired long glass fibers (LGF)/cellulose fibers (CF) concentration; the as-received master-batch pellets from Celanese (60% LGF-filled PP) and International Paper (20% CF-filled PP) were mixed and diluted with neat PP as needed. No extrusion process was used in this innovation. This production method is an efficient one-step injection process that can speed up part of the production. It was found that the hybrid fibers acted as an effective reinforcement. Synergetic effects on the mechanical, thermal, and morphological properties of the hybrid composites were observed. The hybrid composite meets and exceeds all Ford requirements (material and dimensional), as shown in Table 5.2 and Figure 5.10, while helping to reduce CO_2 emissions. No significant dimensional changes were observed. Console key life testing results met all Ford requirements. Cellulose fibers exhibit a high degree of flexibility when compared to glass fibers, allowing them to bend rather than fracture during injection molding, allowing better isotropic behavior in the final part.

In addition to being green, cost effective, and lightweight, hybrid composites flow quickly at relatively low temperatures and pressures. As a result, hybrid composites reduce the energy costs associated with production by 25% and produce an end-product with great structural rigidity and new,

Table 5.2 Hybrid composites *versus* current materials specification for console retainers.

	WRS-M4D1058-A1	Hybrid composites
Filler content	**30–37%**	Cellulose (10%) Glass (15%)
Properties		Average
Young's (tensile) modulus (GPa)	3.4 GPa minimum	4.43
Density (g cm^{-3})	1.14–1.19	1.024
Tensile strength at yield 50 mm min^{-1}	52 min	66
Flexural modulus (GPa)	3.6 GPa minimum	4.64
Heat deflection temperature (°C)	120 minimum	150.8
Notched charpy impact at 23 °C (kJ m^{-2})	0.5 kJ m^{-2} minimum	11.48
Notched charpy impact at 40 °C (kJ m^{-2})	4.5 kJ m^{-2} minimum	12.37
Heat aging	±25% maximum change tensile strength at yield	107.5%
Heat aging	±25% impact strength (notched charpy)	103.1%
Odor	Rating 3 maximum	2/2.5/3
Fogging	70 min	98
Flammability	100 mm per min maximum	42.3

Figure 5.10 An award-winning center console carrier employs a PP composite reinforced with cellulose and glass fibers. Cellulose composites blended with long glass-filled PP offer a completely new and broad array of detailed benefits.

highly marketable performance capabilities. No new design or mold modifications are required. The hybrid material is less abrasive, reducing tool wear to molds. Lastly, the hybrid composites do not cause any complications in processing. There is no issue with thermal degradation and odor as a result of the purity and high thermal stability of the cellulose where these issues might be expected with other wood components and natural fibers.

5.7.3 Rice Hull-filled PP

Ford has also demonstrated its solid commitment to the environment by implementing a robust, industry first, cost effective, lightweight, and sustainable composite for wiring channels as seen in Figure 5.11. The rice hull-filled PP meets all Ford requirements (materials and performance level) while significantly reducing CO_2 emissions. Rice hull-filled polypropylene composites reduced the use of inorganic fillers in wiring channels leading to robust properties and significant weight and cost savings (10% lighter and 10% less costly) while contributing to sustainability of composites. Like cellulose-based hybrid composites and biocarbon-based composites, there is no need for new design or mold modifications. The developed rice hull-filled PP composite also flows very quickly at relatively low temperatures and pressures in comparison to talc-filled PP composites. Overall, rice hulls are used as a reinforcement for injection-molded PP wiring channels, replacing 20% talc-filled PP without loss of properties. The high silica content of rice hulls can provide high resistance to mildew and fungal growth and offers better flame retardance and lower

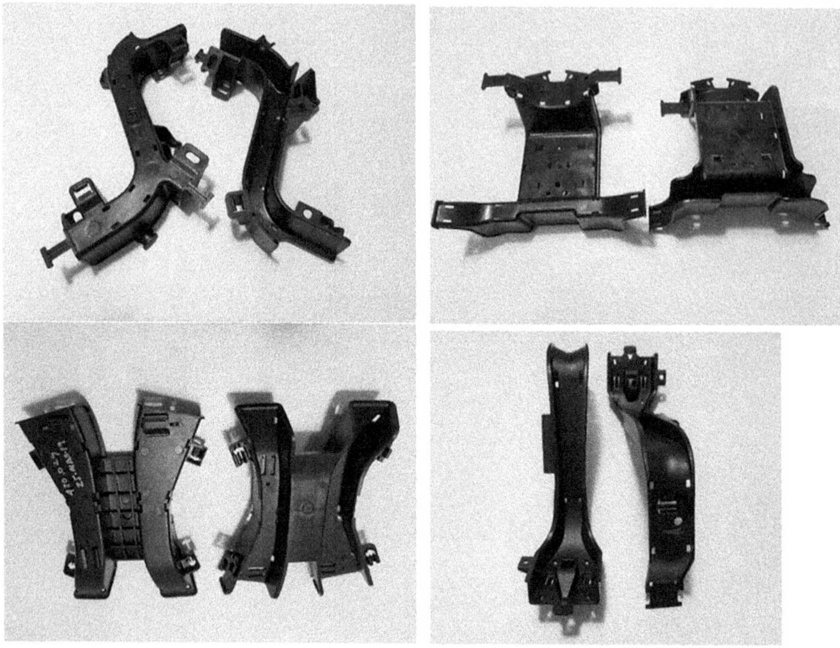

Figure 5.11 10% rice hull-filled PP for injection molded wiring channels, replacing 20% talc-filled PP, without loss of properties.

moisture absorption than natural fiber-filled plastics. Rice hull-filled PP also offers an additional revenue stream to farmers.

5.7.4 Cellulose Filler in Recycled Polyamide 6/Polypropylene Blends

By using combinations of recycled polymers from post-industrial and post-consumer waste and natural reinforcements to replace traditionally unfilled, glass-filled, and talc-filled polymeric components, the environmental impacts, cost, and weight of each vehicle composite can be reduced. Langhorst *et al.* developed composites containing recycled polypropylene (PP), recycled polyamide 6 (PA6), and cellulose with physical properties consistent with unmodified polymers. They also investigated the material properties of PA6/PP blends with the following variables: (a) the addition of a compatibilizer, (b) the addition of cellulose, and (c) the order in which components were mixed.[36]

Abbreviations

CF	Cellulose fibers
LGF	Long glass fibers
PHA	Polyhydroxyalkanoate

PLA	Polylactic acid
PP	Polypropylene
TGA	Thermogravimetric analysis

References

1. P. K. Mallick, in *Fiber-Reinforced Composites: Materials, Manufacturing, and Design*, ed. M. Dekker, 1st edn, 1988, ISBN-13: 978-0824777968.
2. R. F. Gibson, *Principles of Composite Material Mechanics*, McGraw-Hill College, 1st edn, 1994, ISBN-13: 978-0070234512.
3. M. A. Fuqua, S. Huo and C. A. Ulven, Natural fiber reinforced composites, *Polym. Rev.*, 2012, **52**(3), 259–320.
4. M. Zimniewska, M. Wladyka-Przybylak and J. Mankowski, Cellulosic Bast Fibers, Their Structure and Properties Suitable for Composite Applications, in *Cellulose Fibers: Bio- and Nano-Polymer Composites*, ed. S. Kalia, B. Kaith and I. Kaur, Springer, Berlin, Heidelberg, 2011.
5. J. Foulk, D. Akin, R. Dodd and C. Ulven, *Production of flax fibers for biocomposites, Cellulose fibers: Bio-and nano-polymer composites*, Springer, Berlin, Heidelberg, 2011, pp. 61–95.
6. J. A. Foulk, D. Rho, M. M. Alcock, C. A. Ulven and S. Huo, Modifications caused by enzyme-retting and their effect on composite performance, *Adv. Mater. Sci. Eng.*, 2011.
7. A. Amiri and C. A. Ulven, Advanced method for void fraction evaluation of natural fiber composites using micro-Ct technology, *SAMPE Conference Proceedings*, 2016.
8. S. Ahmed and C. A. Ulven, Dynamic in-situ observation on the failure mechanism of flax fiber through scanning electron microscopy, *Fibers*, 2018, **6**(1), 17.
9. M. Ehresmann, A. Amiri and C. Ulven, The effect of different variables on in-plane radial permeability of natural fiber mats, *J. Reinf. Plast. Compos.*, 2018, **37**(19), 1191–1201.
10. S. Mekic, C. A. Ulven, I. S. Akhatov, E. Jerke and E. Kerr-Anderson, Evalutation of in-plane and transverse permeability of flax fiber preforms for biocomposite materials, *J. Biobased Mater. Bioenergy*, 2009, **3**(2), 156–164.
11. S. S. Radkar, A. Amiri and C. A. Ulven, Tensile behavior and diffusion of moisture through flax fibers by desorption method, *Sustainability*, 2019, **11**(13), 3558.
12. S. Huo, V. S. Chevali and C. A. Ulven, Study on interfacial properties of unidirectional flax/vinyl ester composites: resin manipulation on vinyl ester system, *J. Appl. Polym. Sci.*, 2013, **128**(5), 3490–3500.
13. S. Huo, A. Thapa and C. A. Ulven, Effect of surface treatments on interfacial properties of flax fiber-reinforced composites, *Adv. Compos. Mater.*, 2013, **22**(2), 109–121.

14. S. Huo, C. A. Ulven, H. Wang and X. Wang, Chemical and mechanical properties studies of chinese linen flax and its composites, *Polym. Polym. Compos.*, 2013, **21**(5), 275–286.
15. S. Huo, M. Fuqua, V. S. Chevali and C. A. Ulven, Effects of natural fiber surface treatments and matrix modification on mechanical properties of their composites, *SAE Technical Paper*, 2010.
16. C. Taylor, T. Krosbakken, C. A. Ulven, A. Paramarta and D. C. Webster, Structural quality biocomposites of treated flax fiber with epoxidized sucrose soyate resin, *International Conference on Composite Materials*, Montreal, Canada, 2013.
17. R. Whitacre, A. Amiri and C. Ulven, The effects of corn zein protein coupling agent on mechanical properties of flax fiber reinforced composites, *Ind. Crops Prod.*, 2015, **77**, 232–238.
18. A. Amiri, C. A. Ulven and S. Huo, Effect of chemical treatment of flax fiber and resin manipulation on service life of their composites using time-temperature superposition, *Polymers*, 2015, **7**(10), 1965–1978.
19. A. El-Sabbagh, L. Steuernagel, J. Ring and O. Toepfer, Development of natural fiber/engineering plastics composites with flame retardance Properties, *AIP Conf. Proc.*, 2016, **1779**, 030020.
20. Z. C. Yu, M. Townsley, M. Alcock, C. Ulven and S. Potter, Developing Natural Fibre-Based Composite Materials for Canadian Industry Sectors, *J. Biobased Mater. Bioenergy*, 2012, **6**(4), 515–520.
21. M. A. Fuqua, S. Huo, V. S. Chevali and C. A. Ulven, Development of flax fiber/soy-based polyurethane composites for mass transit flooring application SAE, *Int. J. Mater., Mech. Manuf.*, 2010, **3**(1), 230–236.
22. A. Amiri, V. Burkart, A. Yu, D. Webster and C. Ulven, The potential of natural composite materials in structural design, *Sustainable composites for aerospace applications*, Woodhead Publishing, 2018, pp. 269–291.
23. J. Flynn, A. Amiri and C. Ulven, Hybridized carbon and flax fiber composites for tailored performance, *Mater. Des.*, 2016, **102**, 21–29.
24. N. Shah, J. Fehrenbach and C. A. Ulven, Hybridization of Hemp Fiber and Recycled-Carbon Fiber in Polypropylene Composites, *Sustainability*, 2019, **11**(11), 3163.
25. A. Amiri, T. Krosbakken, W. Schoen, D. Theisen and C. A. Ulven, Design and manufacturing of a hybrid flax/carbon fiber composite bicycle frame, *Proc. Inst. Mech. Eng. Part P J. Sport Eng. Technol.*, 2018, **232**(1), 28–38.
26. M. Z. Islam, A. Amiri and C. A. Ulven, Fatigue Behavior Comparison of Inter-Ply and Intra-Ply Hybrid Flax-Carbon Fiber Reinforced Polymer Matrix Composites, *J. Compos. Sci.*, 2021, **5**(7), 184.
27. A. Amiri, Z. Triplett, A. Moreira, N. Brezinka, M. Alcock and C. A. Ulven, Standard density measurement method development for flax fiber, *Ind. Crops Prod.*, 2017, **96**, 196–202.
28. M. Z. Islam and C. A. Ulven, A thermographic and energy based approach to define high cycle fatigue strength of flax fiber reinforced thermoset composites, *Compos. Sci. Technol.*, 2020, **196**, 108233.

29. M. Z. Islam and C. A. Ulven, Influence of Loading Frequency on Thermal and Micro-Mechanical Damage During Fatigue of Flax Fiber Reinforced Composites, *International Manufacturing Science and Engineering Conference*, American Society of Mechanical Engineers, 84256, V001T02A002, 2020.
30. A. Amiri, N. Hosseini and C. A. Ulven, Long-term creep behavior of flax/vinyl ester composites using time-temperature superposition principle, *J. Renewable Mater.*, 2015, **3**(3), 224–233.
31. A. Amiri, A. Yu, D. Webster and C. Ulven, Bio-based resin reinforced with flax fiber as thermorheologically complex materials, *Polymers*, 2016, **8**(4), 153.
32. A. Amiri and C. Ulven, Bio-based composites as thermorheologically complex materials, *Challenges in Mechanics of Time Dependent Materials*, Springer, 2017, vol. 2, pp. 55–63.
33. C. A. Taylor, A. Amiri, D. C. Webster and C. A. Ulven, Long-term behavior of bio-composites for structural applications, *Proceedings of the Composites and Advanced Materials Conference*, Anaheim, CA, USA, 2016, pp. 26–29.
34. C. Taylor, A. Amiri, A. Paramarta, C. Ulven and D. Webster, Development and weatherability of bio-based composites of structural quality using flax fiber and epoxidized sucrose soyate, *Mater. Des.*, 2017, **113**, 17–26.
35. A. Bali, A. Ademagavun and A. Kiziltas, Biocarbon- a renewable and lightweight functional filler for polymer composites, *SPE TPO Automotive Engineered Polyolefins Conference*, Troy, MI, 2019.
36. A. Langhorst, A. Kiziltas, D. Mielewski and E. Lee, Selective dispersion and compatibilizing effect of cellulose filler in recycled PA6/PP blends, *Proceeding of SPE Automotive Composites Conference & Exhibition*, Novi, MI, 2015.

CHAPTER 6

Fermentation of Polyesters (PHA and PLA)

MELISSA MONTALBO-LOMBOY

Experiential Engineering Education Department, Rowan University, 201 Mullica Hill Rd., Glassboro, NJ, 08028, USA
E-mail: montalbolomboy@rowan.edu

6.1 Introduction

Polyesters such as polylactic acid (PLA) and polyhydroxyalkanoates (PHA), as well as other bioplastics, are one of the most widely used and studied bio-based plastics to date. With an increasing awareness of the negative environmental impact of petroleum-based plastics, bio-based plastics, sometimes referred to as bioplastics, became the most promising alternative. Even though the supply of commercial PLA and PHA is well below their petroleum-based counterparts, the supply is growing every year. The production of PLA has been steadily increasing since the opening in 2002 of the largest producer (NatureWorks LLC) of PLA.[1] In 2012, it was predicted that PLA would reach a production capacity of over 800 000 tonnes by the end of 2020.[2] NatureWorks LLC has been the world leader in PLA production but other manufacturers have also emerged in recent years such as Cereplast Inc. (USA), Mitsui Chemicals Inc. (Japan), Purac Biomaterials, Hycail (The Netherlands), Futerro and Galactic (Belgium), FKur Plastics Corporation (Germany), and Snamprogetti (China).[3] In contrast, Metabolix Inc. had been at the forefront of production of a PHA called Mirel since 2002. However, it was reported that Metabolix Inc. exited the bioplastics industry in 2016 when it was sold to a Korean food and bioengineering company (CheilJedang).[4,5] In addition, companies such as Danimer are also increasing capacity of PHA

Green Chemistry Series No. 79
Bioplastics and Biocomposites: A Practical Introduction
Edited by David Grewell

Published by the Royal Society of Chemistry, www.rsc.org

production. Also, there are other new PHA producers such as Full Cycle Bioplastics[6] and Mango Materials,[7] which are both based in California.

Some of the advantages of bioplastic include the use of renewable resources, biodegradability, and reduced carbon footprint. For example, in the US, a commercial PLA called Ingeo, which is produced by NatureWorks LLC, uses renewable materials such as corn, cassava, sugarcane, and sugar beets as raw materials for lactic acid fermentation.[8] These lactic acids are then converted to lactide then to PLA.[8] Currently, Ingeo PLA is used in many applications such as food service products to biomedical fibres to 3D printing polymers.[1] While commercial PHAs produced by Full Cycle Bioplastics and Mango Materials were converted from organic waste such as food waste,[6] agricultural residues, and methane gas from wastewater treatment plants,[7] because they are using a low cost feedstock in their processes, they can be cost-competitive with traditional petroleum plastics.

Typically, polyesters can be produced in many different ways:[9] (1) chemical production, (2) biotechnological production, (3) a combination of chemical and biotechnological production, or (4) isolation from natural resources. The current PLA technology is a combination of chemical and biotechnological methods because it uses biotechnological processes to produce lactic acid then uses chemical processes to polymerize the lactide into polylactic acid.[8] While most of the PHAs are currently produced using mainly biotechnological methods, such as fermentation of PHA accumulating bacteria,[6,7] there are also other studies that explored fermentation of recombinant bacteria and yeast[10–14] and transgenic plants.[15–19] More details can be found in Chapter 7.

Transgenic plants are metabolically engineered plants that can produce PHAs. The first study on PHA transgenic plants was reported by Poirier *et al.*[20] in 1992 where genes from *Alcaligenes eutrophus* bacterium were introduced into *Arabidopsis thaliana* to produce polyhydroxybutyrate (PHB). With the promise that transgenic plants can produce bioplastics, biochemicals, and biofuel, creating multiple product streams and making them cost-competitive, many researchers have studied these processes. Currently, commercial companies such as Yield10 Bioscience Inc. have been developing transgenic switchgrass and oilseed camelina to produce PHBs.[21]

Similar to transgenic plants, the use of recombinant bacteria has a similar concept. The desired genes from PHA accumulating bacteria are spliced and then introduced into a widely studied and easily scaled up bacterium such as *Escherichia coli* (*E. coli*). With the gene incorporated into *E. coli*, it now has the necessary genetic information to produce PHA. *E. coli* has been the preferred bacterium for genetic engineering because it has a high growth rate, high cell density, and uses cheap carbon sources.[22–24] Thus, many studies have used this bacterium for producing PHA. However, this technology also comes with limitations such as high energy demands,[25] low substrate conversions,[26] and expensive downstream processing.[27] Commercialization of PHAs from recombinant bacteria is still in its infancy but this could be one of the solutions for making PHAs cost-competitive in the future. A recent

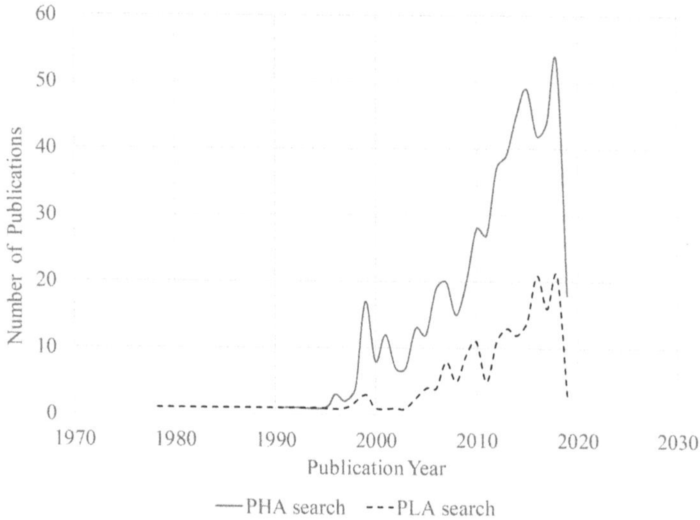

Figure 6.1 The PHA and PLA academic publications from 1978 to part of 2019.

study has thoroughly reviewed a few potential solutions to some of the PHA recombinant bacteria production issues.[28] Even though the fermentation of PHA by recombinant bacteria is very promising, this chapter will only focus on fermentation of wild type and mutant bacterial strains.

In the last two decades, the study of PHA fermentation has been given significant attention. A quick web of science scholarly literature search using "PHA fermentation" and "PLA fermentation" as keywords yielded 544 and 167 studies, respectively. As shown in Figure 6.1, the number of scholarly publications on both PHA and PLA has doubled in the last five years. This growth can be related to the emergence of more PLA and PHA large scale production in the market. Because of the very broad scope of both PLA and PHA fermentation, this chapter will only focus on the fermentation strategies of producing PLA and PHA.

6.2 Fermentation

Fermentation is the chemical conversion of organic materials by micro-organisms to create various essential and high-value products. Although fermentation's definition has evolved in the last century, it is still one of the oldest technologies known to man. It dates back as far as 3000 BC in Egypt and Sumeria where bread and alcoholic beverages were produced using fermentation.[29] Records in ancient Egyptian tombs showed that bread, beer, vinegar, yogurt, cheese, and wine, which are all products of fermentation, were already enjoyed in those times.[30] In 300 BC ancient China, they also used yeast and molds to convert cereal grains to fermented foods.[31] Today, fermentation is a widely used technology and the essential part in the production of many food products, beverages, pharmaceutical products, laundry products, cosmetics products, chemicals, biochemicals, biofuels, and bioplastics.

6.2.1 Fermentation Nutritional Requirements

A microorganism is dependent on its nutritional supplement for growth and reproduction. According to Kampen,[32] microorganisms can be categorized based on their nutritional requirements as follows: (1) photoautotrophic, (2) photoheterotrophic, (3) chemoautotrophic, and (4) chemoheterotrophic. Both photoautotrophs and photoheterotrophs use light as an energy source while both chemoautotrophs and chemoheterotrophs use chemicals as an energy source. Similarly, both photoautotrophs and chemoautotrophs use CO_2 as a carbon source, while photoheterotrophs and chemoheterotrophs use organic compounds as the primary carbon source. Polyester-producing microorganisms have varied classifications and will be discussed thoroughly in a later section of this chapter.

The two essential nutritional requirements for most microorganisms are the carbon and nitrogen sources. Carbohydrates are the most commonly used carbon source in many fermentation processes. Carbohydrates, ranging from complex carbohydrates such as starch and cellulose to simple carbohydrates such as sucrose and glucose, are excellent sources of carbon that cells utilize for metabolic energy. Glucose is found in the majority of fermentation media used. For complex carbohydrates, enzymes are required to hydrolyze them to simple sugars before microorganisms can convert them to metabolic energy. Thus, it is common to pretreat complex carbohydrates to depolymerize the polysaccharides into simple sugars before proceeding to fermentation. It is noted that there have been many studies that combine the enzymatic hydrolysis and fermentation processes in one vessel (simultaneous saccharification and fermentation (SSF)), which saves energy and reduces cost. SSF will be discussed in a later section. Nitrogen is the second most abundant nutrient found in many fermentation media and is the second essential nutrient requirement for microorganism growth. Even though it exists in reduced form, nitrogen occurs in the organic compounds of the cell.[32] Nitrogen is utilized for the cellular synthesis of proteins, DNA, and RNA. Nitrogen is typically supplied in the form of nitrates, nitrites, and protein sources such as peptone, meat extract, or yeast extract.

6.3 Batch Fermentation

Generally, fermentation has three modes: batch, fed-batch, and continuous. Batch fermentations are a closed system where all the ingredients necessary for the microorganism's growth are added at the beginning, and desired products are withdrawn at the end of the fermentation. In small-scale laboratories, the simplest and cheapest form of batch fermentation is shake-flask culture. It is typically done in conical flasks and shaking incubators. However, this method can only control temperature and agitation but not parameters such as pH and aeration, which are also very important in effecting microorganism growth and yield. In most cases, bioreactors are preferred over the shake-flask culture. Batch fermentations are used because they are easy to operate, have a fast turnaround time, and have fewer chances of contamination because of their closed system. However, substrate and product inhibitions can significantly affect the product yield in batch fermentations. In more detail, substrate and product inhibitions occur when an excess amount of the substrate or product is inhibiting the propagation of the microorganism or the desired product.

In batch fermentation, the inoculum or the microorganism will undergo four different phases: lag phase, growth phase, stationary phase, and death phase, as shown in Figure 6.2. The lag phase is where the microorganism has the least amount of growth. This is thought to be the time when the microorganism is adjusting to its new environment. The second phase is the growth phase where the microorganism reproduces exponentially; thus, it is sometimes called the log phase. This phase occurs rapidly; therefore, it is important that all the nutrients, pH, temperature, and aeration (if needed) are at optimum conditions. As the microorganism reaches the stationary phase, the growth slows down and eventually plateaus. As the nutrients or substrate depletes, the microorganism also enters the death phase. Typically, before the death phase occurs, most fermentation is terminated.

6.4 Fed-batch Fermentation

Similar to batch, fed-batch fermentation also withdraws all the products at the end of the fermentation, but the ingredients for the microorganism's growth are

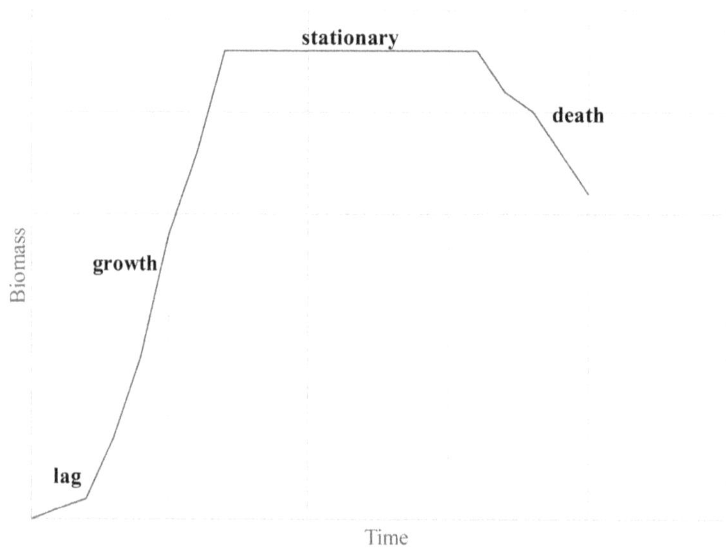

Figure 6.2 The microorganism cell growth curve.

introduced slowly throughout the fermentation period. Fed-batch fermentations were first developed during the early 1900s when yeast producers noticed that yeast cells favour the production of ethanol, rather than reproducing themselves, when there is a high concentration of maltose, which is their primary carbon source.[29] Because yeast is the main product, it is therefore important that an optimum amount of maltose is achieved during the growth phase. To overcome this problem, instead of adding all of the maltose at the beginning, they intermittently added maltose to obtain an optimum level throughout the growth phase.[29] Because the nutrients are added slowly, the fed-batch technique can also be used to minimize substrate and product inhibitions while extending the fermentation log phase and maximizing cell growth. Particularly in microorganisms prone to substrate and product inhibitions, such as baker's yeast and *E. coli*, fed-batch will be a better method than batch.

Depending on the experimental or process design and the desired product, there are various feeding strategies in fed-batch fermentation. The fixed feeding or fixed volume feeding is when the working volume of the fermenter is maintained. This is done by removing some fermented broth and adding the same amount of highly concentrated substrate to compensate for the amount taken out. Depending on the design, the withdrawal of the fermentation broth can be completed right before the feeding or ahead of the time of feeding. One of the advantages of this strategy is the loss of some inhibitory products from the amount taken out. However, a variable feeding strategy occurs when the volume of the fermenter changes as the new substrate is added. The concentrated substrate can be added in large volumes at specified times or added in small volumes at a constant speed. This technique is very typical in ethanol fermentation where some specified volumes of the fresh substrate are added at a specified time. This reduces the substrate inhibition because the substrates are only fed at an optimum tolerance concentration level of the microorganism.

6.5 Continuous Fermentation

Continuous fermentation is a technique that extends the growth phase of the microorganism by feeding new concentrated substrate and withdrawing fermented broth at the same rate, achieving a steady-state system.[29] This is typically employed in large industrial systems because it has the advantage of maintaining the same conditions throughout the growth phase. For example, in batch, many parameters such as pH, temperature, biomass, product, and substrate concentration change over the course of the fermentation. However, these fluctuations are minimized or avoided in continuous fermentations. In some ways, this technique is very similar to the fixed feeding or fixed-volume feeding fed-batch fermentation, except that in continuous fermentation, the feeding and withdrawing are continuously completed; unlike fed-batch, in which feeding is conducted at specific times.

The benefits of continuous fermentation include increased productivity, high cell growth rate, long residence times, and control of microorganism evolution. Technically, the process can be prolonged for a relatively long time, thus reducing time for maintenance.[29] The microorganism evolution can also be directly studied using continuous systems.[29] Because microorganisms respond to changes in their environment, when the operator or scientist wants to "direct the evolution"[29] of the microorganisms, they only have to control select parameters such as temperature, pH, and nutrients and observe how the microorganisms respond to the changes. However, there is also a risk of contamination in these systems because the new substrate is continuously introduced.

6.6 Polyhydroxyalkanoates

Polyhydroxyalkanoates are thermoplastic polyesters that are inherently biodegradable and are produced from microorganisms. Many works of literature have classified PHAs into two categories: short-chain length PHA (scl PHA) or medium chain length PHA (mcl PHA). The difference between the two is dependent on the number of carbon atoms in the branched side chain of the PHA. Typically, scl PHAs such as poly(3-hydroxybutyrate), poly-(4-hydroxybutyrate), and poly(3-hydroxyvalerate) have 3–5 carbon atoms while mcl PHAs such as poly(3-hydroxyhexanoate) and poly(3-hydroxyoctanoate) have 6–14 carbon atoms or more.[33,34] Some studies have shown both mcl and scl in one polymer, such as the copolyester of 3-hydroxybutyric acid and 3-hydroxyalkanoic acid.[35,36]

Polyhydroxybutyrate is a widely produced and the most studied PHA. It was first identified from *Bacillus megaterium* in 1926.[37] It is highly crystalline with excellent thermoplastic properties but poor mechanical properties.[33] The typical molecular weight of PHB ranges from 10 000 to 3 000 000 Da. Some of the recent PHA and PHB studies are listed in Table 6.1.

6.7 PHA Microorganisms

More than 300 species of microorganisms[12,33,36] have already been isolated and identified as capable of producing PHAs. The majority of these microorganisms are Gram-positive or Gram-negative bacteria. The different species of microorganism are identified as either Gram-positive or Gram-negative based on the thickness of their peptidoglycan layer and outer lipid membrane. Gram-positive bacteria have a thick peptidoglycan layer but no outer lipid membrane while Gram-negative bacteria have an outer lipid membrane but thin peptidoglycan layer. Examples of Gram-positive bacteria include *Streptococcus*, *Clostridium*, *Listeria*, *Lactobacillus*, and *Mycobacterium* while Gram-negative bacteria include *Salmonella*, *Pseudomonas*, *Pasteurella*, and *Escherichia*.

Determining whether the microorganism is Gram-positive or Gram-negative is not only important in classification of the type of bacteria but also in determining the fermentation conditions and PHA extraction processes. For example, Gram-positive bacteria are typically more resistant

Table 6.1 Overview of recent PHA studies.

	Ref.	Microorganism	Substrate	Biopolymer	Temp./°C	pH	Mode of fermentation
1	38	Highly osmophilic organism	Hydrolyzed whey, glycerol liquid phase (biodiesel production)	Poly[3(hydroxybutyrate-*co*-hydroxyvalerate)] or poly[3(HB-*co*-HV)]	37	7	Batch
2	39	Mixed culture from activated sludge	Sugarcane molasses	PHA	30 (CSTR)	5 to 7	CSTR (1st stage), SBR (2nd stage), batch (3rd stage)
3	40	*Cupriavidus necator* (DSM 545)	Waste glycerol (biodiesel byproduct), commercial glycerol	Poly(3-hydroxybutyrate) or P(3HB)	34	6.8	Fed-batch
4	41	*Cupriavidus necato*r (MTCC-1954)	*Calophyllum inophyllum* oil	PHA	30	NI[a]	Batch
5	42	Activated sludge	Phenol	PHA	15–23	4 to 9	Batch, sequencing batch reactors (SBR)
6	43	*Burkholderia* sp. AIU M5M02	Mannitol from seaweed	PHB	30	5, 7, 9	Batch
7	44	*Halomonas venusta* KT832796	Glucose	PHA	30	7	Batch, fed-batch
8	45	Aerobic activated sludge augmented with *P. acidivorans*	Sodium acetate solution	PHB	30	7	Sequencing fed-batch reactor, fed-batch
9	46	*Halomonas bluephagenesis* TD01	Glucose, corn steep liquor powder	PHA	37	8.5	2 stage fed-batch
10	47	*Methylocystis hirsuta*	Artificial biogas	PHA	25	NI[a]	Batch
11	48	Mixed culture from oil-contaminated soil	*Parthenium hysterophorus* (herbaceous plant)	P3HB	37	NI[a]	Batch
12	49	*Pandoraea* sp. MA03 from Atlantic rainforest	Crude glycerol from biodiesel industry	P3HB	30	7	Batch

Table 6.1 (*Continued*)

	Ref.	Microorganism	Substrate	Biopolymer	Temp./°C	pH	Mode of fermentation
13	50	*Synechocystis salina* CCALA192	Carbon dioxide	PHB	NI	8.5	Batch
14	51	Activated sludge (abundant microorganism includes *Plasticicumulans*, *Zoogloea*, *Paracoccus*, and *Flavobacterium*)	Mixed volatile fatty acids (25% C-mol based acetic acid, propionic acid, butyric acid, and valeric acid)	PHA	22	7 to 8	Sequencing batch reactors
15	52	Activated sludge (dominated by *Corynebacterium* and *Xantomonadoceae*)	Fermented cheese whey organic acids and ethanol	PHA	23–25	pH not controlled	Sequencing batch reactors
16	53	*Pseudomonas pseudoflava* (NBRC-102513)	Synthetic wastewater	PHA	30	pH not controlled	Batch
17	54	*Bacillus megaterium* KCTC 2194	Acid pre-treated red algae (*Gelidium amansii*)	PHA	30	7	Batch and fed-batch
18	55	*Bacillus cereus* EGU43, *Bacillus thuringiensis* EGU45	VFA from various hydrolyzed biowastes	PHB	37	7.2	Batch
19	56	*R. eutropha* H16 (NCIMB 10442, ATCC 17699)	Oxidized polyethylene wax	PHA	30	NI[a]	Batch
20	57	*R. eutropha* H16 (DSM428) and recombinant *R. eutropha* strain Re2058/pCB113	Waste frying oil and waste animal fats	PHB from *R. eutropha* H16, poly(hydroxybutyrate-*co*-hydroxyhexanoate) from the recombinant strain	30	6.8	Batch

[a] NI means not indicated.

to physical disruptions than Gram-negative bacteria due to their very rigid and less elastic cell wall.[58–60] Thus, the classification of Gram-positive or Gram-negative bacteria will be helpful in determining the bacterial lysis method in the extraction of PHA. As another example, Gram-positive bacteria are usually highly resistant to sodium azide solution while Gram-negative bacteria are less resistant.[61,62] Sodium azide solution is sometimes added as a preservative to reduce contamination in fermentation due to its inhibitory growth effect on many bacteria. For Gram-positive bacteria, which are usually resistant to sodium azide, this antibiotic technique will be effective but not so much with Gram-negative bacteria. Therefore, other preservative techniques must be applied to Gram-negative bacteria.

Many of these microorganisms were isolated from various sources such as marine ecosystems, soil, groundwater, and wastewater. The most successful and the most widely studied among these microorganisms is *Ralstonia eutropha* or *A. eutrophus* (previous name). It has high cell density, excellent tolerance to various substrates, and is a very efficient PHA producer; thus, it is the typical choice for genetic manipulation.[14,26,63,64]

It is widely known that bacteria producing PHA are grouped into two categories based on their environment: those that require nutrient limitation and those that do not. Bacteria that require nutrient limitation include *Ralstonia eutropha*, *Protomonas extorquens*, and *Protomonas oleovorans* while the second group includes *Alcaligenes latus*, a mutant strain *Azotobacter vinelandii*, and a recombinant *E. coli.*[65] The second group of bacteria was reported to produce PHA during the growth phase.[66–68] For the first group, in the presence of excess carbon, PHA is produced by bacteria intracellularly when critical growth conditions occur such as a nitrogen, phosphorus, or magnesium limitation. This mechanism is referred to as feast and famine conditions. This is a condition where the substrate's availability is fed in an alternate pattern. The majority of microorganisms adapt to this type of changing environment to survive. When one of the nutrients is limited, the microorganism uses the abundant substrates to accumulate PHA instead of using them for growth. Another explanation is that during the nutrient limitation phase, there is a reduction in the production of intracellular components such as RNA and enzymes that are necessary for growth; thus, intracellular products such as PHA accumulate.[69–71]

6.7.1 PHA Produced *Via R. eutropha*

Several factors need to be considered when selecting microorganisms for PHA production. This includes cell density, growth rate, ability to consume complex and cheap substrates, and easy recovery. As previously mentioned, *Ralstonia eutropha* (previously known as *A. eutrophus*) is the most extensively studied bacterium to produce PHA. It has been used in industrial-scale production, such as by Imperial Chemical Industries in the UK, who used it to produce PHB from glucose and poly(3-hydroxybutyrate-*co*-3-hydroxy-valerate) from a mixture of glucose and propionic acid.[22] They were able to produce a cell density of over 100 $g L^{-1}$ after 72 hours of fermentation.[72] In China, Tianjin Northern Food Co. Ltd and Zhejiang Tian An Co. Ltd have produced *R. eutropha* of over 200 $g L^{-1}$ with 80% PHB content and 160 $g L^{-1}$ with 75% PHBV content, respectively.[72] Metabolix Inc., USA, has also used *R. eutropha* to produce P3HB4HB from glucose.[72]

Aside from glucose, many other carbon sources have been used by *R. eutropha* for PHA production. Organic acids have been successfully employed as a carbon source. For example, Tsuge *et al.*[73] used L-lactic acid to produce PHB using *A. eutrophus* in fed-batch culture, while Wang and Yu[74] used acetic acids to produce PHB using *R. eutropha* in batch culture. A systematic optimization modelling of mixed acid ratios was investigated by Yang *et al.*[75] to determine the best combination of organic acids to produce the maximum PHA. The study demonstrated that the initial pH of the growth medium gave different cell growth depending on the type of acids and concentration used. Among the three organic acids used (propionate, acetate, and butyrate), it was observed that *R. eutropha* consumed butyrate more rapidly compared to the other acids. The increased butyrate content in the mixture of acids produced more cell dry weight and more PHAs.[75] Many studies have found that plant oils can be a promising substrate as well.[76–79] Additionally, PHB was also produced *via* solid-state fermentation using two agricultural residues (babassu and soy cake) on *R. eutropha.*[80] While supplementing soy cake with sugar cane molasses, they were able to produce 4.9 mg PHB per g medium after 60 hours of fermentation. The supplementation quadrupled the PHB production.

6.7.2 PHA Produced *Via A. latus*

Similar to *R. eutropha*, *Alcaligenes latus* has also gained much attention because of its capability to produce large amounts of PHB on simple glucose medium, fast growth rate, and ability to utilize cheap substrates.[81,82] Chemie Linz AG of Austria (later known as Biotechnologische Forschungsgesellschaft, then became Biomers, Germany) produced PHB from *A. latus* with a final cell dry weight (CDW) of >60 $g L^{-1}$ containing >75% PHA CDW.[72,83,84] A more recent study has modelled various nutrient feeding strategies of *A. latus* for increased production of PHB.[85] The study showed that a dynamic strategy of fed-batch cultivation under pseudo-steady-state produced *A. latus* of up to 39.17 $g L^{-1}$ containing 29.64 $g L^{-1}$ PHB.[85]

One of the substrates used to grow *A. latus* for PHA production was sugar beet juice.[86] In the US, sugar beet production reached approximately 33.88 million tons in 2016.[87] Because of its high sucrose content, it has been considered a promising feedstock for the biofuel industry. It also contains nutrients such as nitrogen, phosphorus, sodium, and potassium; therefore, it is highly suitable for microorganism growth. In Wang *et al.*,[86] sugar beet juice was used to grow *A. latus* (ATCC 29714) for PHB production. The highest PHB produced reached approximately 4.01 ± 0.95 $g L^{-1}$ with a maximum productivity of 0.22 ± 0.01 $g L^{-1} h^{-1}$. Whey[88] and maple sap[89] were two other substrates used to grow *A. latus* for PHB production. Berwig *et al.*[88] pretreated whey with 40% (w/v) citric acid solution to precipitate the protein and only used its supernatant liquid as the growth medium. Their best results showed productivity of 0.11 $g L^{-1} h^{-1}$ and a high PHB yield of 1.08 $g g^{-1}$ ($Y_{Product/Substrate}$). In addition, Canadian researchers[89] studied maple sap as a substrate for PHB production using *A. latus*. Their results showed PHB concentrations of 2.15 $g L^{-1}$ and 3.41 $g L^{-1}$ for sucrose and maple sap, respectively. Interestingly, maple sap (0.34 $g g^{-1}$) gave better PHB yield than pure sucrose (0.23 $g g^{-1}$) in shake-flask culture, which could be attributed to the presence of organic compounds in the maple sap that was used to supplement the bacterial growth.[89]

6.7.3 PHA Produced *Via Methylotrophs*

In many fermentation-based products such as PHA, one of the biggest deterring factors for industrial success is the cost of substrates. Methylotrophic bacteria are heterogeneous prokaryotes that grow on methane or methanol as their carbon source.[90] Methane is a cost-effective substrate, thus it has been explored in the production of PHAs. Additionally, methane is the second largest greenhouse gas contributor next to carbon dioxide; therefore, using it as a substrate for bioplastic will reduce its emission to the atmosphere. Recent reviews by Karthikeyan *et al.*[91] and Strong *et al.*[92] have detailed the potential of methylotrophs for PHA production.

Many studies have already explored the use of methylotrophs for PHA and PHB production. Microorganisms include *Methylobacterium extorquens*,[93–98] *Methylobacterium trichosporium*,[99–102] *Methylobacterium organophilum*,[100,103] *Methylobacterium parvus*,[101,104] *Methylobacterium rhodesianum*,[105] and *Methyloligella halotolerans*.[106] The biosynthesis of PHB in methylotrophs involves three main enzymes, namely β-ketothiolase, acetoacetyl-CoA reductase, and PHB synthase.[90] The first enzyme, β-ketothiolase, is responsible for catalysing the condensation of two molecules of acetyl-CoA while the second enzyme, acetoacetyl-CoA reductase, reversibly reduces acetyl-CoA to hydroxybutyryl-CoA.[90] The third enzyme, PHB synthase, is the main enzyme responsible for the synthesis of PHB.[90] A detailed enzyme and gene biosynthesis process of methylotrophs can be found in Zamakhaeva *et al.*[90]

6.7.4 PHA Produced *Via* Phototrophs

Cyanobacteria, commonly known as blue-green algae, are oxygenic photoautotrophic microorganisms that exist in diverse environments such as freshwater, oceans, soil, and even in rocks. They are the only prokaryote that accumulates PHB through oxygenic photosynthesis.[107] As seen in Figure 6.3, these phototrophs utilize gas as a carbon source such as syngas and carbon dioxide in the presence of light. Cyanobacteria are very promising for synthesizing PHAs because of their minimal nutrient requirement and short generation time.[108] They can also utilize carbon dioxide as a carbon source, which can reduce greenhouse gas emissions.

Many studies have explored the use of cyanobacteria for PHA production. A thermophilic cyanobacteria *Synechococcus* sp. MA19 was isolated from the surface of wet volcanic rock.[107] The study showed 20% (dry cell weight) PHB accumulation when fermented in nitrogen-free inorganic medium aerated with 2% carbon dioxide.[107] In another study, PHA accumulation increased when *Spirulina subsalsa* was grown in two-fold sodium chloride (NaCl) concentration relative to a control.[109] On the other hand, Bhati *et al.*[108] found 19 different genera out of 25 cyanobacterial species examined that produce PHA. Some of the cyanobacteria found were *Anabaena cylindrical*, *Anacystis nidulans*, *Aphanocapsa* sp., *Aphanothece* sp., *Aulosira fertilissima*, *Calothrix* sp., *Cylindrospermum indicum*, *Fischerella muscicola*, *Microcystis* sp., *Nostoc muscorum*, *Oscillatoria limosa*, *Plectonema* sp., *Pleurocapsa* sp., *Rivularia* sp., *Spirulina platensis*, and *Tolypothrix* sp.

Another phototrophic microorganism that accumulates PHAs is *Rhodospirillum rubrum*. *R. rubrum* is a phototrophic, purple, non-sulfur bacterium that utilizes carbon dioxide as a carbon source under anaerobic environments.[110,111] Syngas has been used as a promising substrate for *R. rubrum* growth for PHA production.[112] Synthesis gas or syngas is a product of gasification of solid, carbonaceous fuels.[113–115] It is primarily composed of light gases such as carbon dioxide, carbon monoxide, hydrogen, and methane gas.[112] These products are produced when chemical equilibrium is achieved during gasification.[112] Lately, syngas is produced from the gasification of lignocellulosic materials which are very cheap materials; therefore, syngas has the potential to be an inexpensive substrate for PHA production.[116]

In a study by Do *et al.*,[112] they investigated the growth of *R. rubrum* on synthesis gas to produce carbon monoxide, hydrogen, and PHA. To reduce the cost of the growth medium, they removed MOPS (3-(*N*-morpholino)-propane sulfonic acid) and acetate. This resulted in a 50% decrease in PHA production for acetate removal but had no effect on growth rates or H_2 production for MOPS removal.[112] Although the authors were able to utilize syngas to grow *R. rubrum*, the mass transfer of carbon monoxide into the liquid phase was the rate-limiting step.

R. rubrum was also cultivated in condensed corn solubles for poly-(3-hydroxybutyrate-*co*-3-hydroxyvalerate) production.[117] Condensed corn solubles (CCS) is a co-product of corn ethanol production. The study showed that *R. rubrum* has a chemoheterotrophic preference where it favours fermenting sodium pyruvate when grown anaerobically in the presence of light.[117] Similar findings were also reported by Solaiman and Uffen.[118] The CCS medium showed no significant difference in terms of growth rate or maximum cell density in comparison to another more expensive complex medium.[117] This could be because CCS already contained a sufficient amount of usable carbon for growth.[117] Therefore, CCS is a promising substrate that could offset the cost of PHA production.

6.7.5 PHA Produced *Via* Mixed Culture

The use of mixed culture in PHA production has gained much attention in recent years, as seen in the number of recent studies exploring mixed cultures (see Table 6.1). This is because mixed cultures can use complex, inexpensive substrates that are high in organic nutrients such as food wastewater. With the intention of reducing the overall cost, utilization of waste substrate has made the use of mixed culture promising. Also, unlike pure cultures, the need for sterilization is not required in mixed cultures, making the process easier to handle and reducing the energy use. Additionally, the PHA production is not induced by nutrient limitation because mixed cultures can consume different complex substrates rich in nutrients contrary to their pure culture counterpart.[119] In more detail, in mixed cultures, they can typically use various complex substrates rich in nutrients rather than only one type of substrate such as those in pure cultures. In mixed cultures, microorganisms compete to consume the substrate, while in pure culture, they typically have the same growth rate. In addition, in pure culture, because the concentration of substrates is controlled, nutrients can be easily limited. It is important to note that PHA storage still occurs when growth is prevented regardless of pure and mixed cultures.[119] While there are many advantages that mixed cultures provide, there is still no industrial-scale production at this time.

The first PHB produced from activated sludge bacteria were observed by Deinema in 1972.[120] The study's results showed 0–0.2% PHB content (dry weight basis) for domestic waste and 0.5–2.0% and 0.1% for potato sewage and dairy sewage, respectively. Another study followed in 1974, where PHA was found in the activated sludge of a commercial treatment facility of a grain processing plant and a local wastewater treatment plant.[121] However, they could not produce PHA *in vitro* from the bacteria isolated from the activated sludge.[121] The study suggested that the PHA produced was dependent upon the formation of a precursor which was produced by another bacterial species in the mixed culture.[121] Years later, another study found that PHA could be produced in a co-culture system where two sets of bacterial species were used to produce PHA.[122] In a two-stage fermentation, PHA was produced by *Alcaligenes eutrophus* from xylose *via* L-lacate.[122]

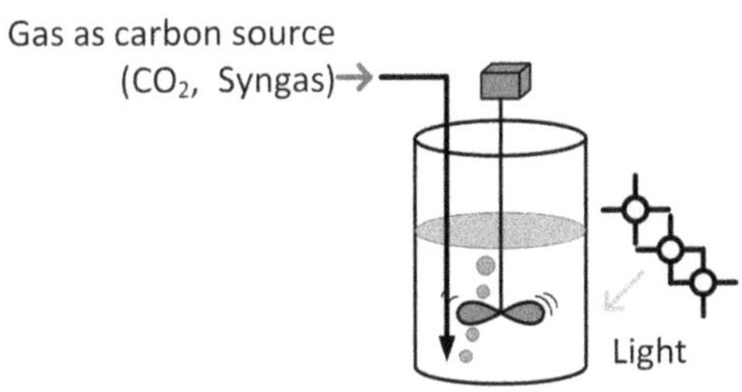

Figure 6.3 Fermentation operation of a phototroph microorganism using gas as a carbon source.

In the first stage, xylose was metabolized by *Lactococcus lactis IO-1* into L-lactic acid and acetic acid. Then, these acids were further converted to PHA by *A. eutrophus* in the second stage of fermentation.[122]

Generally, the production of PHA from a mixed culture is divided into three stages.[52] The first stage involves the conversion of the organic-rich substrate, such as food, domestic, industrial, and agricultural wastes, into organic acids. This is often called the acidogenic fermentation stage.[52] The second stage is the enrichment of the mixed culture where it undergoes feast and famine cycles.[52] This will ensure the utilization of the microorganism's capability to store PHA. The third stage is the PHA production, where organic acids from the first stage are converted into PHA. For further information on the mechanism and metabolism of mixed culture for PHA production, Serafim *et al.*[119] have reviewed this topic well.

One of the benefits of using mixed culture is its ability to use low price and complex substrates such as the organic fraction of municipal solid waste (OFMSW).[123] Typically, municipal solid wastes are collected and landfilled, incinerated, composted, or anaerobically digested. Although the biological treatment (compost and anaerobic digestion) provides additional value from the electricity and compost conversion, plenty is still wasted in landfills. One potential use of OFMSW could be as a substrate for PHA production. A two-stage fermentation was conducted using mixed culture and OFMSW.[123] The first stage was completed in an anaerobic percolation reactor where the liquid digestate obtained from a full-scale anaerobic digester was fed into OFMSW.[123] The liquid percolate was collected, centrifuged, ammonia stripped, diluted to the desired chemical oxygen demand (COD), then fed into a preselected mixed culture of PHA producers.[123] The results showed an average of 223 ± 28 g PHA per kg organic acid fed and 33.22 ± 4.2 g PHA per kg from fresh OFMSW.[123] There are many other promising studies on mixed culture for PHA production.[52,124–133]

6.8 PHA Fermentation Strategies

To produce high cell density PHAs, it is important to optimize the mode of fermentation operations. As detailed in Table 6.1, various fermentation operations have been studied to produce PHA. The strategy for using a particular operation will depend on the substrate, scale, and economics of the operation. In this section, three popular strategies will be discussed, namely batch, fed-batch, and continuous.

6.8.1 PHA Produced in Batch

Batch fermentation is considered the simplest yet the most studied operation for producing PHA. Approximately a third of the PHA published work was conducted using batch fermentations. It is essentially a closed system where all materials are added at the beginning and only removed at the end of fermentation. There is no substrate introduction nor effluent withdrawal at any point during the duration of fermentation.

The majority of the PHA produced in batch fermentation was grown in simple shake-flask cultures or bioreactors. However, other batch fermentation strategies are reported, such as sequencing batch reactors and two-stage batch fermentation. Many shake-flask batch fermentations of PHA involve optimization of inoculum, pH, fermentation media, fermentation time, and temperature.[42,134,135] These studies are particularly beneficial with a new strain of PHA producing microorganism and for exploring its PHA producing capacity at various fermentation conditions. Shake-flask cultures are inexpensive, easy to run, and can be conducted in various volumes, which makes them a good option for gathering preliminary data. In order to study maximum PHA production for scale-up, many researchers move to cell bioreactors which involve more capability for gas, feed, and pH control.

6.8.1.1 PHA Produced in Sequencing Batch Reactors

The sequencing batch reactor (SBR) is a recognized process for treating wastewater using activated sludge. It consists of four operations: fill, treat, settle, and withdraw (see Figure 6.4). The wastewater is introduced into the tank where the activated sludge has already been acclimated. Once it is filled, it operates as a regular activated sludge system where the air is bubbled through the tank and treats the wastewater. After treatment, the air is stopped and the wastewater settles. After a specified period, the treated

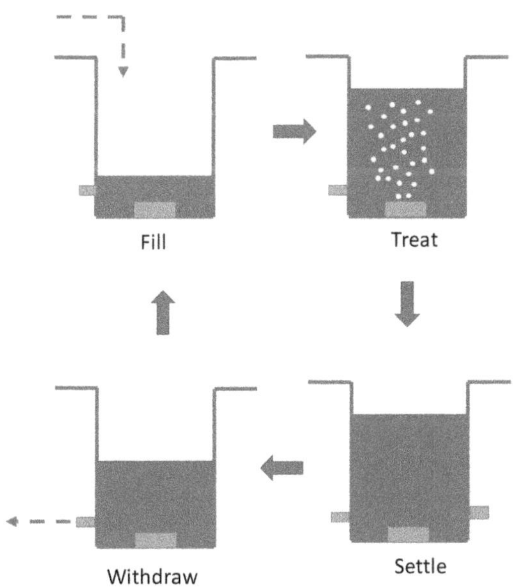

Figure 6.4 Sequencing batch operations of PHA.

wastewater is withdrawn leaving only the activated sludge. The SBR design in treating wastewater has been popular because it has a smaller footprint and is cost-effective because of the use of single reactors instead of two or three in traditional activated sludge systems.[136] Also, it offers flexibility in operations and control. However, it does require larger reactors with many process controls, a higher level of maintenance, and the potential of aeration plugging during operation depending on the aeration system used.[136]

With the thrust to reduce the PHA production cost, many studies have explored using high organic wastewater as a low-cost alternative feedstock and mixed culture as their microorganism. The use of mixed culture can significantly reduce the energy cost of sterilization. Because SBR systems are widely used in wastewater treatment, it has been a popular choice to study PHA produced in mixed cultures and wastewater as a carbon source. A robust investigation towards using municipal wastewater in SBR for producing PHA has been reported by Coats *et al.*[137] The study explored two schemes: a concurrent wastewater treatment and PHA production and a side-stream PHA production in SBR systems. At room temperature and various organic loading rates, the study found that the SBR system could yield 0.23 to 0.31 mg PHA per mg COD and an average COD removal of up to 89%. The study estimated that for 265 000 L per day (1 million gallons per day) of the SBR system, it could yield approximately 11–36 tonne of PHA annually.[137]

In SBR studies, the capacity of the bacteria to store PHA is established by alternating feast and famine conditions. With these systems, the organic loading rates (OLR), solid retention times (SRT), and cycle length are very important parameters to obtain maximum PHA storage. It has been well studied that the mixed culture exhibited good PHA storage at lower OLR (0.3–1.2 g COD per L per day).[138–140] According to Dionisi *et al.*,[141,142] at OLR around 12.75 g COD per L per day, the mixed culture showed a transient response; at OLR of 20 g COD per L per day, the mixed culture shifted from storing PHA to growth; and at 25 g COD per L per day, the mixed culture showed little to no contribution to PHA storage. As for SRT, the PHA storage capacity in SBR decreases when SRT is too low.[143] However, many studies have shown that SRT of one,[144] two,[139] three,[145] and 9.5[138] days can still show a good PHA storage capacity.

In another study,[141] the length of the cycle was varied (one to eight hours) while operating at a fixed OLR (20 g COD per L per day) and HRT (1 day). This study evaluated the PHA production and the competing bacterial growth by varying the length of the cycle in the SBR. This study resulted in better PHA storage capacity at two and four hour cycle lengths while at lower (one hour) or higher cycle lengths (eight hours), the dynamic response of the mixed culture was largely dominated by the growth. With their best conditions, the storage rates reached 500–600 mg COD per g COD per h and productivity of 0.25–0.30 g PHA per L per h, which was according to the study the highest obtained for mixed cultures. A similar study by Valentino *et al.*,[146] with comparable conditions to Dionisi *et al.*,[141] was conducted on a lower OLR (8.5 g COD per L per day) and a cycle length of two to eight hours. They found that their best performance was obtained at a cycle length of two hours with a feed frequency of 12 times per day, although they did note that a cycle length of six hours will simplify the process management under continuous operation.

6.8.2 PHA Produced in Fed-batch

In typical PHA production through fed-batch fermentation, the materials are withdrawn at the end of the fermentation, and the substrate is added until the desired working volume is achieved. Many microorganisms that use fed-batch fermentation to produce PHA have two phases: growth phase and product production phase. The growth phase is the stage where microorganisms can grow exponentially until a necessary biomass concentration is achieved. The product production phase is the stage where excess carbon is fed to the microorganism while restricting other essential nutrients such as nitrogen, phosphorus, and magnesium. This alternating phase is called feast and famine conditions. Because of this changing environment, the microorganism produces PHA instead of reproducing itself.

In fed-batch fermentation of PHA, carbon and nitrogen are typically controlled substrates. One of the popular strategies is "pH-stat" feeding.[73,147–151] The "pH-stat" feeding is a strategy that uses pH to monitor and control carbon or nitrogen level in fermentation. It has been known that the reduction of pH is directly related to biomass growth. Using a nitrogen-based alkali solution such as ammonium nitrate or an organic acid such as lactic acid creates a simple yet effective method to monitor and supply either carbon or nitrogen substrate to the biomass growth. In transitioning the biomass growth to the PHA production phase, the nitrogen-based alkali solution or organic acid is replaced with sodium hydroxide or inorganic acid to allow the carbon and nitrogen to deplete but still maintain the pH of the fermentation.

A pH-stat feeding strategy is beneficial in substrate inhibiting fermentation. Tsuge *et al.*[73] optimized the feeding of L-lactic acid for the production of P(3HB) (poly(3-hydroxybutyrate)) by *A. eutrophus* in a pH-stat fed-batch fermentation. Their preliminary study showed that when organic acids such as lactic acid are used as a carbon source in the fermentation of *A. eutrophus*, they must be kept at low concentration to obtain high cell densities. By using pH-stat feeding, they were able to maintain the pH of the fermentation and maintain the concentration of L-lactate at a minimum. Similarly, in Kobayashi and colleague's[150] study, pH-stat feeding was used to optimize the production of poly(3-hydroxybutyrate-*co*-3-hydroxyvalerate) by *Alcaligenes eutrophus* ATCC17697 from propionic acid. At a carbon to nitrogen ratio of 50, the biomass concentration reached 64 $g L^{-1}$ with 58% P(3HB-*co*-3HV) content. In another study, the growth of *P. oleovorans* was inhibited at 4 $g L^{-1}$ octanoic acid. Thus, Kim[151] utilized the pH-stat system by feeding octanoic acid and ammonium nitrate when the pH rose above 7.1. This allowed him to maintain the carbon to nitrogen ratio of the fermentation broth and obtain the maximum cell density of 63 $g L^{-1}$ with 62% PHA content.

Another popular fed-batch feeding strategy in PHA production is carbon pulse feeding. Many carbon sources have been explored in fed-batch fermentation including glucose,[44,152] oleic acid,[153] waste glycerol,[40,154] soybean oil,[149] sucrose,[155] fructose,[156] molasses,[157] and nonanoic acid.[158] Carbon pulse feeding is a traditional fed-batch fermentation where a carbon source is fed into the culture once the initial carbon source has been depleted, as depicted in Figure 6.5. In Ienczak *et al.*,[159] they reviewed various studies focused on determining fermentation strategy to obtain high cell density in the PHA culture. In their review, they found that to obtain high cell density, carbon pulse feeding should be performed when the carbon source is close to its half-velocity coefficient[160] for the substrate saturation concentration (around 10 $g L^{-1}$) and not when the carbon source has been depleted. The half-velocity coefficient (K_S) is the soluble substrate concentration when the maximum specific substrate utilization rate (μ/μ_{max}) is one-half.[160] For example, in Kim *et al.*,[82] they kept the glucose concentration between 10 and 25 $g L^{-1}$ to obtain high productivity while Shang *et al.*[161] found that keeping the residual glucose concentration at 9 $g L^{-1}$ obtained a high cell density.

Producing high accumulations of PHA, particularly mcl PHAs, has been found to be favourable at high carbon alkanes or alkanoates substrates.[162,163] In Sun and colleagues[158] study, nonanoic acid was used as the carbon source in a fed-batch production of mcl PHA using *Pseudomonas putida* KT2440. Their fermentation strategy followed a two-stage fed-batch fermentation which started with a nutrient-rich growth on glucose substrate followed by a nitrogen-limited PHA accumulation stage on a nonanoic acid substrate. During the PHA accumulation stage, a pulse substrate feeding of 3 $g L^{-1}$ nonanoic acid was used at the beginning, then followed by linear feeding at 0.47 $g L^{-1} h^{-1}$ until the seventeenth hour, then 1.01 $g L^{-1} h^{-1}$ until the end of the fermentation. Their strategy resulted in relatively higher PHA yields (0.6 g PHA per g substrate) compared to other studies. Even though it is not true for other microorganisms, they have concluded that nitrogen limitation was unnecessary in their fermentation with *Pseudomonas putida* KT2440. This study has shown that producing mcl PHA varies greatly depending on the microorganism, carbon sources, and growth conditions.

In an effort to reduce the cost of the main carbon source, da Cruz Pradella *et al.*[149] studied the utilization of soybean oil as a low cost fatty acid alternative to growing *Cupriavidus necator* DSM 545. They initially started with a culture of 10, 20, or 40 $g L^{-1}$ soybean oil concentration in the bioreactor. Once the soybean oil concentration dropped to approximately 5 $g L^{-1}$, a pulse of soybean oil was fed into the bioreactor to restore the initial concentration. Once the pulse feeding began, NH_4OH was exchanged to NaOH to allow the nitrogen from residual NH_4OH to deplete but still maintain the pH. They found that the growth limiting factor was the exhaustion of phosphorus and possibly other minerals such as Cu, Ca, and Fe. The highest productivities were 2.5 $g L^{-1} h^{-1}$, which the authors reported to be the maximum found for edible oil as a carbon source.

6.8.2.1 PHA Produced in Fed-batch with Cell Recycling

Cell recycling in fed-batch systems has been utilized in many commercial settings. Cell recycling is the reuse of the cell culture for the next fermentation cycle. Instead of disposing of the cell culture after one fermentation cycle, it is separated using centrifugation or a filtration process, then added back to the fermentation reactor with fresh medium ready for the next fermentation cycle. It is particularly beneficial in removing or minimizing inhibitory metabolites that are toxic to the microorganism. Additionally, cell recycling can also be a way to reduce the fermentation volume without losing the biomass and creating an opportunity for a fresh medium to be fed into the culture. Fed-batch cell recycling can be achieved in two methods (see Figure 6.5): internally or externally.[159] The internal recycle involves filtration or a sedimentation process within the reactor where cells are retained in the bioreactor while clarified beer is removed. To compensate for the liquid that was removed, fresh medium is also added. With the external recycle, fermented beer is removed from the bioreactor and undergoes a solid–liquid separation process such as centrifugation or a membrane filtration system. The clarified beer or supernatant (centrifugation) is removed while the cells are recovered and recycled back into the bioreactor with the fresh medium. The benefit of the internal cell recycle method is the lowered risk of contamination compared to the external cell recycle method.

One of the notable studies on producing PHA on fed-batch culture with cell recycling was conducted by Ahn and colleagues.[148] They used fed-batch with an external cell recycle strategy employed on a recombinant *E. coli* CGSC 4401 strain using bovine whey solution as the carbon source. To overcome the low solubility of lactose in the feeding solution, fed-batch culture was used to allow continued nutrient feeding once lactose had depleted, thereby also increasing PHB productivity. The starting volume was 1.3 L of a whey solution. The nutrients were fed once the pH exceeded their set value of 6.95,

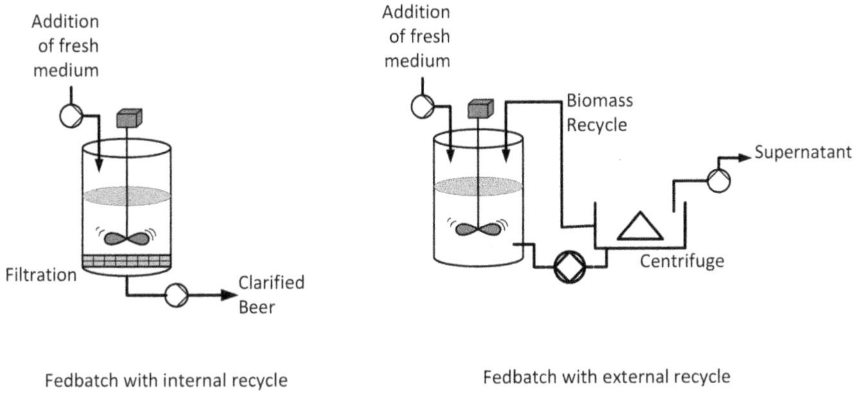

Figure 6.5 Fed-batch fermentations of PHA production showing internal and external recycles.

signifying the depletion of lactose. When the volume reached 2.3 L, a portion of the fermentation beer was pumped to an external crossflow membrane where the permeate (cell-free liquid) was separated while the cells were recycled back to the fermenter. Their effective feeding and recycle regimes resulted in the final cell concentration, PHB concentration, PHB content, and relatively high productivity of 194 g L^{-1}, 168 g L^{-1}, 87%, and 4.6 g PHB per L per h, respectively. Their study showed that high productivity of PHB is possible even for a low solubility substrate such as lactose.

In a similar study by Haas *et al.*,[164] *Cupriavidus necator* DSM 545 was cultivated in fed-batch culture with an external cell recycle using a synthetic medium with glucose as a carbon source. When glucose was completely consumed, the cell recycle was automatically triggered. The fermentation liquid flowed through an external microfiltration membrane unit where the permeate (cell-free liquid) was separated, and cells were recycled back to the reactor. The pH was controlled at 7.0 by adding 12.5% ammonia during the cell growth phase. Once a certain set of optical density was reached, the PHB production phase began by switching to NaOH to control pH. The study's approach had a yield as high as a cell density of 148 g L^{-1} and a high productivity of 3.1 g L^{-1} h^{-1}. It is noteworthy that regardless of the strain (wild-type or recombinant) used, both studies[148,164] showed that high productivity of PHB can be produced using simple carbon sources (glucose and lactose) which can be available in very cheap raw materials such as whey.

6.8.2.2 PHA Produced in Repeated Fed-batch

Repeated fed-batch is a system where a volume of the fermentation broth is removed and replaced with the same volume of fresh medium, as shown in Figure 6.6. This is typically done once the carbon source has been depleted. It is very beneficial for creating high productivity yields and reducing downtimes for cleaning and sterilization.[159] The feeding of nutrients and removal of fermentation broth is very similar to sequencing batch reactors (discussed in an earlier section). However, most studies that refer to the term "sequencing batch reactors" use wastewater as a nutrient source and mixed culture bacteria to produce PHAs.

A repeated fed-batch culture without cell recycle was performed by Khanna and Srivastava[165] on *Ralstonia eutropha* NRRL B14690 using a synthetic medium with fructose as a carbon source. The fermentation was conducted on a 7 L bioreactor with a working volume of 4 L maintained at 30 °C and pH 7.0. When the set residual fructose concentration was achieved, an empty-and-fill process was conducted where 20% fermentation volume was removed and replaced with fresh medium. This process was conducted twice during a 67 hour fermentation period. The study reported an expected drop of dry cell weight during the empty-and-fill process. Despite this, the study resulted in 49 g biomass per L, 25 g PHB per L, and overall productivity of 0.42 g PHB per L per h, which is better than those obtained in batch culture.

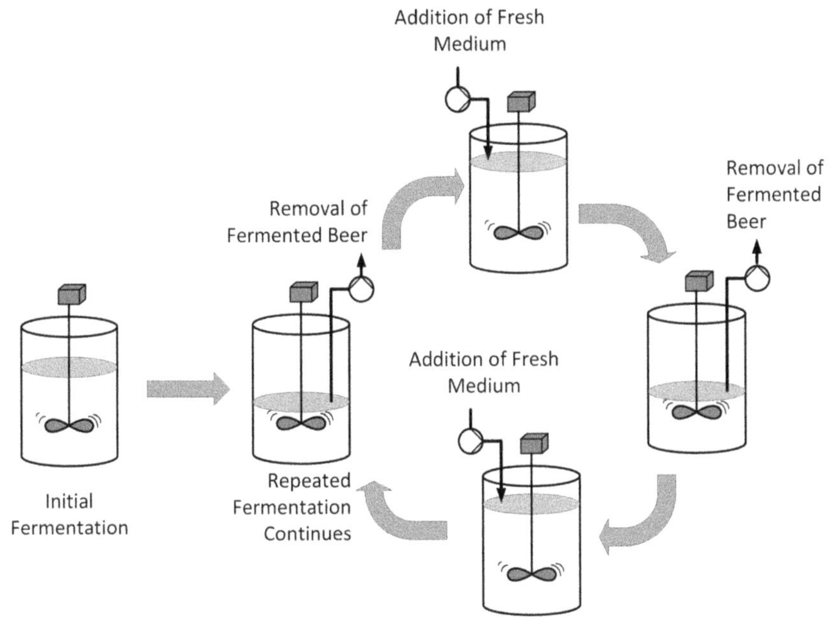

Figure 6.6 Repeated fed-batch fermentation of PHA production.

In Ienczak and colleagues,[166] PHB was produced in repeated fed-batch culture with external cell recycle in *Cupriavidus necator* DSM 545 using a low carbon source concentration medium. One of the strategies to make PHB production cost-effective is the use of low-cost substrates such as whey and lignocellulosic hydrolysates. However, one of their drawbacks is that only a low to moderate concentration can be used. Thus, the authors introduced a low carbon source concentration medium to emulate a cheap substrate. The repeated fed-batch culture was performed in a 5 L bioreactor with an initial working volume of 4 L synthetic medium containing fructose and glucose and maintained at 35 °C and pH 7.0. The cell recycle was conducted by flowing fermentation broth into two crossflow polypropylene external membranes in parallel, where the cell-free permeate was removed, and the cells were recycled. Nutrient media were added after every recycle to maintain the same working volume. There were two and six batch recycles done in the growth phase and the nutrient limiting phase, respectively. The study's strategy successfully resulted in a total biomass of 61.6 $g L^{-1}$, PHB concentration of 42.4 $g L^{-1}$, and PHB productivity of 1.0 $g L^{-1} h^{-1}$. The study showed that the cell recycle can make a significant impact on the PHB productivity.

6.8.3 PHA Produced in Continuous

Many continuous systems are beneficial on a large commercial scale because of their high productivity, high cell growth rate, long residence times, and control on microorganism evolution, as discussed in earlier sections of this chapter. The continuous systems are often interchangeably referred to as "chemostat" systems. Chemostat, as described by Monod,[167] is a system where the cell growth rate reaches a steady-state by varying the dilution rates. In order to achieve this, a fixed volume bioreactor is maintained where fermented beer is continually removed as fresh medium is continuously added,[168] as depicted in Figure 6.7. However, some authors[169–171] may disagree because strictly speaking continuous systems may also include processes such as "pH-stat," "turbido-stat," "redox-stat," or "volume-stat" in addition to "chemo-stat" (chemical environment remaining static). Because continuous is a much broader term than chemostat, this chapter will refer to the term continuous culture systems instead.

Table 6.2 summarizes selected studies of PHA produced in continuous systems in the last two decades, with information on the various biopolymers produced, mode of fermentation, the substrate used, and productivity yield achieved.

An interesting continuous system was investigated by Khanna and Srivastava[178] in the production of PHB on *Wautersia eutropha* using a two-stage continuous culture system. Two 7 L bioreactors were connected in series and fed with a medium containing fructose as the main carbon source. The first stage was cultivated under nitrogen-rich conditions while the second stage was cultivated under conditions suitable for PHB accumulation. The effluent of the first stage and fructose-rich media were fed into the second stage. The first and second stage reactors were maintained at 1.5 and 4 L, respectively. The authors' strategy was to grow the biomass in batch, accumulate PHB in fed-batch, then maintain the biomass and PHB in continuous operation. The results showed a steady-state biomass of 27.7 $g L^{-1}$ and a PHB concentration of 5.5 $g L^{-1}$. The PHB productivity in continuous and fed-batch cultivation yielded 0.55 $g L^{-1} h^{-1}$ and 0.48 $g L^{-1} h^{-1}$, respectively, which were superior to the batch cultivation at 0.18 $g L^{-1} h^{-1}$. Although their productivity results were lower than other continuous culture studies, they demonstrated a novel production technology integrating batch, fed-batch, and continuous process.

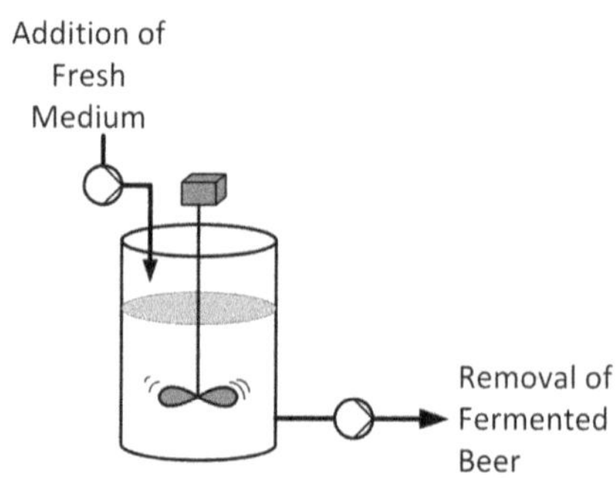

Figure 6.7 Continuous culture system of PHA production.

Table 6.2 Selected studies of PHA produced in continuous systems.

Ref.	Microorganism	Biopolymer	Mode of fermentation	Substrate	Yield
172	*Pseudomonas oleovorans*	MCL poly(3-hydroxy-alkanotaes)	2 stage continuous	Citrate and *n*-octane	1.06 $g L^{-1} h^{-1}$ with 63% cdw PHA
173	*Psuedomonas oleovorans*	MCL poly[(*R*)-3-hydroxy-alkanoates)	Chemostat	Octanoate	0.001–0.147 $g g^{-1} h^{-1}$ with 0.01–0.25 g PHA per g cells
174	*Delftia acidovorans* P4a	Poly(3-hydroxybutyrate-*co*-4-hydroxybutyrate)	2 stage continuous	Acetic acid and γ-butyrolactone (GBL)	[a]1.06 $g L^{-1} h^{-1}$, 0.63 g PHA per g cells
175	*Halomonas campaniensis* LS21	Poly-β-hydroxybutyrate (PHB)	Open and continuous	Mixed substrate (cellulose, starch, lipids, proteins)	[a]0.7 $g g^{-1}$ PHA in CDM, $g L^{-1} h^{-1}$
176	*Ralstonia eutropha* WSH3	Poly(3-hydroxybutrate)	2 stage culture system	Glucose	1.23 $g L^{-1} h^{-1}$ with 0.36 g PHB per g cell
177	*Cupriavidus necator*	Poly([*R*]-3-hydroxybutyrate)	Multi-stage bioreactor	Glucose	1.85 $g L^{-1} h^{-1}$ with 77 ± 7.5% PHB
178	*Wautersia eutropha*	Poly-β-hydroxybutyrate (PHB)	2 stage continuous	Fructose	0.55 $g L^{-1} h^{-1}$, 27.7 $g L^{-1}$ cells with 5.5 $g L^{-1}$ PHB

[a]These values were obtained from Koller,[169] where the original studies were referenced.

Another novel system was reported by Atlic *et al.*[177] in the continuous production of PHB by *Cupriavidus necator* using multi-stage bioreactor operations. The multi-stage system consisted of five continuously stirred tank reactors (CSTR) connected in series where the effluent of the first reactor and a fresh carbon-rich medium were fed into the second reactor, then the effluent of the second reactor and a fresh carbon-rich medium were fed to the third reactor. This scheme was continued until the fifth reactor. The first stage was dedicated to high cell growth while the second to the fifth stages were focused on PHA accumulation where nitrogen limiting conditions were maintained. This unique strategy yielded 81 $g L^{-1}$ cell dry weight, 63 $g L^{-1}$ PHB concentration, 77% PHB content, and 1.85 $g L^{-1} h^{-1}$ of volumetric PHB productivity, which is currently the highest in continuous systems (see Table 6.2). Also, the study compared its results with other continuous studies in the literature as well as fed-batch systems in commercial PHA production. It was evident that the five-stage continuous system was superior to the two-stage continuous systems in terms of cell density, PHB concentrations, and PHB volumetric productivity. In comparing the multi-stage system with the fed-batch system in a commercial setting, the fed-batch system had significantly higher cell dry weight and PHB concentration. Although, it is noted that the PHB content and PHB volumetric productivity of the multi-stage process were comparable to the fed-batch system. Nonetheless, this study's strategy worked effectively for continuous culture systems, which could be very beneficial for future process improvement.

6.8.4 PHA Produced in Solid-state Fermentation

So far, only submerged fermentation has been reviewed which is the production of PHA and other possible bioplastics in an aqueous system. This is because it is the most extensively used method of fermentation for producing PHA. However, over the last two decades, there has been research towards an alternative method called solid-state fermentation (SdSF).[179–182] Solid-state fermentation is the fermentation process where a microorganism grows on wet solid materials with limited free-flowing water between the substrate and the microorganism.[183] The microorganism growth is confined to the surface of the substrate or analogous to biofilm growth,[184] which could be because of the limited substrate mass transfer during fermentation. This emerging technology, which has been extensively reviewed by Sindhu *et al.*,[184] has been deemed beneficial, particularly in agro-industrial substrates. Because these substrates are readily available, abundant, and low cost, it will potentially reduce the overall fermentation cost which is one of the bottlenecks in making PHA production cost-effective.[185]

SdSF is an old fermentation technology that has been used by many cultures. The Japanese used SdSF to ferment "koji" (*Aspergillus oryzae*) in rice to produce an alcoholic beverage called sake, while the French and

Indonesians used SdSF to produce blue cheese and the soybean cake, "tempeh," respectively. This is also the same technique used in traditional bread making where yeast is left to ferment in the dough before baking.

Generally, SdSF's low water content is very favourable to fungi growth while bacteria require a higher amount of water to grow; thus, solid-state fermentation is typically not the ideal environment for their growth. However, in a study by Oliveira *et al.*,[186] they explored the use of bacteria *Ralstonia eutropha* in two solid residues (soy and babassu cake) to produce PHB. The results showed a maximum of 1.2 mg $g_{medium}{}^{-1}$ PHB after 36 hour fermentation for soy cake and 0.7 $mg_{PHB}\,g_{medium}{}^{-1}$ after 84 hour fermentation for babassu cake. When soy cake was supplemented with 2.5% (w/w) glucose and 2.5% (w/w) molasses, the PHB productivity increased from 0.8 mg_{PHB} per $g_{substrate}$ per day with soy cake alone to 1.5 and 2.1 mg_{PHB} per $g_{substrate}$ per day for soy cake supplemented with glucose and molasses, respectively. The reason molasses supplementation contributed more to PHB production than glucose is probably because of the additional nutrients such as vitamins and minerals present in the substrate.[186] The maximum PHB content obtained in this study at 4.9 $mg_{PHB}\,g_{medium}{}^{-1}$ was comparable to those obtained at submerged fermentation using the bacteria *Ralstonia eutropha* in other substrates such as propionate and octanoate.[187,188] Additionally, the same team of authors characterized the PHB produced from SdSF.[189] Their study showed identical properties of PHB produced by SdSF and those grown in submerged fermentation. In fact, PHB grown in SdSF was found to have higher molecular mass, lower polydispersity, and a lower degree of crystallinity. This could show that PHB produced from SdSF provides a broader range of applications.[184]

In another study, Rocha[190] utilized SdSF to grow *C. necator* on sunflower cake, soy bran, and waste biodiesel residue. The study showed that the waste biodiesel residue supplemented with sugarcane molasses improved the PHB production over the non-supplemented substrate. The optimum PHB produced was reported at 2.1 g kg^{-1} substrate after 80 hours of fermentation. Considering that the biodiesel industry has grown over the last decade, this will provide an added value to not only the biodiesel by-product but also a cheap alternative to the expensive cost of PHA production.

In 2013, two more studies using SdSF were conducted using *Bacillus sphaericus* NII 0838 on jackfruit seed hydrolysate substrate[191] and *Bacillus megaterium* MSBN 04 on tapioca industry waste, palm jaggery, and horsegram flour substrate.[192] Ramadas *et al.*[191] introduced an innovative strategy of adding polyurethane foams (PUF) as an inert structural support for bacterial growth on SdSF. One of the downstream processing issues for SdSF was the detachment of bacterial cells from the substrate. The addition of polyurethane foams was considered a potential solution for easy biomass recovery. After optimization using response surface methodology (RSM), they[191] found that the optimized conditions for growing SdSF in PUF were 6.5% inoculum, 1.7% (w/v) $(NH_4)_2SO_4$, and pH 9.0. The study reported a PHB production of 0.169 g g^{-1} PUF. In Sathiyanarayanan *et al.*,[192] using *Bacillus megaterium* MSBN 04 on SdSF, the maximum PHB yield was achieved by the tapioca industry waste at 8.637 mg per g substrate. Their study found that horse-gram flour used as a nitrogen source showed a significant increase in PHB production. Horse-gram belongs to the legume family and is used as a horse feed and sometimes consumed by humans for medicinal purposes. The use of horse-gram in this study has brought new promise for legume plants to be used in PHA production.

SdSF offers several advantages in comparison to submerged fermentation which includes simple substrate handling, minimum water usage, higher substrate capacity, minimum gas transfer issues, and simplified downstream processing.[184] In SdSF, substrates do not require significant pretreatment, such as milling and cooking, to extract bacterial nutrients into the liquid phase. Because SdSF only requires minimum water, there is more space in the reactor for an increased substrate concentration. At the same substrate concentration, SdSF reactors will also have a smaller footprint than submerged fermenters which could lead to cost savings. However, SdSF also has its share of disadvantages including restrictions in the type of microorganism, process monitoring issues, mass transfer limitation, and longer fermentation times.[184] Because the nutrient media is in solid form, the probes used in most submerged fermentations to monitor parameters such as pH and moisture content are not applicable to SdSF. Sampling during fermentation can be a challenge. There are many variabilities between batches which lead to poor reproducibility.[184] Because of the infancy of this technology, there are still many basic process engineering issues that need to be addressed. However, considering all the potential benefits SdSF can offer, this technology could be a promising fermentation alternative for PHA production in the future.

6.9 Polylactic Acid (PLA)

6.9.1 Lactic Acid Bacteria

Lactic acid-producing bacteria have the advantage of being the most known and the most studied among all the lactic acid microorganisms. However, they also have a number of limitations, such as low yield because of the formation of by-products, high and costly essential nutrient requirements, end-product inhibition because of lowering of pH in the fermentation broth, and high risk of contamination which can lead to low chiral purities.[193–195] There are two general types of bacteria that produce lactic acid: *Bacillus* genus and lactic acid bacteria (LAB). LABs are Gram-positive, facultative anaerobic or micro-aerophilic microorganisms that grow at temperatures between 5 and 45 °C and a pH between 4.0 and 4.5.[196] LABs can be isolated in nature from plants, meat, and dairy products. They typically require complex essential nutrients which can make the process expensive. To overcome this expensive process, many studies have explored the use of inexpensive feedstock such as lignocellulosic materials and carbohydrate-rich waste materials to substitute as their essential nutrient. Wang *et al.*[196] and John *et al.*[197] have extensively reviewed the use of renewable materials and agricultural waste residues for lactic acid production. There are also studies that explored the use of carbohydrate-rich materials such as Kodo millet bran,[198] potato starch waste,[199] and municipal food waste[200] for lactic acid production. The limitation of using these inexpensive materials is the additional pretreatment step to extract fermentable sugars from them. Another limitation of LABs is the risk of contamination which is typical for low fermentation temperature conditions. To overcome this limitation, there are studies that investigated promising thermotolerant LAB species.[201–203] LABs can be broadly classified as homofermentative or heterofermentative depending on their end-product. LABs that produce only lactic acid are homofermentative, while LABs that produce lactic acid and other by-products such as ethanol, acetic acid, and carbon dioxide are heterofermentative. The most common LAB are *Lactobacillus*, *Lactococcus*, *Pediococcus*, *Aerococcus*, *Carnabacterium*, *Oenococcus*, *Tetragenococcus*, *Vagococcus*, *Weisella*, *Leuconostoc*, *Streptococcus*, and *Enterococcus.*[196]

Aside from LABs, there are also other bacteria that produce lactic acid, such as *Bacillus coagulans*,[204–206] *Bacillus subtilis*,[207] and *Bacillus* sp.[208,209] Some of the biggest advantages of *Bacillus* species for lactic acid production are their flexibility to grow in thermophilic conditions and the use of an inexpensive complex substrate. For example, Ou *et al.*[205] investigated the use of sweet sorghum juice and bagasse to produce optically pure lactic acid using both native and engineered *Bacillus coagulans* strains. According to the study, they used sweet sorghum because it produces a high amount of soluble sugar among the bioenergy crops. Because *Bacillus coagulans* metabolizes both hexose and pentose sugars at 50–55 °C and a pH of 4.5, it offers better benefits in SSF conditions where complex carbohydrates are used as substrate. The higher temperature minimizes contamination and provides an optimum environment for many commercial fungal cellulases, thus reducing the amount of enzyme required. In some studies, the higher temperatures afforded an open fermentation using nonsterilized media which drives the costs down.[210–212] In Juturu and Wu's[203] paper, they were able to grow *Bacillus coagulans* in hemicellulose hydrolysate obtained by acid hydrolysis of oil palm empty fruit bunch at 50 °C and pH 6.0. Their study also highlighted the use of an inexpensive nitrogen source, dry baker's yeast, to replace the expensive yeast extract. The study's SSF yielded 120 $g L^{-1}$ of L-lactic acid with a productivity of 4.3 $g L^{-1} h^{-1}$, which the authors mentioned to be the highest ever reported from lignocellulose hydrolysates. One trade-off is that it also requires more energy to maintain fermentation at elevated temperatures.

In nature, *E. coli* generally produces a mixture of ethanol and organic acids during fermentation. In recent years, many studies have engineered *E. coli* to improve lactic acid production. Zhou and co-authors[213] reported a short-review of metabolic engineering strategies for D-lactate production of *E. coli*. Because *E. coli* has been extensively studied in metabolic engineering for the last 50 years, it has been an easy choice for a prime prokaryotic genetic model for many industrially important biochemicals.[214] Chen *et al.*[214] conducted an extensive review of the metabolic engineering of *E. coli* in many bio-based chemicals. Some of the notable advantages of using *E. coli* include rapid doubling time and growth rate, high cell density, low production cost, and availability of vast knowledge of its metabolism.[214] Many studies have been conducted to produce lactic acid from *E. coli* using various substrates, including glucose,[215,216] sucrose,[217,218] xylose,[219,220] and glycerol.[221–223] In Abdel-Rahman and co-authors'[194] review, they found that engineered *E. coli* have a much lower productivity, final concentration, and tolerance to lactic acid than those of LABs and *Bacillus* spp. strains. Thus, many investigations are focused on metabolically engineering the *E. coli* to consume specific substrate efficiently.

6.9.2 Fungi and Yeast

Aside from bacteria, many reports have shown that fungi and yeast can also produce lactic acid. Fungi have the advantage of utilizing complex carbohydrate without pre-saccharification because they can produce a hydrolytic enzyme which can break down complex carbohydrates to simple sugars, something that many bacteria are unable to do.[194,224] Also, because of their filamentous nature, fungi are easier to separate from the fermentation broth than bacteria or yeast.[194,224] This reduces the downstream processing step, which also reduces cost and energy usage. In Saito *et al.*,[225] the study investigated the production of lactic acid from xylose and wheat straw by fungi *Rhizopus oryzae*. Their simultaneous saccharification and fermentation resulted in a 22.8 $g L^{-1}$ and 6 $g L^{-1}$ lactic acid concentration for a mixture of glucose and xylose and wheat straw, respectively. The lactic acid results were lower than those obtained from LABs, but the authors pointed out that their study produced lactic acid without the addition of any supplemental material or pH neutralizing agent. Another study[226] investigated the bioconversion of waste office paper to L-lactic acid by the filamentous fungi *Rhizopus oryzae*. The study[226] resulted in 49.1 $g L^{-1}$ of lactic acid with a productivity of 16.3 g per L per day. Although the later study[226] obtained a much higher lactic acid concentration, the authors indicated that the production rate and yield might be inhibited by xylose from the hemicelluloses and an unknown compound present in paper pulp, respectively. Because of its ability to consume complex carbohydrates in its production of lactic acid, the process becomes very attractive, especially for waste materials such as waste office paper[226] and other renewable materials.[224]

Typically, the separation and purification step of the lactic acid is considered a costly process. To overcome this, yeast becomes an attractive alternative because it can tolerate low pH (as low as 1.5) and grow on mineral media, which eases the purification of the acid.[227] The only downside of using yeast is that it is not a natural lactic acid producer but an ethanol producer instead. Thus, studies have explored bioengineering yeast to produce lactic acid instead of ethanol. In Ilmén *et al.*,[228] the production of L-lactic acid by *Candida sonorensis* was bioengineered expressing heterologous bacterial and fungal lactate dehydrogenases. Their study showed glucose conversion to highly pure L-lactic acid at high yield in minimal medium. Many other studies have explored other yeast species such as *Saccharomyces*,[229–231] *Zygosacchromyses*,[232] *Candida*,[228,233] *Pichia*,[234,235] and *Kluyveromyces*.[236]

6.10 PLA Fermentation Strategies

Lactic acid is produced either through fermentation or chemical synthesis. Fermentation has the advantages of using various low-cost feedstocks, the versatility of using a vast array of microorganisms, and technology improvement to reduce energy utilization. Also, fermentation can produce optically pure D(−) or L(+)-lactic acid, while chemical synthesis typically gives a racemate of D/L-lactic acid. As of the time of writing, the majority of the commercial production of PLA uses the fermentation route to produce lactic acid. The decision on the type of fermentation system depends on various factors such as substrate, microorganism, the scale of the production, overall cost, and application of the product. In the next subsections, the various fermentation strategies for producing D-lactic acid or L-lactic acid will be discussed. A list of selected and recent lactic acid fermentation studies is found in Table 6.3. The table shows that many of the PLA studies use glucose as the primary substrate, except for a few that used complex substrates such as corn stover, excess sludge, and broken rice. Also, many of them operate at a mesophilic temperature range and slightly acidic to neutral pH range. The studies in Table 6.3 are further discussed in the next section.

6.10.1 PLA Produced in Batch

The batch system is the most common and widely used PLA production strategy. It is a closed system where substrate, media, and microorganism are added at the beginning of the fermentation, and products are removed at the end of the fermentation. It has a simple operation, produces high lactic acid concentration, and has minimum contamination. Unfortunately, it also comes with disadvantages such as low productivity and substrate/product inhibitions. One of the methods to overcome substrate inhibition is by employing SSF where complex carbon sources such as starch are first enzymatically hydrolyzed to a simple sugar such as glucose then fermented to lactic acid. Because this is done simultaneously in the same fermentation system, the glucose production is gradual, thereby not inhibiting the microorganism. Additionally, the complex carbohydrates are a low-cost carbon source compared to glucose which makes the process more attractive. Using SSF, Nakano *et al.*[237] have produced a high lactic acid content of 96.1% and high productivity of 3.59 $g\,L^{-1}\,h^{-1}$. Similarly, in Zhao *et al.*,[238]

Table 6.3 Selected studies of PLA produced.

Ref.	Microorganism	Substrate	Fermentation	Temp. pH	Lactic acid products
242	*Sporolactobacillus* sp. strain CASD	Glucose	Repeated batch	42 °C, 5.6–5.8	D-Lactic acid 82.8 $g L^{-1}$ – highest conc. for one-reactor system 87.3 $g L^{-1}$ – highest conc. for two-reactor system
240	*Lactobacillus lactis*	Cane sugar and molasses	Shake-flask batch	42 °C, 7.0	D-Lactic acid 81 $g L^{-1}$
243	*Lactobacillus delbrueckii* subsp. *lactis* QU 41	MRS media with 20 $g L^{-1}$ glucose	Batch and continuous with cell recycle	37, 43, 50, and 55 °C, 6.0	D-Lactic acid 20.1 $g L^{-1}$ with >99.99% purity for batch 18.0 $g L^{-1} h^{-1}$ at 0.87 h^{-1} dilution rate for continuous
239	*Lactobacillus* sp., *Lactobacillus amylovorus* DS 20531^T	MRS media with glucose or sucrose or starch	Batch bioreactor	40 °C, 5.5	L-Lactic acid ~8.5 to ~19 $g L^{-1}$ for monoculture ~4.5 $g L^{-1}$ for mixed culture
237	*Lactobacillus delbrueckii*	Broken rice	Simultaneous saccharification and fermentation (SSF)	40 °C, 6.0	D-Lactic acid 79.0 $g L^{-1}$ and 3.59 $g L^{-1} h^{-1}$ using $Ca(OH)_2$ as neutralizing agent
241	*Lactobacillus delbrueckii*	Corn stover	Sequential hydrolysis and fermentation (SHF) on shake-flask culture	37 °C, 6.5	D-Lactic acid 36.3 $g L^{-1}$ (99.8% purity) 0.83 $g g^{-1}$ 1.01 $g L^{-1} h^{-1}$
238	*Pediocococcus acidilactici* DQ2	Corn stover	SSF	48 °C, 5.5	(1/3) D-lactic acid and (2/3) L-lactic acid 101.9 $g L^{-1}$ 77.2% purity 1.06 $g L^{-1} h^{-1}$

212	*Bacillus coagulans*	Excess sludge hydrolysate	Open fermentation (repeated batch)	50 °C, 7.0	L-Lactic acid 98.1 $g\,L^{-1}$ with 99.9% purity 5.45 $g\,L^{-1}\,h^{-1}$ 0.98 $g\,g^{-1}$
244	*Enterococcus mundtii* QU 25	Glucose/xylose mixture	Fed-batch	43 °C, 7.0	L-Lactic acid 129 $g\,L^{-1}$ 5.6 $g\,L^{-1}\,h^{-1}$
245	*Bacillus coagulans* (strain AD)	Corn stover hydrolysate	Continuous	50 °C, 5.5–6.0	Lactic acid 3.69 $g\,L^{-1}\,h^{-1}$ at pH 6.0 1.086 $g\,g^{-1}$
246	*Lactobacillus delbrueckii* Uc-3 (NCIM 5219)	MRS media with 20 $g\,L^{-1}$ glucose	Batch and fed-batch	42 °C, 7.0 (no neutralization)	L-Lactic acid 36.2 $g\,L^{-1}$; 0.5 $g\,L^{-1}\,h^{-1}$ 49.2 $g\,L^{-1}$; 0.512 $g\,L^{-1}\,h^{-1}$ (with adaptive evolution)
198	*Lactobacillus delbrueckii* subsp. *delbrueckii* NBRC3202	Bran of Kodo millet	Shake-flask batch	37 °C, 6.5	D-Lactic acid 25.38 $g\,L^{-1}$ with 97.79% purity 1.18 $g\,g^{-1}$ 0.53 $g\,L^{-1}\,h^{-1}$
225	*Rhizopus oryzae* NBRC 5378	Glucose, xylose, and wheat straw	SSF	30 °C, 3.5	Lactic acid 16.6 $g\,L^{-1}$ – using glucose 14.4 $g\,L^{-1}$ – using xylose 22.8 $g\,L^{-1}$ – using a mixture of glucose & xylose 6 $g\,L^{-1}$ – using wheat straw
228	*Candida sonorensis*	Glucose	2 stage fermentation	30 °C, 5.5	L-Lactic acid 92 $g\,L^{-1}$ 0.94 $g\,g^{-1}$

they were able to produce a high lactic acid titer of 101.9 $g L^{-1}$ and productivity of 1.06 $g L^{-1} h^{-1}$ even at high substrate loading (30%). Another strategy to enhance lactic acid production, particularly for complex carbohydrate, is using a mixed culture of microorganisms. Trontel *et al.*[239] compared the fermentation of mono and mixed cultures of *Lactobacillus* sp. and *Lactobacillus amylovorus* DSM 20531^T using MRS medium with glucose, sucrose, and starch. They found that the mixed culture showed a lower lactic acid concentration than the monoculture. However, the mixed culture of *Lactobacillus* sp. and *L. amylovorus* was able to successfully hydrolyze and ferment complex carbohydrate in the simultaneous saccharification and fermentation of starch.

Shake-flask culture is very widely used as an initial experiment to test various fermentation conditions. It typically grows the microorganism in a conical flask and is agitated using shaking incubators. The biggest advantage of this process is that it can easily be done with minimal equipment requirement compared to bioreactor fermentation. Because of the flexibility in the fermentation volume, replication and sterilization can be easily achieved. Like any batch process, it still suffers from feedback inhibition from substrate and products. Also, there is limited automation, therefore it is more laborious during sampling than bioreactor fermentation. Depending on the frequency, it can be prone to contamination during sampling. However, shake-flask culture can still yield a high concentration of D-lactic acid, as shown by Joshi *et al.*[240] and Zhang and Vadlani.[241]

In fermentations that produce organic acid as a product, the increase in acidity can inhibit microorganism growth. Therefore, many lactic acid fermentations use a neutralizing agent to control the pH. In Nakano *et al.*,[237] they investigated the effects of using various neutralizing agents ($Ca(OH)_2$, NH_4OH, and NaOH) on lactic acid production. Interestingly, among the three neutralizing agents investigated, $Ca(OH)_2$ produced the highest lactic acid concentration, product yield per substrate consumption ratio, and productivity. The study discussed that this could be because of the calcium lactate being easily recovered upon precipitation and that its low molarity has enhanced the growth of LABs, thus increasing the lactic acid production.

6.10.2 PLA Produced in Fed-batch

In order to overcome substrate inhibition, many studies use fed-batch systems. Similar to batch systems, fed-batch systems are a closed system fermentation where products are not withdrawn until the end of the fermentation. However, with fed-batch systems, substrates are fed into the system gradually to limit the substrate concentration throughout the fermentation process. Regardless of the substrate feeding strategy, whether continuously, sequentially, or pulsed, it still produced a higher lactic acid concentration than conventional batch.[194] Abdul-Rahman *et al.*[244] investigated various neutralizing agents using fed-batch fermentations feeding on a glucose/xylose mixture substrate and using the *E. mundtii* QU 25 NITE BP-965 strain. To produce a higher lactic acid concentration, the study found that glucose and xylose concentration had to be maintained below 25 $g L^{-1}$ and above 10 $g L^{-1}$, respectively.[244] There are many advantages to using fed-batch systems, especially in a full-scale application, but the end-product inhibition can still pose a problem. It is also difficult to conduct optimization studies with fed-batch systems.

6.10.3 PLA Produced in Continuous Fermentation

Substrate and product inhibition are major drawbacks of batch fermentation; therefore, more continuous fermentation can be conducted to improve the productivity of D-lactic acid production. Continuous fermentation is beneficial at producing high lactic acid productivity, controlled growth rates, and cost/energy savings because of the minimum cleaning or maintenance. A possible downside includes that the substrate is not completely consumed. Tashiro and co-authors[243] have shown that high D-lactic acid productivity can be achieved using continuous fermentation with cell recycle. Continuous fermentation has also been used with a complex carbohydrate substrate, such as corn stover hydrolysate.[245] Ahring *et al.*[245] have shown that the pH and dilution rate did not create a significant effect on the lactic acid yield. However, they found that there was a 100% xylose conversion at a lower dilution rate.

6.10.4 PLA Produced in Repeated Batch

In recent years, the repeated batch process has become popular. Similar to the repeated batch process described in PHA fermentation, it requires less time for sterilization, maintenance, and inoculum propagation. The savings in time and labour translate to savings in cost and energy as well. Because it intermittently removes fermentation broth, this process also overcomes end-product inhibition. In Zhao *et al.*,[242] the use of multiple reactors repeated batch systems was investigated using *Sporalactobacillus* sp. strain CASD and glucose as substrate. They found that the two-reactor fermentation system showed a higher lactic acid concentration than the conventional one-reactor system[242] (see Table 6.3). They also found that the D-lactic acid production became cell-growth-dependent which shifted from being a cell-maintenance-dependent fermentation.[242] In Ma *et al.*,[212] excess sludge hydrolysate and unsterilized open fermentation were the strategy used as an alternative to an expensive substrate and energy-intensive fermentation. The study found a high lactic acid concentration of 97 $g L^{-1}$ with 99% purity can be achieved with three repeated batch cycles.

6.11 Summary

Fermentation is known to be the heart of polyester production; however, it is also a costly process.

The major factors that will make this process economically viable are advancements in the use of substrate, microorganisms, and the fermentation operation strategies. There are many studies that focused on a cheap alternative substrate, such as municipal and industrial wastewater and agricultural residues. Additionally, the introduction of recombinant microorganisms has shown great promise in terms of productivity. As it goes with any process, the technology matures with time. With the increase in PLA and PHA fermentation strategies presented in recent years, it is likely that the fermentation process will become cost-effective in the near future.

Abbreviations

CCS	Condensed corn solubles
CDW	Cell dry weight
COD	Chemical oxygen demand
CSTR	Continuously stirred tank reactors
LAB	Lactic acid bacteria
mcl PHA	Medium chain length PHA
MOPS	3-(*N*-Morpholino)propane sulfonic acid
OFMSW	Organic fraction of municipal solid waste
OLR	Organic loading rates
PHA	Polyhydroxyalkanoates
PHB	Polyhydroxybutyrate
PLA	Polylactic acid
PUF	Polyurethane foams
RSM	Response surface methodology
scl PHA	Short-chain length PHA
SBR	Sequencing batch reactor
SRT	Solid retention times
SSF	Simultaneous saccharification and fermentation
SdSF	Solid-state fermentation

References

1. Natureworks, Ingeo in Use, 2018.
2. Nova-Institute, Growth in PLA bioplastics: a production capacity of over 800 000 tonnes expected by 2020, 6th June, 2012.
3. M. Jamshidian, E. A. Tehrany, M. Imran, M. Jacquot and S. Desobry, *Compr. Rev. Food Sci. Food Saf.*, 2010, **9**, 552–571.
4. Metabolix, *Metabolix, Inc. Announces Name Change to Yield10 Bioscience, Inc.*, 2017.
5. P. Technology, *Metabolix Sells Bioplastics Business*, 2016.
6. FullCycleBioplastics, A Better Approach to Plastics, 2018.
7. MangoMaterials, Mango Materials, 2018.
8. Natureworks, About NatureWorks, 2018.
9. A. Steinbüchel and B. Füchtenbusch, *Trends Biotechnol.*, 1998, **16**, 419–427.

10. H. Chung, J. E. Yang, J. Y. Ha, T. U. Chae, J. H. Shin, M. Gustavsson and S. Y. Lee, *Curr. Opin. Biotechnol.*, 2015, **36**, 73–84.
11. C. G. Chae, Y. J. Kim, S. J. Lee, Y. H. Oh, J. E. Yang, J. C. Joo, K. H. Kang, Y.-A. Jang, H. Lee and A.-R. Park, *Biotechnol. Bioprocess Eng.*, 2016, **21**, 169–174.
12. Y. K. Leong, P. L. Show, C. W. Ooi, T. C. Ling and J. C.-W. Lan, *J. Biotechnol.*, 2014, **180**, 52–65.
13. M. Ganesh, A. Senthamarai, S. Shanmughapriya and K. Natarajaseenivasan, *Bioresour. Technol.*, 2015, **192**, 677–681.
14. S. J. Park, Y. A. Jang, W. Noh, Y. H. Oh, H. Lee, Y. David, M. G. Baylon, J. Shin, J. E. Yang and S. Y. Choi, *Biotechnol. Bioeng.*, 2015, **112**, 638–643.
15. L. A. Petrasovits, L. Zhao, R. B. McQualter, K. D. Snell, M. N. Somleva, N. A. Patterson, L. K. Nielsen and S. M. Brumbley, *Plant Biotechnol. J.*, 2012, **10**, 569–578.
16. T. Yokoo, K. i. Matsumoto, T. Ooba, K. Morimoto and S. Taguchi, *Biosci., Biotechnol., Biochem.*, 2015, **79**, 986–988.
17. R. Mozes-Koch, E. Tanne, A. Brodezki, R. Yehuda, O. Gover, H. D. Rabinowitch and I. Sela, *J. Biol. Eng.*, 2017, **11**, 44.
18. K. i. Matsumoto, K. Morimoto, A. Gohda, H. Shimada and S. Taguchi, *J. Biosci. Bioeng.*, 2011, **111**, 485–488.
19. T. Yoshizumi, M. Yamada, M. Higuchi-Takeuchi, S. Taguchi, M. Matsui and K. Numata, *Plant Biotechnol.*, 2017, **34**, 39–43.
20. Y. Poirier, D. E. Dennis, K. Klomparens and C. Somerville, *Science*, 1992, **256**, 520–523.
21. Yield10Bioscience, Yield10 bioscience agricultural company – Corporate Overview, 2018.
22. S. Khanna and A. K. Srivastava, *Process Biochem.*, 2005, **40**, 607–619.
23. H. Li, S. Pattathil, M. B. Foston, S. Y. Ding, R. Kumar, X. Gao, A. Mittal, J. M. Yarbrough, M. E. Himmel, A. J. Ragauskas, M. G. Hahn and C. E. Wyman, *Biotechnol. Biofuels*, 2014, 7, 50.
24. S. Fidler and D. Dennis, *FEMS Microbiol. Rev.*, 1992, **9**, 231–235.
25. T. Li, X. B. Chen, J. C. Chen, Q. Wu and G. Q. Chen, *Biotechnol. J.*, 2014, **9**, 1503–1511.
26. Z.-J. Li, Z.-Y. Shi, J. Jian, Y.-Y. Guo, Q. Wu and G.-Q. Chen, *Metab. Eng.*, 2010, **12**, 352–359.
27. X.-R. Jiang and G.-Q. Chen, *Biotechnol. Adv.*, 2016, **34**, 435–440.
28. G.-Q. Chen and X.-R. Jiang, *Synth. Syst. Biotechnol.*, 2017, **2**, 192–197.
29. S. Macauley-Patrick and B. Finn, in *Practical fermentation technology*, ed. B. McNeil and L. M. Harvey, 2008, ch. 4, pp. 69–95.
30. E. El-Mansi, C. F. Bryce, A. R. Allman and A. L. Demain, in *Fermentation Microbiology and Biotechnology*, ed. E. Mansi, C. Bryce, B. Dahhou, S. Sanchez, A. Demain and A. Allman, CRC Press, 2011.
31. W. Shurtleff and A. Aoyagi, *A Brief History of Fermentation, East and West*, 2004.

32. W. H. Kampen, in *Fermentation and Biochemical Engineering Handbook*, ed. H. C. Vogel and C. M. Todaro, William Andrew Publishing, Boston, 3rd edn, 2014, pp. 37–57.
33. A. Anjum, M. Zuber, K. M. Zia, A. Noreen, M. N. Anjum and S. Tabasum, *Int. J. Biol. Macromol.*, 2016, **89**, 161–174.
34. S. Taguchi and Y. Doi, *Macromol. Biosci.*, 2004, **4**, 145–156.
35. M. Kato, H. Bao, C.-K. Kang, T. Fukui and Y. Doi, *Appl. Microbiol. Biotechnol.*, 1996, **45**, 363–370.
36. S. H. Lee, D. H. Oh, W. S. Ahn, Y. Lee, J. I. Choi and S. Y. Lee, *Biotechnol. Bioeng.*, 2000, **67**, 240–244.
37. M. Lemoigne, *Bull. Soc. Chem. Biol.*, 1926, **8**, 770–782.
38. M. Koller, R. Bona, G. Braunegg, C. Hermann, P. Horvat, M. Kroutil, J. Martinz, J. Neto, L. Pereira and P. Varila, *Biomacromolecules*, 2005, **6**, 561–565.
39. M. G. E. Albuquerque, M. Eiroa, C. Torres, B. R. Nunes and M. A. M. Reis, *J. Biotechnol.*, 2007, **130**, 411–421.
40. J. M. B. T. Cavalheiro, M. C. M. D. de Almeida, C. Grandfils and M. M. R. da Fonseca, *Process Biochem.*, 2009, **44**, 509–515.
41. A. Arumugam, S. G. Senthamizhan, V. Ponnusami and S. Sudalai, *Int. J. Biol. Macromol.*, 2018, **112**, 598–607.
42. Y. Zhang, A. Wusiman, X. Liu, C. Wan, D.-J. Lee and J. Tay, *J. Biotechnol.*, 2018, **267**, 36–44.
43. M. Yamada, A. Yukita, Y. Hanazumi, Y. Yamahata, H. Moriya, M. Miyazaki, T. Yamashita and H. Shimoi, *Fish. Sci.*, 2018, **84**, 405–412.
44. A. Stanley, H. N. Punil Kumar, S. Mutturi and S. V. N. Vijayendra, *Appl. Biochem. Biotechnol.*, 2018, **184**, 935–952.
45. L. Marang, M. C. van Loosdrecht and R. Kleerebezem, *New Biotechnol.*, 2018, **41**, 55–61.
46. Y. Ren, C. Ling, I. Hajnal, Q. Wu and G.-Q. Chen, *Appl. Microbiol. Biotechnol.*, 2018, **102**, 4499–4510.
47. J. C. López, E. Arnáiz, L. Merchán, R. Lebrero and R. Muñoz, *Chem. Eng. J.*, 2018, **333**, 529–536.
48. P. C. Sabapathy, S. Devaraj and P. Kathirvel, *Curr. Sci.*, 2017, **112**, 2106–2111.
49. F. C. de Paula, S. Kakazu, C. B. C. de Paula, J. G. C. Gomez and J. Contiero, *J. King Saud Univ., Sci.*, 2017, **29**, 166–173.
50. C. Troschl, K. Meixner and B. Drosg, *Bioengineering*, 2017, **4**, 26.
51. X. Wang, A. Oehmen, E. B. Freitas, G. Carvalho and M. A. M. Reis, *Water Res.*, 2017, **112**, 269–278.
52. C. S. S. Oliveira, C. E. Silva, G. Carvalho and M. A. Reis, *New Biotechnol.*, 2017, **37**, 69–79.
53. M. Venkateswar Reddy, Y. Mawatari, R. Onodera, Y. Nakamura, Y. Yajima and Y.-C. Chang, *Bioresour. Technol.*, 2017, **234**, 99–105.
54. B. Alkotaini, H. Koo and B. S. Kim, *Korean J. Chem. Eng.*, 2016, **33**, 1669–1673.
55. P. Kumar, S. Ray and V. C. Kalia, *Bioresour. Technol.*, 2016, **200**, 413–419.

56. I. Radecka, V. Irorere, G. Jiang, D. Hill, C. Williams, G. Adamus, M. Kwiecień, A. A. Marek, J. Zawadiak and B. Johnston, *Materials*, 2016, **9**, 367.
57. S. L. Riedel, S. Jahns, S. Koenig, M. C. E. Bock, C. J. Brigham, J. Bader and U. Stahl, *J. Biotechnol.*, 2015, **214**, 119–127.
58. M. R. Salton, *Microbiology*, 1953, **9**, 512–523.
59. M. Mahalanabis, H. Al-Muayad, M. D. Kulinski, D. Altman and C. M. Klapperich, *Lab Chip*, 2009, **9**, 2811–2817.
60. A. P. J. Middelberg, *Biotechnol. Adv.*, 1995, **13**, 491–551.
61. M. L. Snyder and H. C. Lichstein, *J. Infect. Dis.*, 1940, **67**, 113–115.
62. H. C. Lichstein and M. H. Soule, *J. Bacteriol.*, 1944, **47**, 221.
63. S. J. Park, S. Y. Lee and Y. Lee, *Appl. Biochem. Biotechnol.*, 2004, **113**, 373–379.
64. C. Phithakrotchanakoon, V. Champreda, S.-I. Aiba, K. Pootanakit and S. Tanapongpipat, *J. Polym. Environ.*, 2015, **23**, 38–44.
65. E. Galindo, C. Peña, C. Núñez, D. Segura and G. Espín, *Microb. Cell Fact.*, 2007, **6**, 7.
66. K.-S. Chou, H. W. R. Chang-Ho and P. R. Goodrich, *Biotechnol. Lett.*, 1997, **19**, 7–10.
67. W. J. Page, *FEMS Microbiol. Lett.*, 1992, **103**, 149–157.
68. G.-Q. Chen and W. J. Page, *Biotechnol. Tech.*, 1997, **11**, 347–350.
69. C. Krishna and M. C. Van Loosdrecht, *Water Res.*, 1999, **33**, 2374–2382.
70. A. J. Anderson and E. A. Dawes, *Microbiol. Rev.*, 1990, **54**, 450–472.
71. J. M. Dias, P. C. Lemos, L. S. Serafim, C. Oliveira, M. Eiroa, M. G. Albuquerque, A. M. Ramos, R. Oliveira and M. A. Reis, *Macromol. Biosci.*, 2006, **6**, 885–906.
72. G. Q. Chen, *Chem. Soc. Rev.*, 2009, **38**, 2434–2446.
73. T. Tsuge, K. Tanaka, M. Shimoda and A. Ishizaki, *J. Biosci. Bioeng.*, 1999, **88**, 404–409.
74. J. Wang and J. Yu, *J. Ind. Microbiol. Biotechnol.*, 2001, **26**, 121–126.
75. Y.-H. Yang, C. J. Brigham, C. F. Budde, P. Boccazzi, L. B. Willis, M. A. Hassan, Z. A. M. Yusof, C. Rha and A. J. Sinskey, *Appl. Microbiol. Biotechnol.*, 2010, **87**, 2037–2045.
76. G. Eggink, H. van der Wal, G. N. M. Huijberts and P. de Waard, *Ind. Crops Prod.*, 1992, **1**, 157–163.
77. T. Fukui and Y. Doi, *Appl. Microbiol. Biotechnol.*, 1998, **49**, 333–336.
78. M. R. López-Cuellar, J. Alba-Flores, J. N. G. Rodríguez and F. Pérez-Guevara, *Int. J. Biol. Macromol.*, 2011, **48**, 74–80.
79. P. Kahar, T. Tsuge, K. Taguchi and Y. Doi, *Polym. Degrad. Stab.*, 2004, **83**, 79–86.
80. F. C. Oliveira, D. M. G. Freire and L. R. Castilho, *Biotechnol. Lett.*, 2004, **26**, 1851–1855.
81. F. Wang and S. Y. Lee, *Appl. Environ. Microbiol.*, 1997, **63**, 3703–3706.
82. B. S. Kim, S. C. Lee, S. Y. Lee, H. N. Chang, Y. K. Chang and S. I. Woo, *Biotechnol. Bioeng.*, 1994, **43**, 892–898.
83. O. Hrabak, *FEMS Microbiol. Lett.*, 1992, **103**, 251–255.

84. G.-Q. Chen, K.-H. König and R. M. Lafferty, *Antonie van Leeuwenhoek*, 1991, **60**, 61–66.
85. G. Gahlawat and A. K. Srivastava, *Appl. Biochem. Biotechnol.*, 2017, **183**, 530–542.
86. B. Wang, R. R. Sharma-Shivappa, J. W. Olson and S. A. Khan, *Ind. Crops Prod.*, 2013, **43**, 802–811.
87. T. S. P. Statista, Sugar beet production in the U.S. from 2000 to 2017 (in 1000 tons), 2018.
88. K. H. Berwig, C. Baldasso and A. Dettmer, *Bioresour. Technol.*, 2016, **218**, 31–37.
89. A. Yezza, A. Halasz, W. Levadoux and J. Hawari, *Appl. Microbiol. Biotechnol.*, 2007, **77**, 269–274.
90. S. Zamakhaeva, D. Fedorov and Y. A. Trotsenko, *Appl. Biochem. Microbiol.*, 2017, **53**, 389–400.
91. O. P. Karthikeyan, K. Chidambarampadmavathy, S. Cirés and K. Heimann, *Crit. Rev. Environ. Sci. Technol.*, 2015, **45**, 1579–1610.
92. P. J. Strong, B. Laycock, S. N. S. Mahamud, P. D. Jensen, P. A. Lant, G. Tyson and S. Pratt, *Microorganisms*, 2016, **4**, 11.
93. D. Bourque, Y. Pomerleau and D. Groleau, *Appl. Microbiol. Biotechnol.*, 1995, **44**, 367–376.
94. N. Korotkova, L. Chistoserdova and M. E. Lidstrom, *J. Bacteriol.*, 2002, **184**, 6174–6181.
95. S. Ueda, S. Matsumoto, A. Takagi and T. Yamane, *Appl. Environ. Microbiol.*, 1992, **58**, 3574–3579.
96. B. Taidi, A. J. Anderson, E. A. Dawes and D. Byrom, *Appl. Microbiol. Biotechnol.*, 1994, **40**, 786–790.
97. N. Korotkova and M. E. Lidstrom, *J. Bacteriol.*, 2001, **183**, 1038–1046.
98. I. Orita, K. Nishikawa, S. Nakamura and T. Fukui, *Appl. Microbiol. Biotechnol.*, 2014, **98**, 3715–3725.
99. Y. Zhang, J. Xin, L. Chen, H. Song and C. Xia, *J. Nat. Gas Chem.*, 2008, **17**, 103–109.
100. C. Zúñiga, M. Morales, S. Le Borgne and S. Revah, *J. Hazard. Mater.*, 2011, **190**, 876–882.
101. K. H. Rostkowski, A. R. Pfluger and C. S. Criddle, *Bioresour. Technol.*, 2013, **132**, 71–77.
102. J. Xin, Y. Zhang, J. Dong, H. Song and C. Xia, *Afr. J. Biotechnol.*, 2011, **10**, 7078–7087.
103. S. W. Kim, P. Kim, H. S. Lee and J. H. Kim, *Biotechnol. Lett.*, 1996, **18**, 25–30.
104. E. R. Sundstrom and C. S. Criddle, *Appl. Environ. Microbiol.*, 2015, **81**, 4767–4773.
105. W. Babel, *FEMS Microbiol. Lett.*, 1992, **103**, 141–148.
106. M. Poroshina, N. Doronina, V. Ezhov and Y. A. Trotsenko, *Appl. Biochem. Microbiol.*, 2014, **50**, 253–258.
107. M. Miyake, M. Erata and Y. Asada, *J. Ferment. Bioeng.*, 1996, **82**, 512–514.

108. R. Bhati, S. Samantaray, L. Sharma and N. Mallick, *Biotechnol. J.*, 2010, **5**, 1181–1185.
109. A. Shrivastav, S. K. Mishra and S. Mishra, *Int. J. Biol. Macromol.*, 2010, **46**, 255–260.
110. R. Kerby, P. Ludden and G. Roberts, *J. Bacteriol.*, 1995, **177**, 2241–2244.
111. R. Kerby, S. Hong, S. Ensign, L. Coppoc, P. Ludden and G. Roberts, *J. Bacteriol.*, 1992, **174**, 5284–5294.
112. Y. S. Do, J. Smeenk, K. M. Broer, C. J. Kisting, R. Brown, T. J. Heindel, T. A. Bobik and A. A. DiSpirito, *Biotechnol. Bioeng.*, 2007, **97**, 279–286.
113. P. McKendry, *Bioresour. Technol.*, 2002, **83**, 55–63.
114. G. Guéhenneux, P. Baussand, M. Brothier, C. Poletiko and G. Boissonnet, *Fuel*, 2005, **84**, 733–739.
115. P. McKendry, *Bioresour. Technol.*, 2002, **83**, 47–54.
116. G. Maschio, A. Lucchesi and G. Stoppato, *Bioresour. Technol.*, 1994, **48**, 119–126.
117. R. Smith, T. West and W. Gibbons, *J. Appl. Microbiol.*, 2008, **104**, 1488–1494.
118. D. Solaiman and R. L. Uffen, *J. Bacteriol.*, 1982, **152**, 1175–1187.
119. L. S. Serafim, P. C. Lemos, M. G. Albuquerque and M. A. Reis, *Appl. Microbiol. Biotechnol.*, 2008, **81**, 615–628.
120. M. H. Deinema, *Appl. Microbiol.*, 1972, **24**, 857–858.
121. L. L. Wallen and W. K. Rohwedder, *Environ. Sci. Technol.*, 1974, **8**, 576–579.
122. K. Tanaka, K. Katamune and A. Ishizaki, *Can. J. Microbiol.*, 1995, **41**, 257–261.
123. B. Colombo, F. Favini, B. Scaglia, T. P. Sciarria, G. D'Imporzano, M. Pognani, A. Alekseeva, G. Eisele, C. Cosentino and F. Adani, *Biotechnol. Biofuels*, 2017, **10**, 201.
124. J. L. Waller, P. G. Green and F. J. Loge, *Bioresour. Technol.*, 2012, **120**, 285–289.
125. F. Valentino, C. Riccardi, S. Campanari, D. Pomata and M. Majone, *Bioresour. Technol.*, 2015, **192**, 304–311.
126. L. S. Serafim, P. C. Lemos, R. F. Oliveira and M. A. M. Reis, *Biotechnol. Bioeng.*, 2004, **87**, 145–160.
127. M. Reis, M. Albuquerque, M. Villano and M. Majone, in *Comprehensive biotechnology*, ed. F. Fava, S. Agathos and M. Young, Elsevier, Amsterdam, 2011, DOI: 10.1016/b978-0-08-088504-9.00464-5.
128. F. Monlau, C. Sambusiti, A. Barakat, M. Quemeneur, E. Trably, J. P. Steyer and H. Carrere, *Biotechnol. Adv.*, 2014, **32**, 934–951.
129. A. F. Duque, C. S. S. Oliveira, I. T. D. Carmo, A. R. Gouveia, F. Pardelha, A. M. Ramos and M. A. M. Reis, *New Biotechnol.*, 2013, **31**, 276–288.
130. B. Colombo, T. Pepè Sciarria, M. Reis, B. Scaglia and F. Adani, *Bioresour. Technol.*, 2016, **218**, 692–699.
131. Z. Chen, L. Huang, Q. Wen, H. Zhang and Z. Guo, *J. Environ. Sci.*, 2017, **52**, 76–84.
132. Z. Chen, L. Huang, Q. Wen and Z. Guo, *J. Biotechnol.*, 2015, **209**, 68–75.

133. M. H. Bushara, F. Alkoaik, A. Abasaeed and R. Fulleros, *Open J. Appl. Sci.*, 2018, **8**, 150.
134. S. O. Kulkarni, P. P. Kanekar, S. S. Nilegaonkar, S. S. Sarnaik and J. P. Jog, *Bioresour. Technol.*, 2010, **101**, 9765–9771.
135. R. Sindhu, N. Silviya, P. Binod and A. Pandey, *Biochem. Eng. J.*, 2013, **78**, 67–72.
136. U. E. P. A. (EPA), Wastewater Treatment Fact Sheet Sequencing Batch Reactors, 1999.
137. E. R. Coats, K. E. VandeVoort, J. L. Darby and F. J. Loge, *J. Environ. Eng.*, 2010, **137**, 46–54.
138. J. J. Beun, F. Paletta, M. C. M. Van Loosdrecht and J. J. Heijnen, *Biotechnol. Bioeng.*, 2000, **67**, 379–389.
139. J. J. Beun, K. Dircks, M. C. M. Van Loosdrecht and J. J. Heijnen, *Water Res.*, 2002, **36**, 1167–1180.
140. L. S. Serafim, P. C. Lemos, R. Oliveira and M. A. Reis, *Biotechnol. Bioeng.*, 2004, **87**, 145–160.
141. D. Dionisi, M. Majone, G. Vallini, S. D. Gregorio and M. Beccari, *Biotechnol. Prog.*, 2007, **23**, 1064–1073.
142. D. Dionisi, M. Majone, G. Vallini, S. Di Gregorio and M. Beccari, *J. Biotechnol. Bioeng.*, 2006, **93**, 76–88.
143. F. Valentino, M. Beccari, S. Fraraccio, G. Zanaroli and M. Majone, *New Biotechnol.*, 2014, **31**, 264–275.
144. K. Johnson, Y. Jiang, R. Kleerebezem, G. Muyzer and M. C. van Loosdrecht, *J. Biomacromol.*, 2009, **10**, 670–676.
145. M. Beccari, M. Majone, P. Massanisso and R. J. W. R. Ramadori, *J. Water Res.*, 1998, **32**, 3403–3413.
146. F. Valentino, M. Beccari, S. Fraraccio, G. Zanaroli and M. Majone, *New Biotechnol.*, 2014, **31**, 264–275.
147. W. S. Ahn, S. J. Park and S. Y. Lee, *Appl. Environ. Microbiol.*, 2000, **66**, 3624–3627.
148. W. S. Ahn, S. J. Park and S. Y. Lee, *Biotechnol. Lett.*, 2001, **23**, 235–240.
149. J. G. da Cruz Pradella, J. L. Ienczak, C. R. Delgado and M. K. Taciro, *Biotechnol. Lett.*, 2012, **34**, 1003–1007.
150. G. Kobayashi, K. Tanaka, H. Itoh, T. Tsuge, K. Sonomoto and A. Ishizaki, *J. Biotechnol. Lett.*, 2000, **22**, 1067–1069.
151. B. S. Kim, *J. Biotechnol. Lett.*, 2002, **24**, 125–130.
152. S. Munir and N. Jamil, *J. Basic Microbiol.*, 2018, **58**, 247–254.
153. S. Y. Lee, H. H. Wong, J. I. Choi, S. H. Lee, S. C. Lee and C. S. Han, *J. Biotechnol. Bioeng.*, 2000, **68**, 466–470.
154. J. M. Cavalheiro, R. S. Raposo, M. C. M. de Almeida, M. T. Cesário, C. Sevrin, C. Grandfils and M. Da Fonseca, *Bioresour. Technol.*, 2012, **111**, 391–397.
155. J. G. da Cruz Pradella, M. K. Taciro and A. Y. P. Mateus, *Bioresour. Technol.*, 2010, **101**, 8355–8360.

156. J. Ienczak, L. Quines, A. De Melo, M. Brandellero, C. Mendes, W. Schmidell and G. Aragão, *Braz. J. Chem. Eng.*, 2011, **28**, 585–596.
157. S. Kulpreecha, A. Boonruangthavorn, B. Meksiriporn and N. Thongchul, *J. Biosci. Bioeng.*, 2009, **107**, 240–245.
158. Z. Sun, J. A. Ramsay, M. Guay and B. A. Ramsay, *Appl. Microbiol. Biotechnol.*, 2007, **74**, 69–77.
159. J. L. Ienczak, W. Schmidell and G. M. F. de Aragão, *J. Ind. Microbiol. Biotechnol.*, 2013, **40**, 275–286.
160. J. Monod, *Annu. Rev. Microbiol.*, 1949, **3**, 371–394.
161. L. Shang, M. Jiang and H. N. Chang, *J. Biotechnol. Lett.*, 2003, **25**, 1415–1419.
162. H. O. Kang, C. W. Chung, H. W. Kim, Y. B. Kim and Y. H. Rhee, *J. Antonie van Leeuwenhoek*, 2001, **80**, 185–191.
163. R. A. Gross, C. DeMello, R. W. Lenz, H. Brandl and R. C. Fuller, *J. Macromol.*, 1989, **22**, 1106–1115.
164. C. Haas, T. El-Najjar, N. Virgolini, M. Smerilli and M. Neureiter, *J. New Biotechnol.*, 2017, **37**, 117–122.
165. S. Khanna and A. K. Srivastava, *Biotechnol. Lett.*, 2005, **27**, 1401–1403.
166. J. L. Ienczak, M. Schmidt, L. K. Quines, K. Zanfonato, J. G. da Cruz Pradella, W. Schmidell and G. M. F. de Aragao, *Appl. Biochem. Biotechnol.*, 2016, **178**, 408–417.
167. J. Monod, *Ann. Inst. Pasteur*, 1950, **79**, 390–410.
168. N. Ziv, N. J. Brandt and D. Gresham, *J. Visualized Exp.*, 2013, e50168.
169. M. Koller, *Fermentation*, 2018, **4**, 30.
170. M. Koller and A. Muhr, *Chem. Biochem. Eng. Q.*, 2014, **28**, 65–77.
171. M. Koller and G. J. B. Braunegg, *Bioengineering*, 2015, **2**, 94–121.
172. K. Jung, W. Hazenberg, M. Prieto and B. Witholt, *Biotechnol. Bioeng.*, 2001, **72**, 19–24.
173. R. Durner, B. Witholt and T. Egli, *Appl. Environ. Microbiol.*, 2000, **66**, 3408–3414.
174. G. Mothes and J. Ackermann, *Eng. Life Sci.*, 2005, **5**, 58–62.
175. H. Yue, C. Ling, T. Yang, X. Chen, Y. Chen, H. Deng, Q. Wu, J. Chen and G.-Q. Chen, *Biotechnol. Biofuels*, 2014, **7**, 108.
176. G. Du, J. Chen, J. Yu and S. Lun, *J. Biotechnol.*, 2001, **88**, 59–65.
177. A. Atlić, M. Koller, D. Scherzer, C. Kutschera, E. Grillo-Fernandes, P. Horvat, E. Chiellini and G. Braunegg, *Appl. Microbiol. Biotechnol.*, 2011, **91**, 295–304.
178. S. Khanna and A. K. Srivastava, *J. Chem. Technol. Biotechnol.*, 2008, **83**, 799–805.
179. A. Pandey, C. R. Soccol and D. Mitchell, *Process Biochem.*, 2000, **35**, 1153–1169.
180. D. A. Mitchell, M. Berovic and N. Krieger, *Biotechnol. Annu. Rev.*, 2002, **8**, 183–225.
181. D. A. Mitchell, M. Berovič and N. Krieger, in *Solid-State Fermentation Bioreactors*, Springer, 2006, pp. 1–12.

182. A. Pandey, *Biochem. Eng. J.*, 2003, **13**, 81–84.
183. S. Bhargav, B. P. Panda, M. Ali and S. Javed, *Chem. Biochem. Eng. Q.*, 2008, **22**, 49–70.
184. R. Sindhu, A. Pandey and P. Binod, *Chem. Biochem. Eng. Q.*, 2015, **29**, 173–181.
185. B. S. Kim, *Enzyme Microb. Technol.*, 2000, **27**, 774–777.
186. F. Oliveira, D. Freire and L. Castilho, *Biotechnol. Lett.*, 2004, **26**, 1851–1855.
187. Y.-M. Jung and Y.-H. Lee, *J. Biosci. Bioeng.*, 2000, **90**, 266–270.
188. C. Reddy, R. Ghai and V. C. Kalia, *Bioresour. Technol.*, 2003, **87**, 137–146.
189. F. C. Oliveira, M. L. Dias, L. R. Castilho and D. M. Freire, *Bioresour. Technol.*, 2007, **98**, 633–638.
190. P. Rocha Jr, PhD thesis, Universidade Federal do Rio de Janeiro, Rio de Janeiro, 2005.
191. N. V. Ramadas, R. Sindhu, P. Binod and A. Pandey, *Ann. Microbiol.*, 2013, **63**, 1265–1274.
192. G. Sathiyanarayanan, G. S. Kiran, J. Selvin and G. Saibaba, *Int. J. Biol. Macromol.*, 2013, **60**, 253–261.
193. S. P. Chahal and J. N. Starr, *Ullmann's encyclopedia of industrial chemistry*, 2000, pp. 1–8.
194. M. A. Abdel-Rahman, Y. Tashiro and K. Sonomoto, *Biotechnol. Adv.*, 2013, **31**, 877–902.
195. S. Jantasee, K. Marlene, N. Mungma and M. Siebenhofer, *J. Chem. Technol. Biotechnol.*, 2017, **92**, 2885–2893.
196. Y. Wang, Y. Tashiro and K. Sonomoto, *J. Biosci. Bioeng.*, 2015, **119**, 10–18.
197. R. P. John, K. M. Nampoothiri and A. Pandey, *Appl. Microbiol. Biotechnol.*, 2007, **74**, 524–534.
198. R. Balakrishnan, S. R. R. Tadi, S. Sivaprakasam and S. Rajaram, *Ind. Crops Prod.*, 2018, **111**, 731–742.
199. S. P. N. Ayudthaya, A. H. Weijer, A. H. Gelder, A. J. Stams, W. M. Vos and C. M. Plugge, *Biotechnol. Biofuels*, 2018, **11**, 13.
200. K. Sakai, M. Taniguchi, S. Miura, H. Ohara, T. Matsumoto and Y. Shirai, *J. Ind. Ecol.*, 2003, **7**, 63–74.
201. B. Liu, M. Yang, B. Qi, X. Chen, Z. Su and Y. Wan, *Biochem. Eng. J.*, 2010, **52**, 212–219.
202. K. Ma, G. Hu, L. Pan, Z. Wang, Y. Zhou, Y. Wang, Z. Ruan and M. He, *Bioresour. Technol.*, 2016, **219**, 114–122.
203. V. Juturu and J. C. Wu, *Biotechnol. Appl. Biochem.*, 2018, **65**, 145–149.
204. E. C. van der Pol, G. Eggink and R. A. Weusthuis, *Biotechnol. Biofuels*, 2016, **9**, 248.
205. M. S. Ou, D. Awasthi, I. Nieves, L. Wang, J. Erickson, W. Vermerris, L. Ingram and K. Shanmugam, *Bioenergy Res.*, 2016, **9**, 123–131.
206. D. Pleissner, A.-K. Neu, K. Mehlmann, R. Schneider, G. I. Puerta-Quintero and J. Venus, *Bioresour. Technol.*, 2016, **218**, 167–173.
207. D. Awasthi, L. Wang, M. S. Rhee, Q. Wang, D. Chauliac, L. O. Ingram and K. T. Shanmugam, *Biotechnol. Bioeng.*, 2018, **115**, 453–463.

208. S. Thitiprasert, K. Kodama, S. Tanasupawat, P. Prasitchoke, T. Rampai, B. Prasirtsak, V. Tolieng, J. Piluk, S. Assabumrungrat and N. Thongchul, *Bioprocess Biosyst. Eng.*, 2017, **40**, 1787–1799.
209. N. Assavasirijinda, D. Ge, B. Yu, Y. Xue and Y. Ma, *Microb. Cell Fact.*, 2016, **15**, 3.
210. J. Ouyang, C. Cai, H. Chen, T. Jiang and Z. Zheng, *Appl. Biochem. Biotechnol.*, 2012, **168**, 2387–2397.
211. Y. Zhang, X. Chen, B. Qi, J. Luo, F. Shen, Y. Su, R. Khan and Y. Wan, *Bioresour. Technol.*, 2014, **163**, 160–166.
212. K. Ma, T. Maeda, H. You and Y. Shirai, *Bioresour. Technol.*, 2014, **151**, 28–35.
213. L. Zhou, W. j. Cui, Z. m. Liu and Z. m. Zhou, *J. Chem. Technol. Biotechnol.*, 2016, **91**, 576–584.
214. X. Chen, L. Zhou, K. Tian, A. Kumar, S. Singh, B. A. Prior and Z. Wang, *Biotechnol. Adv.*, 2013, **31**, 1200–1223.
215. Y. Zhu, M. Eiteman, K. DeWitt and E. Altman, *Appl. Environ. Microbiol.*, 2007, **73**, 456–464.
216. S. Zhou, L. Yomano, K. Shanmugam and L. Ingram, *Biotechnol. Lett.*, 2005, **27**, 1891–1896.
217. V. Shukla, S. Zhou, L. Yomano, K. Shanmugam, J. Preston and L. Ingram, *Biotechnol. Lett.*, 2004, **26**, 689–693.
218. Y. Wang, T. Tian, J. Zhao, J. Wang, T. Yan, L. Xu, Z. Liu, E. Garza, A. Iverson and R. Manow, *Biotechnol. Lett.*, 2012, **34**, 2069–2075.
219. M. A. Eiteman, S. A. Lee, R. Altman and E. Altman, *Biotechnol. Bioeng.*, 2009, **102**, 822–827.
220. H. Lu, X. Zhao, Y. Wang, X. Ding, J. Wang, E. Garza, R. Manow, A. Iverson and S. Zhou, *BMC Biotechnol.*, 2016, **16**, 19.
221. S. Mazumdar, J. M. Clomburg and R. Gonzalez, *Appl. Environ. Microbiol.*, 2010, **76**, 4327–4336.
222. S. Mazumdar, M. D. Blankschien, J. M. Clomburg and R. Gonzalez, *Microb. Cell Fact.*, 2013, **12**, 7.
223. J. A. Posada, C. A. Cardona and R. Gonzalez, *Appl. Biochem. Biotechnol.*, 2012, **166**, 680–699.
224. Z. Y. Zhang, B. Jin and J. M. Kelly, *Biochem. Eng. J.*, 2007, **35**, 251–263.
225. K. Saito, Y. Hasa and H. Abe, *J. Biosci. Bioeng.*, 2012, **114**, 166–169.
226. E. Y. Park, P. N. Anh and N. Okuda, *Bioresour. Technol.*, 2004, **93**, 77–83.
227. M. Sauer, D. Porro, D. Mattanovich and P. Branduardi, *Biotechnol. Genet. Eng. Rev.*, 2010, **27**, 229–256.
228. M. Ilmén, K. Koivuranta, L. Ruohonen, V. Rajgarhia, P. Suominen and M. Penttilä, *Microb. Cell Fact.*, 2013, **12**, 53.
229. A. Pacheco, G. Talaia, J. Sá-Pessoa, D. Bessa, M. J. Gonçalves, R. Moreira, S. Paiva, M. Casal and O. Queirós, *FEMS Yeast Res.*, 2012, **12**, 375–381.
230. T. L. Turner, G.-C. Zhang, S. R. Kim, V. Subramaniam, D. Steffen, C. D. Skory, J. Y. Jang, B. J. Yu and Y.-S. Jin, *Appl. Microbiol. Biotechnol.*, 2015, **99**, 8023–8033.

231. T. Mimitsuka, K. Sawai, K. Kobayashi, T. Tsukada, N. Takeuchi, K. Yamada, H. Ogino and T. Yonehara, *J. Biosci. Bioeng.*, 2015, **119**, 65–71.
232. P. Branduardi, M. Sauer, L. De Gioia, G. Zampella, M. Valli, D. Mattanovich and D. Porro, *Microb. Cell Fact.*, 2006, **5**, 4.
233. H. Tamakawa, S. Ikushima and S. Yoshida, *J. Biosci. Bioeng.*, 2012, **113**, 73–75.
234. P. B. A. de Lima, K. C. L. Mulder, N. T. M. Melo, L. S. Carvalho, G. S. Menino, E. Mulinari, V. H. de Castro, T. F. dos Reis, G. H. Goldman and B. S. Magalhães, *Microb. Cell Fact.*, 2016, **15**, 158.
235. H. J. Park, J. H. Bae, H. J. Ko, S. H. Lee, B. H. Sung, J. I. Han and J. H. Sohn, *Biotechnol. Bioeng.*, 2018, **115**, 2232–2242.
236. M. M. Bianchi, L. Brambilla, F. Protani, C.-L. Liu, J. Lievense and D. Porro, *Appl. Environ. Microbiol.*, 2001, **67**, 5621–5625.
237. S. Nakano, C. U. Ugwu and Y. Tokiwa, *Bioresour. Technol.*, 2012, **104**, 791–794.
238. K. Zhao, Q. Qiao, D. Chu, H. Gu, T. H. Dao, J. Zhang and J. Bao, *Bioresour. Technol.*, 2013, **135**, 481–489.
239. A. Trontel, A. Batusic, I. Gusic, A. Slavica, B. Santek and S. Novak, *Food Technol. Biotechnol.*, 2011, **49**, 75.
240. D. Joshi, M. Singhvi, J. Khire and D. Gokhale, *Biotechnol. Lett.*, 2010, **32**, 517–520.
241. Y. Zhang and P. V. Vadlani, *Bioprocess Biosyst. Eng.*, 2013, **36**, 1897–1904.
242. B. Zhao, L. Wang, F. Li, D. Hua, C. Ma, Y. Ma and P. Xu, *Bioresour. Technol.*, 2010, **101**, 6499–6505.
243. Y. Tashiro, W. Kaneko, Y. Sun, K. Shibata, K. Inokuma, T. Zendo and K. Sonomoto, *Appl. Microbiol. Biotechnol.*, 2011, **89**, 1741–1750.
244. M. A. Abdel-Rahman, Y. Xiao, Y. Tashiro, Y. Wang, T. Zendo, K. Sakai and K. Sonomoto, *J. Biosci. Bioeng.*, 2015, **119**, 153–158.
245. B. K. Ahring, J. J. Traverso, N. Murali and K. Srinivas, *Biochem. Eng. J.*, 2016, **109**, 162–169.
246. M. Singhvi, T. Zendo, D. Gokhale and K. Sonomoto, *Appl. Microbiol. Biotechnol.*, 2018, 1–11.

CHAPTER 7

Bio-based Polyesters

D. GREWELL,[a] TIANYI SUN,[b] THIDARAT TOSUKHOWONG,[b] JED RANDALL*[c] AND MOLLY MORSE[d]

[a] College of Engineering and Engineering Technology, Northern Illinois University, 590 Garden Rd, Dekalb, IL 60115, USA; [b] GC Innovation America, Houston, Texas, USA; [c] NatureWorks LLC, Plymouth, MN, USA; [d] Mango Materials, Redwood City, CA, USA
*Email: Jed_Randall@natureworksllc.com

7.1 Polyesters

Polyesters are polymerized esters, which are typically produced through a reaction between an acid and alcohol. During this reaction at least one –OH group and one –COOH group is replaced with a –COO– group. The final chemical structure is seen in Figure 7.1 in which R and R′ are organic structures such as alkanes or aromatic rings.

In nature, esters often exhibit a fragmentation pattern. However, the plastics industry has developed many polymerized esters to produce a range of plastics. One of the most common is polyethylene terephthalate (PET) which is used for many applications including the bottles for carbonated beverages because of its good vapor barrier properties, toughness, strength, transparency, and costs. Recently there have been a number of bio-based polyesters that are commercially available. Depending on the type and formulation, they have a wide range of thermal mechanical properties and may or may not be degradable through natural processes. For example, Coca-Cola has developed an infrastructure to produce a partially bio-based PET that is a "drop in" for petrochemical-based PET. While the current technology only allows for a cost-effective approach to replacing about one third of plastics with a bio-based feedstock, the ethylene portion of the plastic, there is hope

Green Chemistry Series No. 79
Bioplastics and Biocomposites: A Practical Introduction
Edited by David Grewell

Published by the Royal Society of Chemistry, www.rsc.org

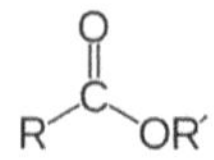

Figure 7.1 Fundamental chemical structure of ester functional group.

that the entire molecule, including the terephthalic acid portion, can be bio-based in the future. It is important to note that, as reviewed in more detail in Chapter 3, a bio-based PET drop in, while partially bio-based, is not readily biodegradable. The balance of this chapter reviews some of the common bio-based polyesters. Additionally, because many bio-based polyesters are produced from fermentation processes, Chapter 6 provides additional information on various fermentation methods.

7.2 Polylactic Acids

7.2.1 Introduction

One of the most promising and exciting new materials in the world of thermoplastics and polymers continues to be polylactide (PLA) or poly(lactic acid) as it is referred to by many. As we near the twenty years milestone for the first large commercial scale plant to produce PLA, the NatureWorks 150 kta Ingeo™ biopolymer facility in Blair, NE, and over ninety years since Carothers first described synthesizing PLA in 1932,[1] it may seem the phrase "new to the world" polymer is becoming dated. However, in an industry where over 350 million tonnes per year of plastics are produced annually,[2] just a tiny fraction is classified as biopolymer (0.6%), and an even smaller fraction is PLA (0.08%).[3] Rapid growth in PLA production and use over the years has only recently exposed many consumers, entrepreneurs, scientists, and developers to PLA products and technology; so, for the experience of most, PLA is indeed very new to the world.

Industrial and academic research on PLA and related topics has grown exponentially with publications exceeding 2000 per year over the past five years, contributed to by researchers throughout the world. Over the years, the published literature has been compiled into several comprehensive chapters and dedicated books which detail many aspects of PLA production, performance properties, and uses.[4–10] As the world looks to enable a more stable and prosperous future, the concept of circular economy has been embraced, demanding more sustainable solutions for products, pollution, and waste.[11] Commercial PLA is sourced from renewable materials, and it has attractive end-of-life options that fit into the circular economy concept such as compostability and recycling through mechanical and chemical means. The process and resources required to produce today's PLA have been well studied, and results demonstrate favorable improvements in the environmental profile/impact *vs.* conventional thermoplastic materials.[12] The processing and performance properties of PLA currently meet many commercial needs and fit for use demands. A few examples of excellent performance advantages in both properties and processing are high modulus, excellent grease resistance, high strength-to-weight ratio, high 3D-printing quality, controlled hydrolysis, flavor and aroma barriers, tunable surface energy, and flexible processability with many fabrication techniques without sacrificing suitability for food contact. While the world calls for solutions to reduce consumption of natural resources and plastics waste, indicators signal that PLA could significantly contribute to meeting 21st century needs. Still, it feels as though this new-to-the-world polymer is just beyond infancy and that we have just scratched the surface on the role that PLA and its derivatives, renewables, and degradables will play in the future renewable age.

PLA represents a family of structures derived from lactic acid (2-hydroxy-propionic acid). Lactic acid has a chiral carbon resulting in two stereo-isomers classified as L(−)-lactic and D(+)-lactic acid which will be referred to throughout this section as simply L- and D-. Through fermentation of sugars such as dextrose (dextrorotatory form of glucose) and sucrose, highly pure L- and D-lactic acids are available, which have enabled polymerization of both highly pure L- and D-PLA (PLLA and PDLA) as well as many variations on the stereo structure. The intermediate lactide, a cyclic lactone, dimer ester of two lactic acid molecules, is often prepared which provides a second layer of flexibility to polymerize various stereo structures. Across this range of structures, PLA properties can vary from highly crystalline to completely amorphous which changes many mechanical and chemical properties. PLAs are generally rigid at room temperature with a glass transition ranging from 45 to 55 °C when amorphous. Crystal melting points typically range from about 125 to 175 °C with some special examples up to 220–230 °C (*i.e.*, stereocomplex PLA). Stereo structure, molecular weight, and molecular architecture in combination with polymer processing techniques and functional additives are used to control polymer properties and processing behavior. PLA is processed into rigid thermoforms, biaxially oriented and blown films, many types of nonwoven and woven fibers, 3D printing filaments, foams, coatings, injection and profile moldings, and other types of products. End uses vary across a wide range of single use items, durable goods, and controlled hydrolysis products.

7.2.2 Synthesis and Production

As technology and infrastructure advance, the source of carbonaceous material to produce lactic acid and PLA is expected to change with time. To conceptualize where PLA comes from today, a schematic from NatureWorks is shown in Figure 7.2.[13]

Present technology uses the first generation of industrial agricultural feedstocks. Companies have commercialized PLA products made by fermentation of sugars from corn, sugar cane, and cassava, see Figure 7.3. Smaller demonstrations have been made by companies and academics using sugar beets, potato processing waste, wheat starch, and milk whey, to name

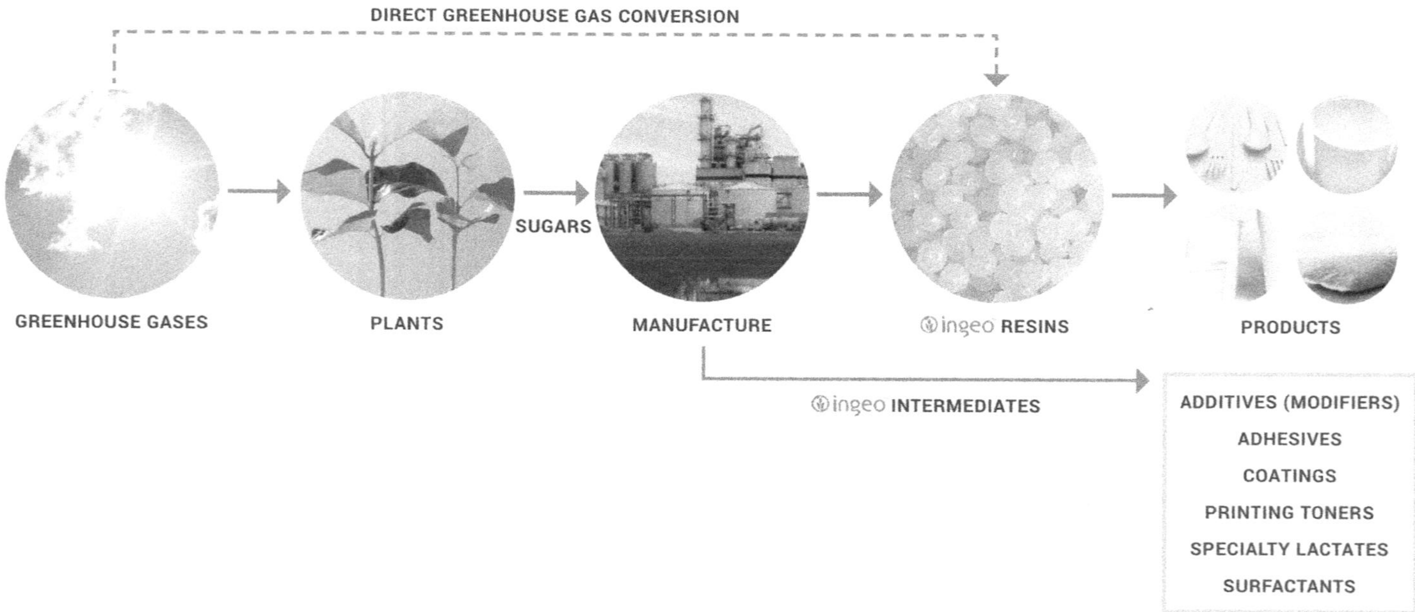

Figure 7.2 NatureWorks literature, Turning greenhouse gas into Ingeo-branded biomaterials.

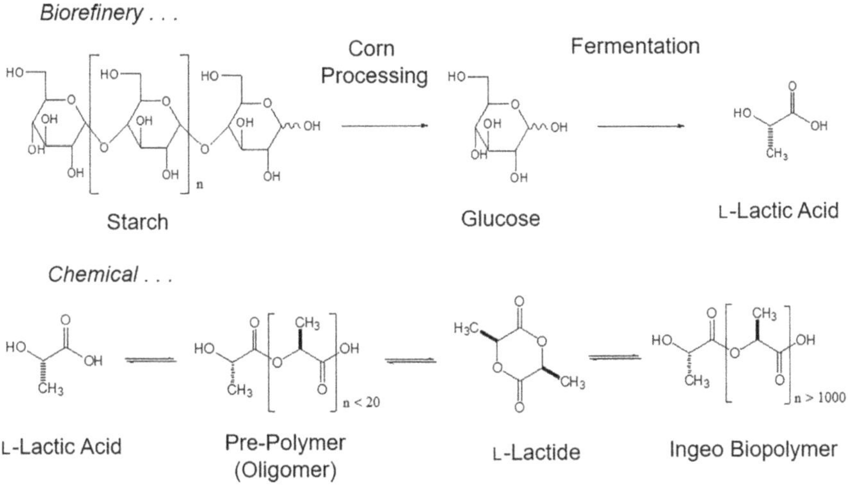

Figure 7.3 The chemistry of making Ingeo PLA from agricultural feedstock.

a few. Research and development programs look to diversify the feedstocks that can be utilized for manufacturing PLA and precursor lactic acid including direct greenhouse gas conversion technologies.[14,15]

High molecular weight PLA synthesis can be accomplished through condensation esterification of lactic acid under very specific conditions as side reactions and water removal become problematic in achieving high molecular weights. Direct polymerization (DP) has been demonstrated in solvent with solvent/water azeotrope removal and in solid state polymerizations.[16–24] One of the side reactions that occurs is isomerization of lactic acid resulting in loss of stereo purity. A second reaction consists of the hydroxyl chain end "back biting" within the polymer chain to generate a stable cyclic lactide. Lactide lacks the hydroxyl and carboxylic acid groups of lactic acid and is thus less reactive and easier to separate and handle in anhydrous environments. Exploitation of the lactide formation reaction leads to lactide products which can be purified to a high degree and used as monomers in ring opening polymerization (ROP) to make polylactide which is very similar to poly(lactic acid) products from DP. ROP products can have higher stereo sequencing control *versus* DP.

Starting with starch from a biorefinery (*e.g.*, corn wet mill), simple sugars are created to feed a highly efficient fermentation.[25] In practice, the naturally occurring and most common form of lactic acid produced is the stereo pure L-form. Following fermentation, a chemical pathway is used to derive low molecular weight PLA polymers (prepolymer), lactides, and high molecular weight polymers. In general, six-carbon (C_6) sugars have very favorable fermentation yield when producing the three-carbon (C_3) lactic acid.

Lactic acid enantiomers racemize to 1 : 1 chemical equilibrium with time and temperature, but this is a slow process. In practice, low amounts of D- are formed during prepolymer and lactide production steps. Lactide is available

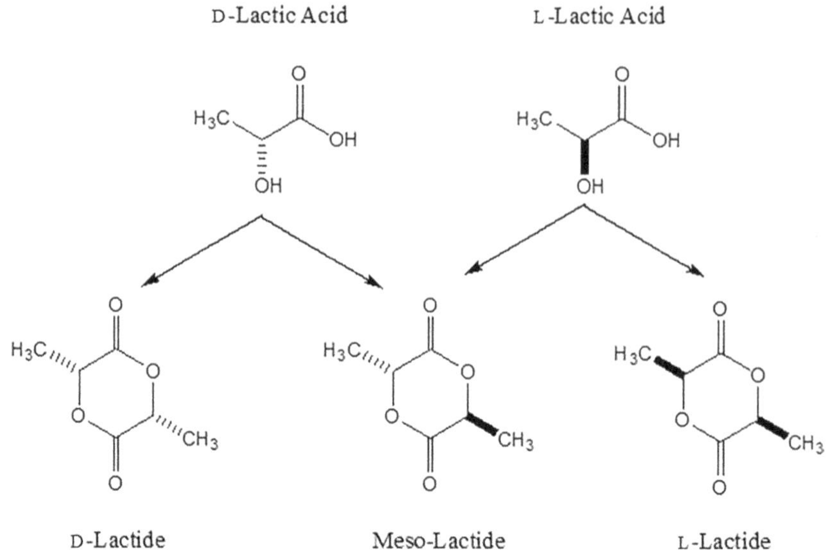

Figure 7.4 Lactic acid and lactide types.

as L-lactide, D-lactide, and *meso*-lactide, see Figure 7.4. A 1 : 1 mixture of L-lactide and D-lactide form a stereo crystal structure known as racemic lactide or *rac*-lactide that has an especially high melting point (120 °C *vs.* 98 °C for L-lactide or D-lactide and 56 °C for *meso*-lactide).

Physical property differences between the lactides, such as volatility and crystallization, enable separation and purification. ROP is initiated with alcohols and catalyzed with metals (aluminum, zinc, tin, and lanthanides) and proceeds *via* anionic, cationic, or coordination mechanisms dependent on the catalyst and initiator selection. Often non-organometallic Sn(II) is used for polymerization. Stereo sequencing of PLA has been shown to vary with catalyst type and monomer feed composition.[26–28]

The most common polymer structures are mainly PLLA with small amounts of random D-lactide and *meso*-lactide incorporated which cause disruptions or "kinks" in PLA's natural helical conformation.[29] The kinks act as "defects" in the crystal arrangement which result then in depression of the melting point, reduction in the level of attainable crystallinity, reduction in the rate of crystallization, delay of strain hardening during drawing/orientation between T_g and T_m, and reduction in modulus above T_g. For products made from D-fermentation, relationships of stereo purity hold true, with L-being the opposite minor isomer in PDLA. Highly disrupted polymers of different architectures can be made from *meso*-lactide, *rac*-lactide, or various combinations of lactides.

General properties for PLA are shown in Table 7.1. Of note, PLA has stiffness most like general purpose polystyrene (GPPS), T_g just below that of PET, and melting point about the same as polypropylene (PP) homopolymer. The oxygen transmission rate of PLA is much lower than GPPS and PP, while

Table 7.1 General properties of PLA compared with commodity polymers.

	PLA	GPPS	PET	PP
Tensile strength,[a] MPa	53.1	45.5	58.6	35.9
Elongation at break, %	4.1	1.4	5.5	350
Tensile modulus, GPa	3.45	3.03	3.45	1.31
Izod impact, $J\,m^{-1}$	16	21.4	26.7	48.1
T_g, °C	60	102	74	−20
Peak melting point, °C	170	NA	270	165
Density, $g\,cc^{-1}$	1.24	1.05	1.35	0.9
OTR[b]	38–42	260	3–6.1	150–800
WVTR	18–22	8.5	1–2.8	0.5–0.7
CO_2	201	—	15–25	150–650

[a] Tensile properties measured on 3.2 mm thick injection molded tensile bars.
[b] Transmission rates: oxygen and CO_2 mL per mil per 100 in.^{2} per day; water vapor g per mil per 100 in.^{2} per day.

about 10× that of PET. The water vapor transmission rate (WVTR) of PLA is higher than the three conventional polymers shown.

Often in the literature, the properties of PLA are cited with significant shortcomings, with brittleness, low heat distortion temperature, poor durability, and low melt strength commonly referenced. As described above, PLA is a family of polymers with a large range of properties available through various strategies. PLA properties available today are much advanced from 20–30 years ago when the literature first started to proliferate. It is beyond the scope of this chapter section to describe the many variations and technologies used with PLA to obtain excellent properties. However, the following sections highlight some of the more relevant and practical work that can be used by researchers and product developers. As technology development of PLA continues, the toolbox for product development is expected to continue to grow.

7.2.3 Heat Resistance

Amorphous PLA's upper use temperature is typically limited by T_g (45–60 °C) for rigid articles. An amorphous PLA can be used as a coextruded layer for low temperature heat seal or thermal or chemical bonding. Coatings made with amorphous PLA on paperboard or other substrates are often used above the glass transition temperature, for example in a coated paper cup for hot drinks. Many low temperature use items for food serviceware and packaging use predominantly amorphous PLA.

Blends with more heat resistant polymers usually have an upper use temperature limited by the second polymer or the mixture properties. PC/PLA,[30] PMMA/PLA,[31] ABS/PLA,[32] PBS/PLA, PP/PLA, and PBAT/PLA are all examples cited in the literature, and many can be found in commercial products. Products range from durable goods made by injection or profile molding to single-use compostable items made by thermoforming, injection molding, and blown film.

Crystalline PLA exists as both homocrystal (HC) and stereocrystal (SC). HC is made with predominantly PLLA both in pure form and with small amounts of D-lactide or *meso*-lactide to provide crystallization disruption. Melting points of HC products range from ~125 to 175 °C.[33] Degree of crystallinity can vary from low to very high (~75%). Several crystal types are possible with the α phase being most common. The α phase is a left-handed 10^3 helix for PLLA packed into an orthorhombic unit cell with parameters $a = 1.07$ nm, $b = 0.645$ nm, and $c = 2.78$ nm.[34] Less common HC types are α′ and β phases. PDLA has the opposite helix with the same α phase crystal parameters. SC is made from a mixture of highly pure PLLA and PDLA either blended from separate polymerizations or from single polymerizations of block copolymers of the two stereo types. Blends of SC can be asymmetric (meaning unequal weight ratio) or symmetric; they may be unbalanced (different molecular weight (MW) or stereo regularity) or balanced. SC has a high melting point of ~220 °C which varies with the stereo purity of the components. The stereocomplex γ crystal phase helix is 3^1, formed by two antiparallel helices packed in an orthorhombic unit cell with $a = 0.995$ nm, $b = 0.625$ nm, and $c = 0.88$ nm. Wide angle X-ray scattering (WAXS) is used to distinguish the crystal types.

Below T_g, PLA is glassy and stiff. When heated, the rubbery plateau region above T_g of crystalline PLA can bear load with the use temperature limited by load, modulus above T_g, onset of melting, degree of crystallinity, and amorphous phase orientation. With naturally slow crystallization kinetics, almost all PLA types are quenchable, meaning they can be quenched in the amorphous state if very rapid cooling is applied. Crystallinity must be attained by orientation and/or thermal history between T_g and T_m (see Figure 7.5).[35]

Composite reinforcement can be used to increase modulus above T_g through a combination of crystallinity and reinforcing agents such as minerals, fibers, or fillers. Copolymers with higher T_g monomers have been cited in the literature to increase the use temperature *versus* that of amorphous PLA.[36]

Crystallization kinetics were mentioned as naturally slow but progress has been made to accelerate the bulk crystallization rate by the addition of heterogeneous nucleating agents, thermal treatments, use of plasticizers, dissolved gas, and shear. Optimum crystallization temperatures are usually from 105–130 °C.[37] In extensional flows, crystallization can be fast as chain alignment/orientation is forced through mechanical means. In particular, oriented films[38] and fibers take advantage of stress induced crystallization (SIC).[39] Heat set can be applied under tension or in some cases allowing free shrinkage to "lock in" crystal structure enabling higher heat performance and low shrinkage below T_m. Oriented PLA also tends to be much tougher, possessing both high modulus and much higher ductility.

7.2.4 Brittleness and Toughness

Toughness describes a material's ability to absorb energy without fracturing. The types of tests used to determine toughness vary greatly and reflect very

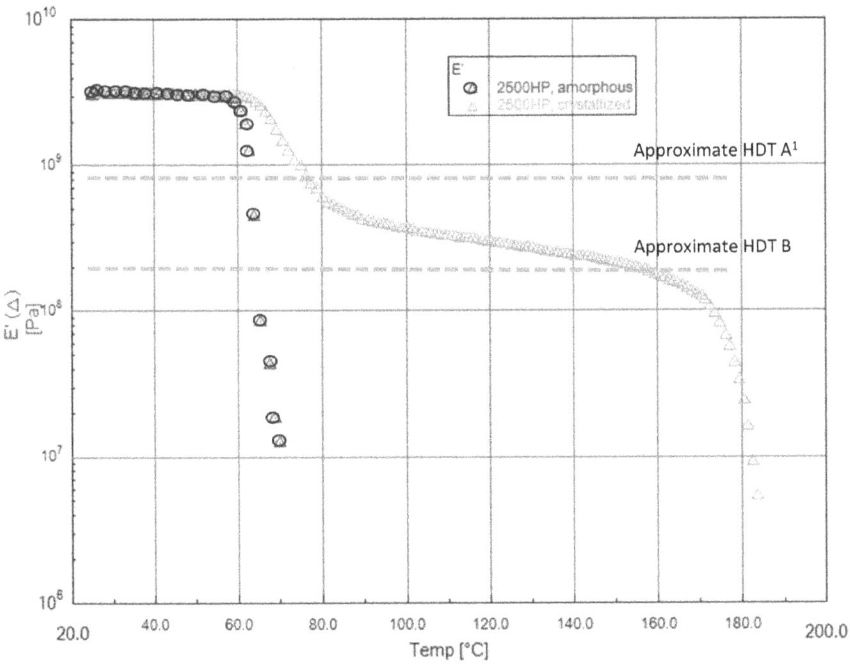

Figure 7.5 Elastic modulus, E', by dynamic mechanical analysis of crystalline and amorphous Ingeo™ 2500HP (high molecular weight PLLA) from NatureWorks using three-point bend geometry (ASTM D5023) with HDT estimates using the method of Takemori, see ref. 35.

different strength/toughness needs of the myriad of polymer applications. Integration of the stress–strain curve during tensile deformation is a common measure referred to as "tensile toughness." Impact strength is another measure of toughness commonly used in polymers where tests typically involve striking a sample to determine energy to break or energy where 50% of specimens fail. Specimens may be impact tested with a notch, to concentrate stress, or without a notch. Injection molded articles, flat sheets, and films are tested using specific methods to determine failure energy. Brittle materials quickly fracture upon mechanical loading as the material is unable to absorb energy. Ductile materials sustain plastic deformation before fracture and absorb significant energy. Strain rate, force concentration, and temperature are all extremely important for characterizing ductility.

Most PLAs will act brittle when amorphous and unoriented with no other additives present. Molecular weight improves tensile toughness to some degree, with elongation to break of most high molecular weight polymers reported at 4–6% and low notched Izod of about 0.3–0.4 ft lb per in. (16–21 J m^{-1}). Similar to the glassy polymer polystyrene (PS), below T_g PLA will craze and shear under stress, and large crazes can lead to crack propagation and breaks.

Many options to improve the toughness of PLA have been reported in the literature. Strategies include adding traditional impact modifiers, creating alloys with other polymers, adding reinforcing agents, copolymerizing with energy absorbing block copolymer segments, plasticization, orientation, and crystallization.

PLA can be impact modified with inclusion of an energy absorbing minor phase. With PS systems, polybutadiene rubber is incorporated during polymerization to make high impact polystyrene (HIPS) and the minor phase works to absorb energy and terminate cracks resulting in much smaller craze dimensions of higher craze density. Impact modified grades of PLA can work similarly where a minor phase is introduced to absorb energy and terminate cracks. As of the writing of this text, commercial grades of PLA are not polymerized incorporating rubbery phases, but many materials can be added to form blends and compounds. Commercial masterbatches are also available from several companies where impact modifiers are prepared in a concentrate to be blended with virgin or neat PLA during fabrication or compounding.

Options for ductility modification of PLA often start with product requirements and a first pass of selection is typically driven by food contact and/or compostability criteria and needs. The list of available modifiers with potential to meet industrial compostability standards when used at concentrations exceeding 1 wt% in the final product is fairly small with materials such as polybutylene succinate (PBS) and polybutylene succinate co-adipate (PBSA) derivatives, polybutylene adipate-*co*-terephthalate (PBAT), polycaprolactone (PCL), epoxidized natural rubber (ENR),[40] and polyhydroxyalkanoates (PHA). Blends with PBSA were shown to significantly improve PLA tensile toughness to about 70 MJ m^{-3} and also exhibited high elongation to break at 20 wt% PBSA.[41] Notched Izod impact typical of commercial HIPS is reached at about 50 wt% PBSA in PLA.[42] Similar impact strength is reached with about 15 wt% of rubbery PHA copolymer.[43]

Crystallization of PLA in the absence of orientation can lead to some small improvements in toughness. Ingeo 2500HP literature shows notched Izod increase from 0.35 to 0.75 ft lb per in., an increase of 114%. At the same time, the crystalline material exhibits substantially higher flex modulus and heat distortion temperature improvement from 54 to 144 °C. Impact modification of crystalline versions of PLA were shown to improve at much lower concentrations of impact modifier than quenched, amorphous versions of the exact same polymer type.

Orientation driven crystallization, discussed earlier as SIC, has been shown to dramatically improve PLA toughness qualities. NatureWorks literature shows elongation to break improvements for Ingeo 4032D biaxially oriented films (high MW, 1.5% D-) from 6% elongation to 180% and 100% in machine and transverse directions. Accompanying the tensile improvement, dart impact improved from 2.44 to 18.7 J. These improvements were shown to also hold at −15 °C. Fibers experience high orientation during draw from the melt or during secondary draw steps. The toughness of fibers is often compared in terms of strength/tenacity, or elongation to break, and typical PLA staple fibers have about 10–70% elongation to break. Orientation during thermoforming can also result in some toughness improvements.

Traditional impact modifiers are one of the most important methods for toughness modification of PLA. Modifiers induce plastic deformation before crack initiation/propagation occurs. Selection of the modifier type and level is dictated by the balance of end-use application, fabrication process, and type of impact performance required. Considerations for optics, organoleptics, compostability, and regulatory clearances are critical. Composting standards (ASTM D6400) allow non-toxic, non-degradable additives to be used below 1 wt% with a rule of no more than five of such different materials. Soft thermoplastics, such as ethylene based Biomax Strong 120 and polyvinyl acetate based Vinnex, are effective at improving PLA toughness. Core–shell rubbers have been used very effectively with some materials maintaining high clarity. Polymethylmethacrylate (PMMA) shells are favorable as PMMA has been shown to be miscible with PLA. Core polymers of butyl-acrylate, silicone-acrylate, and butadiene have been shown to work well for improving impact performance. Table 7.2, prepared by Owusu,[42] compares commercially available core–shell impact modifier types.

Figure 7.6 shows a chart of notched Izod impact strength *versus* impact modifier loading for crystalline PLAs Ingeo 2500HP and 2003D (circles with a darker outline) (0.5 and 4.5% D- and 58 and 48% crystallinity respectively).

The highly crystalline 2500HP grade increases in impact strength with much lower dosing of the impact modifier KaneAce Eco-100. Impact strengths in the range of HIPS are reached at about 1 wt% dosing, and the Izod range of poly(acrylonitrile butadiene styrene) (ABS) is achieved at around 5 wt% dosing, see Figure 7.7. Non-crystallized 2500HP achieved less than 1 ft lb per in. (53 J m^{-1}) notched Izod impact over the entire range of dosing. In non-crystallized PLA studies, it was shown that the range of notched Izod impact exhibited by ABS could be achieved with about 15 wt% impact modifier using core–shell products Metablen S-2001, Blendex 3160, and Biostrength 150.

7.2.5 Controlled Hydrolysis

It is well known that PLA is susceptible to hydrolysis and can undergo molecular weight reduction over time. It is also known that the rate of PLA hydrolysis is catalyzed by acids and proceeds by an autocatalytic mechanism,

Table 7.2 Commercially sourced core–shell impact modifiers.

Product	Manufacturer	Chemistry
KaneAce Eco-100	Kaneka	Acrylate rubber core/PMMA shell
Blendex 3160	Galata chemicals	Butadiene core/styrene acrylonitrile shell
Blendex 960A	Galata chemicals	Acrylate core/SAN shell
Biostrength 150	Arkema	Butadiene core/PMMA shell
Biostrength 280	Arkema	Acrylate core/PMMA shell
BPM 500/515	Dow chemical	Acrylic core/PMMA shell
BPM 520	Dow chemical	Butadiene core/PMMA shell
Metablen S2001	Mitsubishi	Silicone core/PMMA shell
Kane Ace 731	Kaneka	Butadiene core/PMMA shell

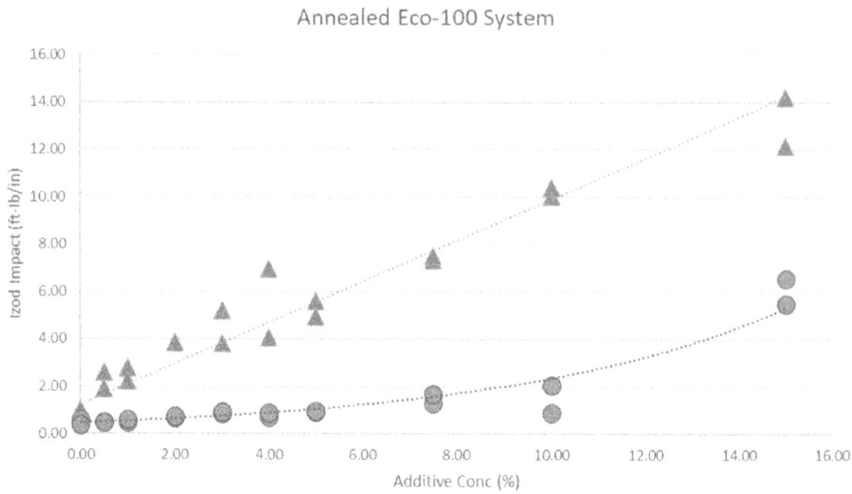

Figure 7.6 Comparison of impact improvement of crystalline Ingeo 2003D (4.5% D-) and 2500HP (0.5% D-). The upper (triangles) and lower (circles) are crystalline Ingeo 2500HP and 2003D, respectively.

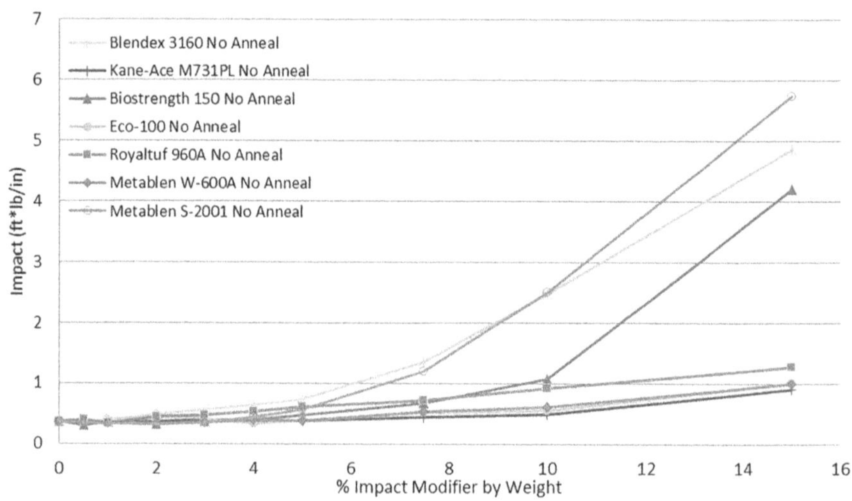

Figure 7.7 Notched Izod impact strength of amorphous Ingeo 2003D with various core–shell impact modifiers.

whereby the increasing acid concentration that results as each ester group is hydrolyzed further catalyzes the rate of hydrolysis. An expression describing acid catalyzed hydrolytic degradation of PLA is:

$$\mathrm{d[Ester]/d}t = k_\mathrm{h}[\mathrm{H_2O}][\mathrm{Ester}]([-\mathrm{COOH}]K_\mathrm{a})^{1/2}$$

Temperature strongly affects the rate of hydrolysis with a significant increase in hydrolysis kinetics above T_g.[8] This property allows for industrial compostability

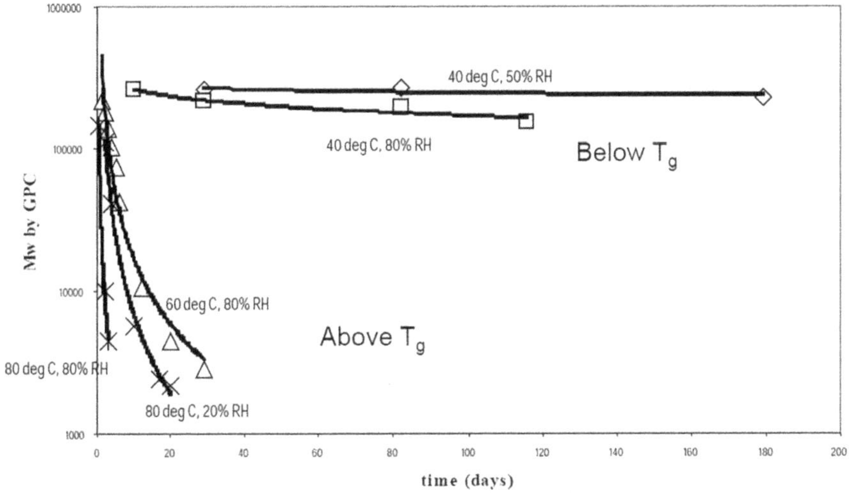

Figure 7.8 Molecular weight loss for amorphous 10% PDLA stored at different temperatures and relative humidities.

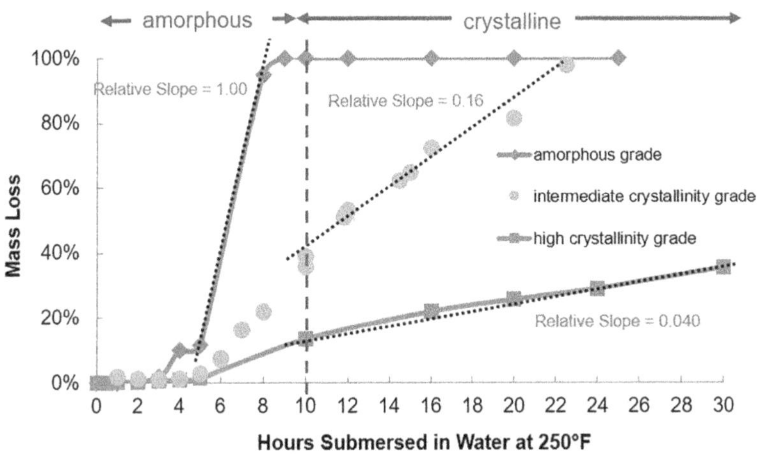

Figure 7.9 Differences in the rate of mass loss for amorphous and crystalline Ingeo grades.

as hydrolysis proceeds quickly in warm and moist composting environments where existing microbes metabolically transform the carbon present in low MW lactic acid oligomers to primarily carbon dioxide, see Figure 7.8. The time for the molecular weight of PLA to decrease by 50% was shown to be 20–50 years for unmodified PLAs at room temperature and 50% humidity.

Highly crystalline PLLA of high stereo purity tends to be more stable to hydrolysis than amorphous and stereo disrupted PLA, see Figure 7.9. The amorphous phase was reported to hydrolyze >20× faster than crystalline phase when PLA was submersed in water at 121 °C.[44]

Strategies to control hydrolysis generally include the selection of the PLA type followed by various means to control acid catalysis and water transport. For most situations, water can quickly diffuse in and out of PLA and the polymer will undergo autocatalytic bulk hydrolysis, not surface erosion. Mechanical properties decrease rapidly for polymers that degrade by bulk erosion, and the PLA will embrittle and fracture as molecular weight decreases substantially. Hydrophilic and hydrophobic additives can be used to modify the rate of water transport to some degree. Acid catalysis can be intentionally increased or decreased by the use of functional additives and/or acid end capping. Some of the most effective acid catalysts may have delayed onset to provide a "trigger" for hydrolysis. Hydrolysis stabilizers such as carbodiimides can react with acids produced during PLA hydrolysis to effectively delay the onset of acid catalysis until they are consumed, after which the acid-catalyzed hydrolysis of PLA proceeds. The degradation rate of PLA can be "tuned" to provide the desired mass loss performance for different applications. A chart from NatureWorks describing the range of time various Ingeo tunable hydrolysis strategies can deliver is shown in Figure 7.10.

At each temperature in Figure 7.10, options exist to provide a range of $t_{1/2}$ values (*e.g.*, at 140 °F, different formulations deliver $t_{1/2}$ values from 7 to 1143 days).

7.2.6 Melt Strength

The rheology of linear PLA has been described in the literature to have molecular weight dependence on zero shear viscosity typical of that of many thermoplastics.[45,46] Shear thinning is observed but not as strong of a feature as many other materials. Strain hardening occurs during extensional flow. The entanglement molecular weight is about 8700 g mol^{-1}, and the plateau modulus is about 5×10^5 Pa.[47] In general, at typical processing temperatures of 180–230 °C, linear PLA does not exhibit high melt strength. Similar to polymers such as PP and PET, branching is used to increase the melt strength of PLA. Polymerization initiated by multifunctional alcohols[48] and/or use of hydroxy lactones or bicyclic lactones[49] as co-monomers can lead to branching during synthesis. Branching has also been achieved using reactive extrusion with branching along the PLA backbone using free radical initiated chemistry[50] and branching using the available PLA end groups using epoxy,[51] isocyanate, phosphite,[52] and carbodiimide chemistry.

7.2.7 Summary

As the world looks to use more sustainable, high-performing materials, PLA has proven to have many features that are attractive to the market for applications including food serviceware, packaging, coatings, fibers, films, foams, powders, 3D printing filament, injection molded articles, and profile extrusion molded articles. As more companies and researchers invest

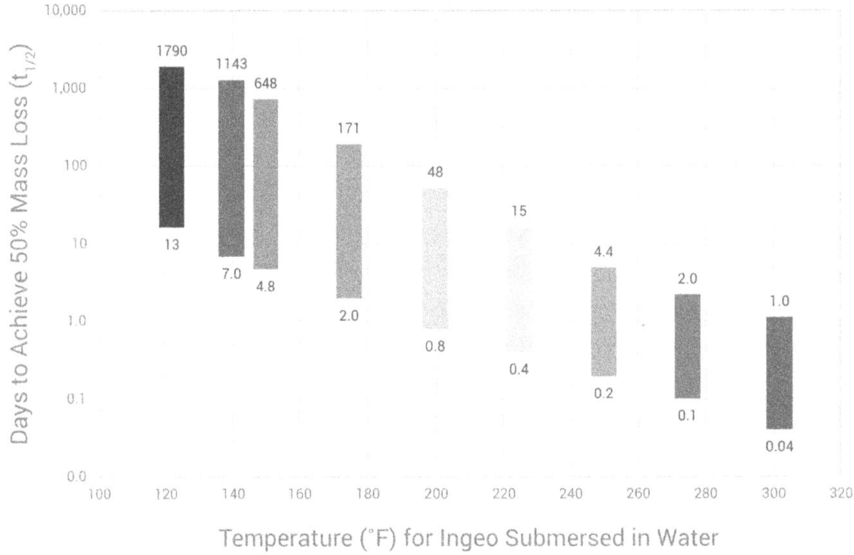

Figure 7.10 Range of mass loss behavior using current commercial options to modify Ingeo biopolymer hydrolysis rate.

resources into further developments, significant expansion of properties and products is expected to continue accompanied by significant market growth.

7.3 Polyhydroxyalkanoates

Polyhydroxyalkanoates are a promising family of biopolymers that can be a sustainable alternative to conventional, petroleum-based plastics. They are completely biodegradable under aerobic and anerobic conditions,[53] are derived from renewable resources,[54] and can have properties similar to polypropylene.[53] There is significant motivation to find alternatives to conventional plastics because the majority of plastics are derived from fossil hydrocarbons and most are not biodegradable. In fact, it has been estimated that 79% of all plastics ever produced have accumulated in landfills or the natural environment.[55] PHAs are naturally-occurring polyesters synthesized by bacteria and stored as intracellular carbon and energy reserves,[53,54] see Figure 7.11.

As detailed in Chapter 3 "Overview of Bioplastics," not all bio-based plastics are "biodegradable" as summarized in Figure 7.3. As we strive towards a circular bioeconomy, it will be important to adopt plastics that are both bio-based and biodegradable. Bio-based biopolymers are produced from rapidly renewable, non-fossil carbon sources[56] while biodegradable biopolymers may be broken down by naturally occurring microorganisms. PHAs can be both bio-based and biodegradable and thus offer an advantage over other biomaterials that may be bio-based or biodegradable but not both, see Figure 7.12. PHAs can be produced from many sustainable feedstocks

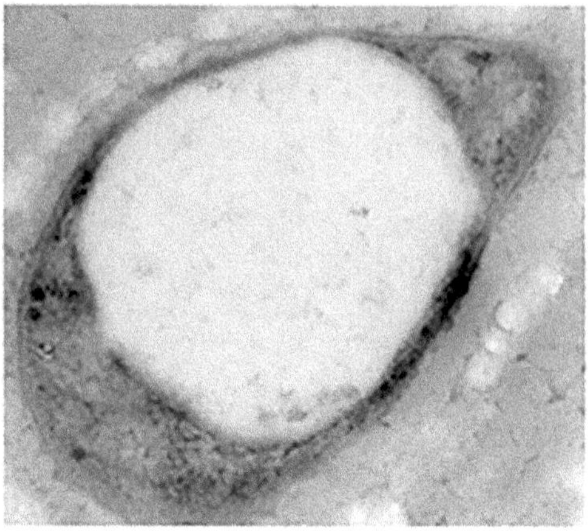

Figure 7.11 Scanning electron micrograph of PHA granule produced inside a bacterial cell.

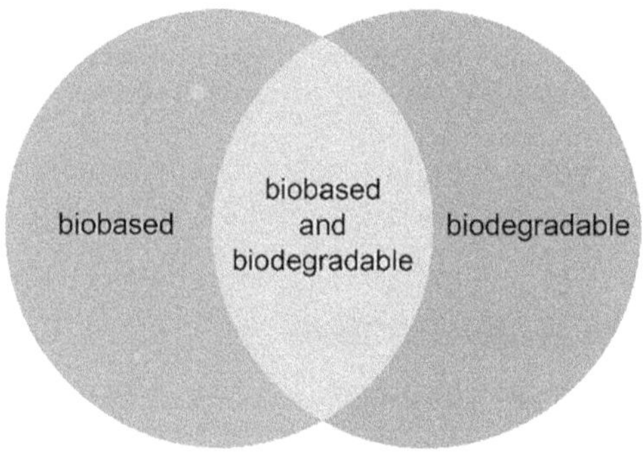

Figure 7.12 Types of biopolymers – bio-based, biodegradable, or both (European Bioplastics 2016).

including waste biogas, solid organic waste, sugars, starches, or cellulose and can biodegrade in a variety of industrial and natural environments.

A unique characteristic of PHAs is that in some production processes, the carbon released during PHA biodegradation can be recovered for use as feedstock in PHA production, such as when using methane or volatile fatty acids, creating the potential for promoting closed-loop life cycles. These qualities have made PHAs into the fastest-growing biopolymers in the world.[57]

7.3.1 Production of PHAs

PHAs can be produced by a variety of organisms from a variety of feedstocks.[54,58] They are most commonly synthesized by prokaryotes which utilize PHAs for intracellular carbon storage and energy reserves. Nearly a hundred different PHAs have been identified,[59] and there are numerous factors, including monomer composition, formulation, and processing, that

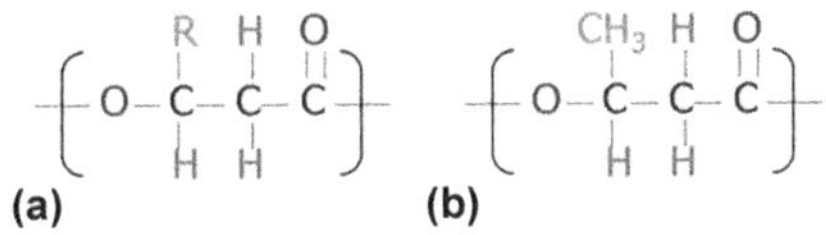

Figure 7.13 (a) Chemical structure of a generic PHA. (b) Chemical structure of PHB.

influence the physical, material, and end-of-life properties of the end polymer. Polyhydroxybutyrate (PHB), see Figure 7.13, is the most common biopolymer in the PHA family and the most extensively studied.[53]

Various carbon sources have been used to produce PHAs with different chain lengths based on the attached R groups (see Table 7.1).[58,60] PHAs created from monomers of three to five carbon atoms are called short-chain-length PHAs (scl-PHAs) and those created from monomer units of 7–14 carbon atoms are called medium-chain-length PHAs (mcl-PHAs). However, six carbon monomers can be incorporated into both classes of PHAs. Most commercial production has focused on P3HB based scl-PHAs with scl co-monomers.

A wide variety of bacterial strains are capable of accumulating PHAs (see Table 7.3).[58] Although many bacterial strains can produce PHAs naturally, genetically modified bacteria have been created to take advantage of different carbon sources.

Although PHA is not yet commercially available on large scales, there are several companies working to scale production through large-scale bacterial fermentation. The use of genetically modified plants for PHA production has been explored; however, significant advances must occur for this to be fully commercially viable.

Industrial production requires bacterial synthesis of PHA in large-scale fermentation vessels designed to maximize growth and PHA production rates (see Chapter 6 Fermentation for additional information). It is important to control the rate of carbon/nutrient and gas feeding. Typically, this is accomplished through a two-stage process where the first stage supports rapid population growth and the second stage is a period of PHA production, often triggered by nutrient limitation or simply by excess carbon substrate. This process may be completed as a batch, sequencing batch, or chemostat reactor.

PHA is naturally accumulated in the form of granules inside bacterial cell walls and a series of downstream processing steps must then be employed to lyse the cells and purify and dry the PHA. Purification may be accomplished by various methods where the cells are lysed and the PHA granules are separated from the cell mass. Purification may also be accomplished by dissolving the PHA in a solvent such as chloroform, filtering to remove the biomass, and precipitating the polymer. Although P3HB has limited solubility in solvents, other PHAs have a wider solubility range. Patents by Metabolix have highlighted the use of solvent-based methods for PHA recovery.[79]

Table 7.3 Overview of microorganisms capable of producing PHA biopolymers from various substrates.

Genus	Biopolymer	Substrate	Ref.
Azotobacter	P3HB	Glucose	Lasemi *et al.*, 2012[61]
Cupriavidus	P3HB, 3HB, 3HV	CO_2, glucose, corn oil, oleic acid, olive oil, palm oil, butyrate, lactic acid, propionic acid, potato starch, saccharified waste	Ishizaki and Tanaka, 1991;[62] Kim *et al.*, 1994;[63] Fukui and Doi, 1998;[64] Chakraborty *et al.*, 2009;[65] Haas *et al.*, 2008[66]
Hydrogenophaga	P3HB3HV	Hydrolyzed whey and valerate	Koller *et al.*, 2007[67]
Methylobacterium	P3HB	Methanol	Bourque *et al.*, 1995[68]
Methylocystis	P3HB3HV	Methane	Wendlandt *et al.*, 1998[69]
Paracoccus	P3HV	*n*-Pentanol	Yamane *et al.*, 1996[70]
Pseudomonas	mcl-PHA, p3HB, scl-mcl-PHA	Cane molasses, fructose, sucrose, terephthalic acid from polyethylene terephthalate, 1,3-butanediol, octanoate, styrene	Tripathi *et al.*, 2012;[71] Kenny *et al.*, 2008;[72] Lee *et al.*, 1995[73]
Thermus	scl-mcl PHA	Whey	Pantazaki *et al.*, 2009[74]
Bacillus	P3HB, scl-mcl-PHA, 3HB, 3HV, 3HHx	Citric acid, glucose, glycerol, succinic acid, octanoic acid, acetate, *n*-alkanoate, propionate, sucrose, valerate	Shahid *et al.*, 2013[75]
Corynebacterium	P3HB, mcl-PHA, 3HB, 3HV	Acetic acid, citric acid, glucose, glycerol, succinic acid	Shahid *et al.*, 2013;[75] Haywood *et al.*, 1991[76]
Nocardia	3HB, 3HV	Acetate, succinate	Haywood *et al.*, 1991[76]
Haloferax	P3HB3HV	Vinasse, crude glycerol from biodiesel production	Bhattacharyya *et al.*, 2012;[77] Hermann-Krauss *et al.*, 2013[78]

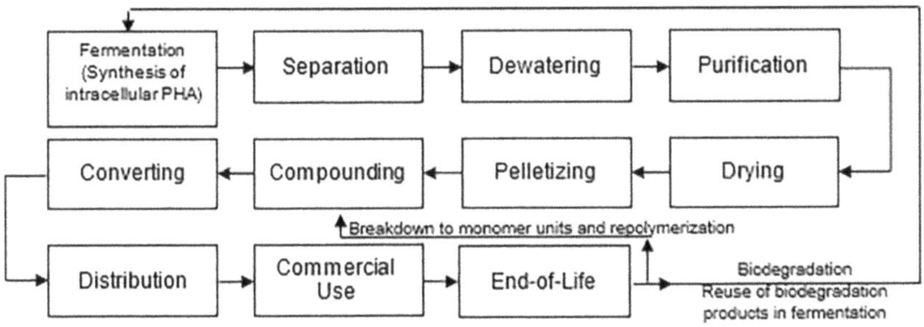

Figure 7.14 The industrial process of PHA production.

After the PHA granule and residual biomass have been separated, a series of de-watering, purification, rinsing, and drying steps are required to produce purified PHA powder that can be pelletized. The pellets may be compounded with additives to improve properties before being processed into commercial goods, as discussed in the next section. The basic steps for the industrial production, use, and disposal of PHA are shown in Figure 7.14.

7.4 Processing and Properties of PHA

7.4.1 Properties

PHAs have material properties (see Table 7.4[80–82]) that could allow them to be used in a variety of consumer applications. PHAs can compare favorably to petroleum-based plastics, with similar crystallinity, glass transition temperature, melting temperature, elastic modulus, and tensile strength to polypropylene. PHAs also can have better water resistance and heat resistance compared to other biodegradable polymers such as polylactic acid and starch.

Mechanical and thermal properties can vary significantly based on the monomer composition and can be enhanced by blending with other polymers and additives. A copolymer of P3HB-*co*-3HV with 14% HV polymer has a melt temperature of 150 °C allowing for a wider processing window to avoid thermal degradation of PHA. However, lower melting PHB copolymers also have reduced heat distortion temperatures; a 14% HV polymer lowers the heat distortion temperature to 112 °C, 45 °C lower than the PHB homopolymer.[83] In general, co-monomers in P3HB based-PHAs decrease melt temperature, glass transition temperature, and crystallinity while increasing toughness and ductility. Co-monomers also decrease heat distortion temperature and tensile strength relative to pure P3HB. The degree by which the mentioned properties can be enhanced depends on the structure of the co-monomer. For example, linear monomers such as 4-HB tend to enhance elongation to break at low molar fractions and, while 3HV copolymers exhibit some unusual properties because of the isodimorphic crystal structure, can promote crystallizing for weeks after melt.[80,84]

Table 7.4 Thermal properties of PhB.

Material property	Value	Units
Crystallinity	60–80	%
Decomposition temperature	200	°C
Density	1.25	$g\,cm^{-3}$
Elastic modulus	3.5	GPa
Elongation at break	6	%
Flexural modulus	3.4	GPa
Max. flexural stress	54	MPa
Glass transition temperature	4	°C
Melting temperature	180	°C
Molecular weight	180	°C
Tensile strength	40	MPa

The ability to customize and tune the material properties of PHAs gives the material versatility.

7.4.2 Processing

scl-PHAs are thermoplastics that can be processed on extrusion and product manufacturing equipment typically used with conventional plastics. scl-PHAs require special attention when processing because of their relatively low thermal degradation temperatures that often result in narrow processing windows. Processing near the melt temperature of 180 °C leads to approximately five chain scissions per ten million monomer units per second; this rate increases four- and tenfold at 190 °C and 200 °C, respectively.[81] Thus, processing leads to molecular weight degradation as well as thermomechanical properties. Therefore, care must be taken to avoid excessive heating, both by controlling melt temperature and minimizing shear stress. Processing work at Metabolix led to the recommendation of reverse temperature profiles for extrusion of PHAs,[85] where early zones are the hottest and temperature setpoints gradually decrease as the polymer moves toward the die.

scl-PHAs undergo crystallization at a slow rate, leading to large crystalline spherulites that create potential fracture planes and cause increased brittleness. Nucleation and annealing of the polymer can increase crystallization rates and reduce spherulite size dramatically, thus improving polymer properties. Mineral fillers such as talc and calcium carbonate have been used effectively to aid crystallization. Boron nitride has long been the gold standard of nucleating agents for PHAs with a 1% addition increasing the crystallization temperature by 18 degrees.[86] More recently, heterocyclic nitrogen compounds have been found to induce an increase in crystallization temperatures greater than boron nitride.[87,88]

In certain applications such as blow molding, the use of PHAs can be limited by low melt viscosity, a challenge that is exacerbated by thermal degradation. Reactive processing with organic peroxides has been shown to increase melt strength by promoting branching of polymer chains.[89]

Peroxides can also be used to compatibilize PHA with non-miscible polymers, such as PLA or cellulose.[90,91] Processing with branching agents has great promise to broaden possible applications for scl-PHAs.

7.4.3 PHA as a Circular Material

Efforts to reduce plastic pollution have sparked interest in developing circular materials that can be regenerated after use. The carbon in biopolymers such as PHAs can be recovered using traditional waste management processes such as recycling, composting, or take-back programs, making it an interesting candidate for future development.

7.4.4 Recycling

Similar to conventional plastics, PHAs are thermoplastics that can ultimately be recycled and reformed into new products. Recycling may be an attractive alternative to biodegradation because it requires less energy compared to depolymerizing and re-synthesizing the polymer. Because the molecular weight of PHA can degrade during thermal processing, repeated recycling may require applications that use lower-MW PHA or require techniques to increase the MW during processing.

As PHAs become more widely adopted, the increased volumes may incentivize material recovery facilities to source-separate them along with other biopolymers such as PLA. To accomplish this source separation, current infrastructure at some recycling facilities may need to be upgraded with the addition of a UV-detector, or the biopolymer resin could contain a marker that would facilitate separation and isolation of different resin types.

7.4.5 Biodegradation

PHAs can degrade in natural and industrial environments under both aerobic and anaerobic conditions. PHAs will degrade under enzymatic attack or upon exposure to strong acids or bases. However, commercial products made from PHA will not degrade during everyday use. Because PHAs are produced microbially, many microorganisms possess enzymes to break them down intra- or extra-cellularly. PHAs can biodegrade in industrial waste facilities such as aerobic composting, anaerobic digestion, or modern landfills. They have also been shown to biodegrade in natural environments including soils,[92] fresh water and oceans,[93] and living tissue.[94]

There are numerous tests and certifications that a material or product can undergo to confirm biodegradability. For instance, the ASTM D6400 standard defines a biodegradable polymer as one "in which the degradation results from the action of naturally occurring microorganisms such as bacteria, fungi, and algae."[95] In order to be labeled as complying to ASTM D6400, the product must pass this test. All polymers that are termed "compostable" are biodegradable, but the reverse is not always true. Environmental conditions such as medium pH, moisture and oxygen content, and temperature can significantly affect the biodegradation of biopolymers.[96]

Biodegradation begins when microorganisms colonize and attach to the surface of PHAs. The hydrolysis step occurs first where secreted enzymes (hydrolases) and water are used to break down the biopolymer into smaller monomer units through the addition of water. These monomers are broken down into fatty acids by fermentation. Acetogenesis then takes place which converts the fatty acids to acetate or hydrogen. The acetate is broken down into methane using acetate-utilizing methanogens while the hydrogen is converted to methane using hydrogen-utilizing methanogens.[97,98] This stepwise conversion of PHB is shown in Figure 7.15.

Aerobic biodegradation is the biological breakdown of a substance by microorganisms in environments where oxygen is present. Aerobic environments can be found in numerous natural ecosystems as well as in industrial facilities such as compost facilities. Aerobic biodegradation tests of different PHAs have been conducted in numerous conditions including soil, salt water, fresh water, canal water, and household compost[99] as well as in soil[100] and earth worm compost,[101] and by *Comomonas*[102] and *P. lemoignei.*[103]

Preliminary investigations studied adding 3-hydroxyvalerate (HV) as a comonomer and revealed that as more HV is added to PHB more weight is lost during degradation by extracellular PHB depolymerase from *Emehcellopsis minima W2.* It has also been established that PHB-3HV degrades quicker in soil than the homopolymer,[100] and there was more biofilm growth on P(HB-*co*-8HV) during biodegradation using fertile garden soil than plain PHB.[104]

Anaerobic (without oxygen) environments are present in estuaries, soils, sediments, and in industrial facilities such as modern landfills, agricultural facilities, and water treatment plants. Complete weight loss of PHB thin films was described for samples biodegraded in anaerobic sludge.[105] Anaerobic biodegradation of PHB has also been investigated using anaerobic sewage sludge[106] as well as PHB depolymerase from *Alcaligenes faecalis.* Anaerobic biodegradation of PHB was reported to occur in sewage sludge[83] with PHB powder and plate biodegrading more thoroughly than a common

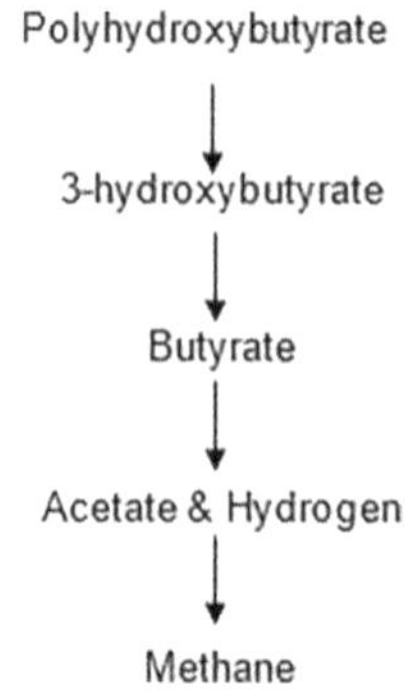

Figure 7.15 Stepwise decomposition of PHB.

standard cellulose powder which is known to be biodegradable under anaerobic conditions.[107] In addition, natural additives have been shown to modify the characteristics and degradation rate of the PHB without impairing its final biodegradation in anaerobic environments.[107]

PHA and its copolymers have also been shown to biodegrade in both saltwater and freshwater ecosystems. Bacteria are the main PHA degrading agents in water ecosystems unlike in soil environments where fungi are the most significant degrading agents.[108] Copolymer PHBV (polyhydroxyalkanoate-type polymer) was shown to biodegrade in a continuum of environments common in tropical coastal waters. Both PHBV and P(3HB-*co*-4HB) were shown to biodegrade in seawater at a rate almost independent of the copolymer compositions.[80] Furthermore, PHA films were deemed highly biodegradable in seawater using ASTM standard method 6691.[109] PHAs were also shown to degrade at different depths of a saltwater lake.[108] In natural waters, PHB and PHBV were both shown to effectively biodegrade under real-life conditions.[110]

7.4.6 Closed-loop Production of PHAs

It is possible that PHAs can be produced in a partly closed-loop lifecycle where the same feedstocks used to produce PHA can be regenerated upon its degradation. For example, methane or volatile fatty acids can be used as feedstock for PHA; upon biodegradation of commercial, PHA-based products at anaerobic waste management facilities, the resulting acids or methane could be captured to serve as carbon feedstock in the production of new PHAs. The natural recycling of carbon in this manner ensures that value is generated from waste and is a step towards a sustainable bioeconomy.

7.5 PHA Applications

As more commercial brands adopt PHAs for their bio-based and biodegradable qualities, they have quickly become one of the fastest growing biopolymers in the world.[57] It is estimated that global production of PHAs will increase more than six-fold to 160 000 tonnes in 2024 from 25 000 tonnes in 2019 because of a ramp-up of commercial production.[111] With physical and mechanical properties similar to polypropylene, PHAs could ultimately substitute numerous polymer applications.[112]

PHAs can cover a wide range of mechanical properties, allowing them to be utilized for many different applications ranging from rigid packaging to flexible thin films. Successful initial applications should be tailored to the properties inherent to PHA while capitalizing on the sustainability advantages of a bio-based and biodegradable material.

As the largest market for plastics, packaging is an attractive application for PHAs. PHAs are suited to protect both food and non-food items before breaking down into environmentally harmless components upon disposal. PHAs have desirable barrier properties compared to other bio-based polymers including low gas permeability and could be an attractive biodegradable replacement for paperboard coatings.

In response to the growing demand for biodegradable alternatives to persistent petroleum-based plastics, PHA producers are scaling up production capacity to increase supply on the market. Today, PHA can be found in a number of diverse consumer products, including beach toys (Zoë b Organic), denitrification pellets (Bluepha), and 3D printing filament (ColorFabb). PHA is often used in its pure form or in a blend with other biopolymers to improve material properties of the end product. Many PHA products are now currently under development including textiles (Mango Materials), cosmetics packaging (Mango Materials), straws (Danimer Scientific), food packaging (Danimer Scientific), furniture (Newlight Technologies), automotive products (Kaneka), bottles (Cove), and single-use cutlery (RWDC Industries).

7.6 PHA Conclusions

While PHAs have been studied for many years, advances in biotechnology, industrial microbiology, and polymer processing, coupled with a drive towards increased sustainability, are finally propelling them to the forefront of the bioeconomy. With many companies and academics working to address the challenges posed by using PHAs as petrochemical-based plastic replacements and with a growing need for bio-based and biodegradable alternatives to plastics, they are starting to gain market traction and may now fulfill their commercial potential.

7.7 Other Polyester

While this chapter has primarily focused on PLA, PBS, and PHAs, it is important to note there are other common bio-based polyesters. For example, PCL is a common material, its chemical structure seen in Figure 7.16.

While polycaprolactone has a relatively low melting temperature (approximately 60 °C) and has a tensile strength typically between 10 and 16 MPa, one of its unique properties is the ability to be absorbed by the human body. Thus, it has been used in medical applications as absorbable medical sutures, for drug delivery, and in implants. The rate of absorbance can be tuned by compounding it with polyglycolic acid, polylactic acid, and polydioxanone.

Another common bio-based polyester is polybutylene adipate terephthalate (PBAT), see Figure 7.17. The polyester is highly amorphous, has relatively good thermal stability (230 °C) and can be highly flexible with relatively low strength between 1 and 2 MPa. Because of its relative ease of

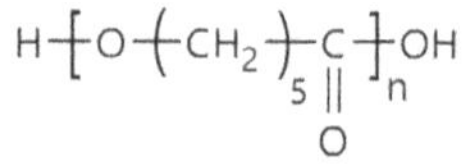

Figure 7.16 Chemical structure of PCL.

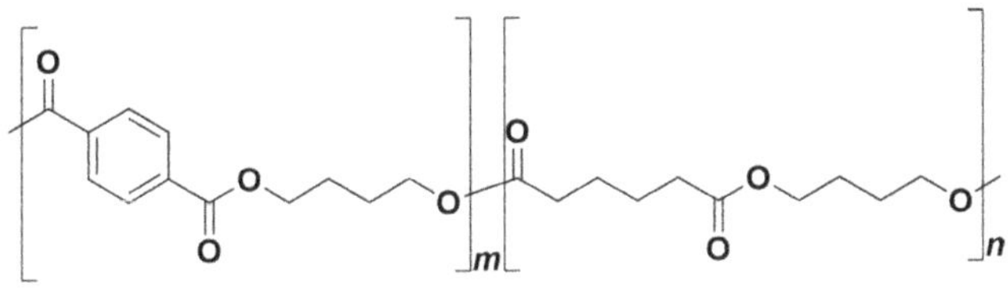

Figure 7.17 Chemical structure of PBAT.

degradability and flexibility, it is often used for film bags and food packages, including wrapping as well as coating of paper food packaging. In addition, it is often compounded with other bio-based plastics to enhance the mechanical properties.

One of the largest bio-based polyesters is bio-based polyethylene terephthalate. This is the plastic most commonly used to produce carbonated beverage drink bottles. The bio-based version is currently up to one third bio-based, has the exact chemical structure of petrochemical-based PET, and thus is a one-to-one drop in.

7.8 Introduction to PBS

7.8.1 Summary

PBS is a biodegradable aliphatic polyester formed from the reaction of succinic acid and 1,4-butanediol (BDO). The molecular structure is shown in Figure 7.18. The polymer is semi-crystalline with physical properties that are comparable to those of polypropylene or polyethylene (PE) due to similarity in mechanical and processing properties. The typical melting point is around 112–116 °C and the glass transition temperature is about −33 °C. In addition to biodegradability in a broad range of environments, including in soil and seawater, PBS is also found to have good compatibility with many other polymers and monomers leading to increasing interest in using PBS as a blend or composite component for various applications.

Due to these properties, PBS is gaining increasing attention and has become one of the bioplastic materials with good commercialization potential, mainly pursued by Asian companies. Showa Denko was among the first companies to build PBS capacities in the early 1990s and marketed it under the "Bionolle™" trade name, which was unfortunately terminated in 2017. Major Chinese producers include Kingfa S&T, Anqing Hexing Chemical, and Xinfu Pharma, most of which were designed as swing plants between PBS and PBAT.[113] Other major Asian manufacturers include Mitsubishi Chemical, PTTMCC Biochem, SK Chemicals, and Ire Chemical. Their PBS trade names are GS Pla™, BioPBS™, SkyGreen™, and EnPol™, respectively. PBS target applications include disposable goods in food packaging, medical devices, and agriculture, as well as durable goods in the automotive and construction industries.

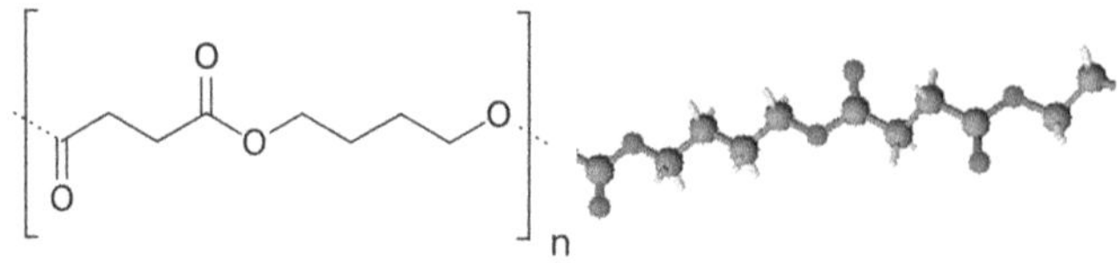

Figure 7.18 Structure of PBS.

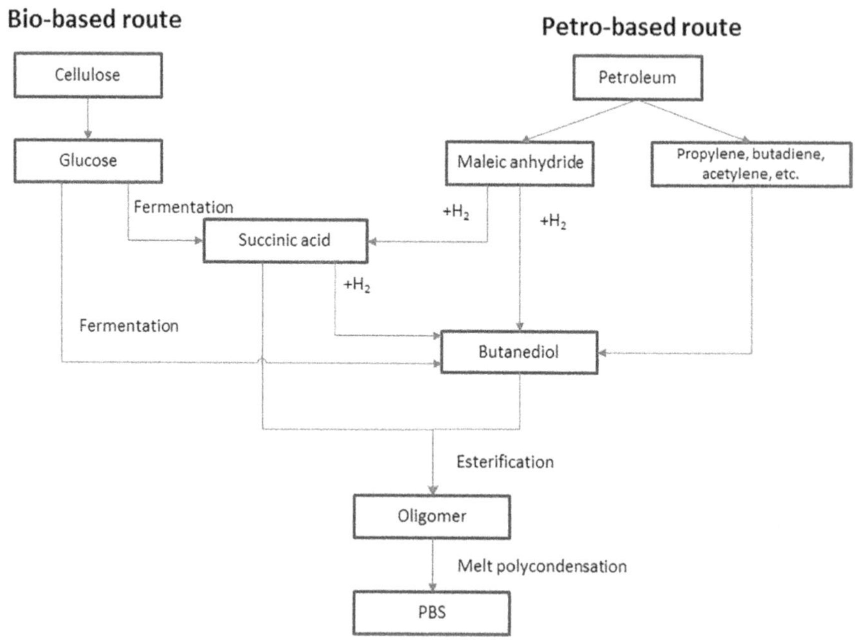

Figure 7.19 Bio-based and petro-based PBS synthesis routes.

In the following sections, we will provide an overview of PBS synthesis, properties, processing, and applications. For further reading on PBS, ref. 114, 122, 123 and 129 are all very useful resources for more detail.

7.8.2 Synthesis

PBS was first synthesized in the 1860s and systematically studied around the 1930s.[115] There are two main routes for PBS synthesis: a transesterification process from succinate diesters and a direct esterification process starting from the succinic acid. Transesterification involves reaction of dimethyl succinate with BDO in the presence of a catalyst. This reaction can take place at 180 °C.[116] Direct esterification of succinic acid with BDO is the most common way to produce PBS (see Figure 7.19). First, an excess of 1,4-BDO is esterified with the succinic acid to generate PBS oligomers. Then, these oligomers go through the polycondensation process to form a high molecular weight polymer. Depending on the source of the BDO and succinic acid, PBS can be 100% bio-based.

While the basic reaction between succinic acid and 1,4-BDO is straightforward, to obtain practical manufacturing of PBS is less so. A common issue encountered at the commercial scale is to achieve a high enough molecular weight to allow for useful processability in a reasonable time. There are several different approaches to solve this problem. One major approach is using chain extenders such as isocyanate chain extenders at late stages of the polymerization. However, in food-contact applications, isocyanate cannot be used. PTT MCC Inc., for example, use their proprietary chain extender that is allowable in food contact application. There are also studies looking into enzymatic polymerization and even direct polymerization under microwave irradiation,[117,118] but none is close to commercialization scale yet.

7.8.3 Properties

As mentioned above, PBS is a flexible and soft semi-crystalline polyester with good heat resistance and biodegradability. Its key properties are summarized in Table 7.5 below. Conventional PP, low-density polyethylene (LDPE), and biodegradable plastic polylactide (PLA) are also included in Table 7.5[119–121,124] for comparison.

7.8.3.1 Mechanical Properties

The density of PBS is between 1.2 g cm^{-3} to 1.3 g cm^{-3}, making it heavier than PP and LDPE but lighter than PET and polyvinyl chloride (PVC). Its mechanical properties are similar to those of polyolefin such as PP and LDPE (see Table 7.5). This similarity is partly because PBS shares a similar crystallization behavior to polyolefin: the thickness of the lamellae depends on the crystallization temperature; thickening of the lamellae occurs by annealing as well as a relatively high degree of crystallinity (typically in the range from 34% to 45% for PBS).[122,123] Compared with other biopolymers such as PLA, PBS is much more flexible and less stiff, with elongation at break in the range from Mitsubishi GS Pla FZ91's 210% to Showa Bionolle's 700% and Young's modulus of only 0.7 GPa. As a result, PBS is viewed as a complementary plastic to the mechanical properties of PLA and is often used in compounds and blends.

Table 7.5 Summary of PBS properties.

Properties	PBS	PLA	PP	LDPE
Density (g cm^{-3})	1.26	1.24	0.9	0.92
Glass-transition temperature (°C)	−33	55	−5	−120
Young's modulus (GPa)	~0.71	3.5	0.9–1.5	0.1–0.3
Tensile strength (MPa)	36	66	33	10
Elongation at break (%)	210–700	4	415	300
Melting point (°C)	112–116	170–180	163	110
Decomposition temperature (°C)	360–410	350–400	328	290
Heat deflection temperature (°C @ 0.45 MPa)	95	55	110	49

7.8.3.2 Thermal Properties

PBS has a low-glass transition temperature (T_g) at −33 °C with a high heat deflection temperature (HDT), 95 °C, and a wide processing window. The low T_g marks another similarity between PBS and polyolefin. This makes the material feasible for applications that require higher temperature stability than amorphous PLA can offer, such as in packaging of hot food and beverages. Depending on the molecular weight and the thermal history, PBS homopolymers have a melting point of 112–116 °C and a degradation temperature between 360 and 410 °C. The big difference between the melting point and the degradation temperature directly contributes to PBS's wide processing window.

7.8.3.3 Barrier Properties

Moisture and oxygen barrier properties are crucial parameters, especially for packaging applications. As can be seen from Table 7.6,[124] PBS's WVTR is slightly better than PBAT and PLA but one order of magnitude poorer than that of LDPE. However, PBS's oxygen transmission rate (OTR) is the best among PLA, PBAT, as well as LDPE, and thus can serve as a medium oxygen barrier.

It is worth noting that gas barrier properties needed for some food packaging applications can be very stringent. For example, coffee bean packaging typically needs WVTR of <8 g per m^2 per day per atm and an OTR of <55 cc per m^2 per day per atm.[125] Multi-layer packaging, additional coating, compounds, and composites are common approaches. For example, clays are often used as a filler to reduce WVTR, as demonstrated in ref. 126–128. Formulation of bio-based compostable packaging is a research area that needs further investigation.

7.8.4 Degradability

There are three major PBS degradation mechanisms: non-enzymatic hydrolytic degradation, biodegradation in natural environments, and enzymatic degradation. In non-enzymatic hydrolytic degradation, PBS reacts with water and its molecular weight decreases steadily with time as a result

Table 7.6 PBS barrier properties (20 μm films).

	Gas transmission rate[a]	
Material	WVTR (g per m^2 per day per atm)	OTR (cc per m^2 per day per atm)
PBS (BioPBS FZ91)	520	460
PLA	680	1900
PBAT	900	>2000
LDPE	26	8850

[a] Testing standard: ASTM E96/E96m-10@38 °C/90% RH (WVTR); ASTM D3985-05@23 °C/0% RH (OTR).

Table 7.7 Relationship between polymer structures and biodegradability.

Biodegradability test	PBS	PBSA
In hot compost	Normal	Rapid
In moist compost	Normal	Rapid
In the sea	Slow	Rapid
In water with activated sludge	Slow	Slow

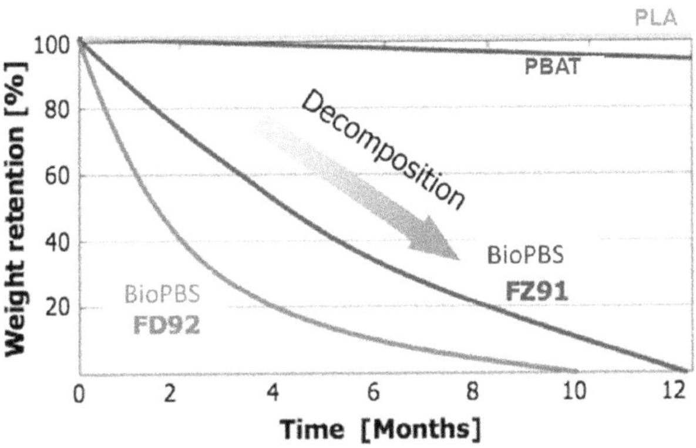

Figure 7.20 Weight retention of 200 μm sheet sample in soil at 30 °C/50% RH.

of random chain scission. pH is one particularly important condition that affects the rate considerably; the higher the pH, the greater the hydrolytic degradation rate.[122,129]

PBS, together with its copolymer PBSA, exhibits good biodegradability in a broad range of environments including in soil and seawater, as shown in Table 7.7.[130] It decomposes naturally into water and CO_2 at ambient temperature. PBSA, in particular, has a fast degradation rate in a home composting environment without the need of industrial composting facilities. Its degradation rate is also much faster than that of PLA and PBAT, as shown in Figure 7.20.[124] This has given PBS some advantages in applications where end-of-life is a key consideration.

Many lipase enzymes, such as *Rhizopus delemar*, *Rhizopus arrhizus*, and *Mucor meihei*, can accelerate the degradation of PBS.[131–133] Some enzymes, such as *Cryptococcus* sp., can also accelerate the degradation of other biopolymers including PLA and PCL, see Figure 7.21.[134,135]

7.8.5 Copolymers/Blends/Composites

Since proper blending can improve PBS's various attributes and PBS is compatible with many different bioplastics and natural fibers, PBS is often compounded or made into composites with other materials. Table 7.8[123] summarizes several example compounding/blending compositions and their improved attributes. One particular combination worth mentioning is

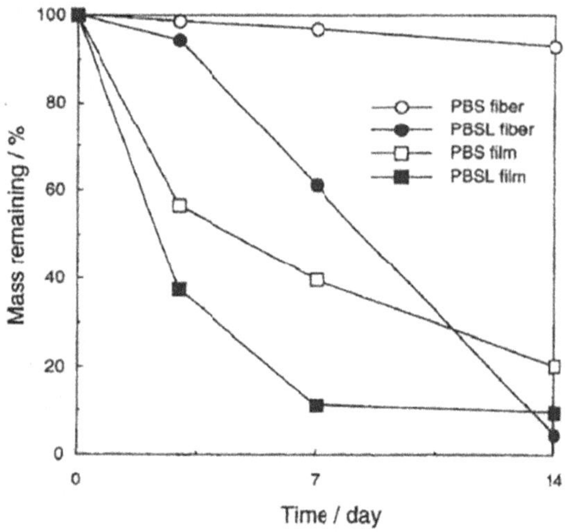

Figure 7.21 Biodegradation of PBS and PBSL incubated with Lipase PS1 at 50 °C and pH 6.0.

Table 7.8 PBS compounds and improved properties.

PBS compounds/blends	Improved properties
PBSA	Crystallinity, tensile strength, biodegradation (compared to PBS)
PBS/PLA	Tensile strength, elastic modulus, elongation at break, ductility (compared to neat PLA)
PBSA/PLA	Impact strength, ductility
PBS/talcum	Cost, heat distortion temperature, creep resistance
PBS/PBAT	Tensile strength, toughness
PBS/carbon nanotube	Tensile modulus, conductivity, storage moduli
PBS/PLA/$CaSO_4$-whiskers	Crystallization, thermal resistance, impact resistance, elongation at break

PBS/PLA. PBS is soft and flexible while PLA is stiff and brittle (Table 7.5). Such a huge difference in mechanical properties creates a large design space for PLA/PBS compounds leading to an increasing number of investigations on PBS/PLA compounds and applications.[136–141]

7.8.6 Processing

PBS can be processed by most conventional plastics processing. Typically, PBS with molecular weight less than 100 000 can be processed with injection molding, extrusion, and thermoforming, and PBS with higher molecular weight or long-chain is favored in film-blowing and casting.[128] Below is a sample parameter (see Table 7.9[143]) for sheet extrusion of PBS based on SK Chemical's SkyGreen™.[142]

Other than making individual parts, PBS can also serve as a coating layer, such as the liner for paper cups. This is a particularly useful area for PBS

Table 7.9 Processing data for sheet extrusion of PBS.

Extruder	45 mm single screw, $L/D = 30/1$, 100 rpm, vacuum = 40 mbar
Extruder temperature	Zone 1: 180 °C; zone 2: 245 °C; zone 3: 265 °C; zone 4: 260 °C; zone 5: 240 °C
Melt pressure	83 bar
Die	Zone 1: 247 °C; zone 2: 249 °C
Roll	1: 45 °C; 2: 54 °C; 3: 63 °C

Table 7.10 Comparison of PBS and PLA in paper coating processes.

	PBS	PLA
Line speed	Fast – same as LDPE	Slower
Processability	Stable, excellent drawability	High draw resonance
Machine	Can use same screw as LDPE (metering screw)	Prefers barrier screw
Coating thickness	Thinner – as low as 18 gsm	Thicker
Adhesion to paper	Excellent. No delamination.	Good
Heat seal strength	Excellent	Poorer
Heal seal temperature	120–130 °C	>160 °C
Foldable	Able to fold without breaking	Cannot fold
Biodegradation at room temperature	Yes – complete biodegradation within 1 year	No – requires composting conditions (60 °C or 140 °F)
Food contact at 100 °C	Suitable	HDT = 60 °C

because PBS has good adhesion, processability, and would not affect the recyclability or compostability of paper. Table 7.10 is a comparison of the processing conditions for coating PBS (PTT MCC's BioPBS) and PLA on paper.[124]

Many factors can affect the thermal stability of PBS during processing: water content, content of residual carboxylic acid terminals, and molecular structure. Thus, PBS must be dried sufficiently to prevent excessive hydrolysis which can compromise the physical properties. A typical optimum drying condition can be found in Table 7.11.[143]

7.8.7 Applications

With increasing concerns regarding plastic wastes, PBS has demonstrated potential as an attractive alternative in many applications. Based on PBS relevant patents and products available on the market, we summarize PBS example applications in Table 7.12 and explain in more detail in the following paragraphs.

Table 7.11 Drying conditions for PBS processing.

Items	Value
Drying temperature	65–70 °C
Drying time	6–12 h
Air flow of dry air	>0.065 m^3 per min per kg per h
Dew point of dry air	<−30 °C
Residual moisture content	<0.05%
Drying hopper capacity	6–12 times extruder output per hour
Height/diameter ratio	>2 : 1, 3 : 1 is better for plug flow
Insulation	Insulate well to improve energy efficiency
Monitoring	Inlet air temperature and dew point

Table 7.12 PBS application examples.

Industry	Example applications
Agriculture	Mulch film, fishing gear, seedling pot, protection nets, fertilizer coating
Construction	Insulation board, parts for tubes/artificial turfs/wall papers
Food packaging	Disposable utensils, containers, cling films, coated paper cups and trays
General packaging	Trash bags, wrapping films, envelope windows
Medical devices	Coatings/parts for stents and scaffolds
Textile	Layers in absorption sheets, hygiene products
Other	Hot melt adhesives, toys, golf tees

In an agricultural industry, PBS adds biodegradability to plastic films, parts, and tools, and leaves no toxic residues in the soil. For example, SABIC developed PLA/PBS compounds as fertilizer coatings with controllable release time.[144] Agricultural mulch films are a major application for PBS,[145] where PBS mulch films can be left in the soil after harvesting season and hence significantly reduce labor costs associated with mulch film removal. The Polish Institute of Biopolymers and Chemical Fibers investigated biodegradable seedling pots blending PBS with potassium sulfate fertilizer[146] where PBS could maintain shape and structure for three to four weeks before degradation in soil in five to seven weeks. The timing is suitable for free growth of the roots as well as the whole plant. PBS is of particular interest in the fishing sector. Roquette Frères proposed PBS/silica composites for fishing gear use.[147] This invention keeps the strength, enables high inflatability, and adds biodegradability to fishing nets. S-EnPol, a subsidiary of Lotte Fine Chemical Co., Ltd, produces and markets its PBS resins specifically for the fishing market under the trade name of EnPol™.

In the construction industry, PBS can reduce carbon footprint and avoid noxious gases in case of fire. It has been proposed to be used in various roles, including as a major composition in resin boards and wallpapers, as a component in the adhesive layer for protective films or as a coating for a part. For example, LG Hausys proposed a biodegradable wallpaper composition using

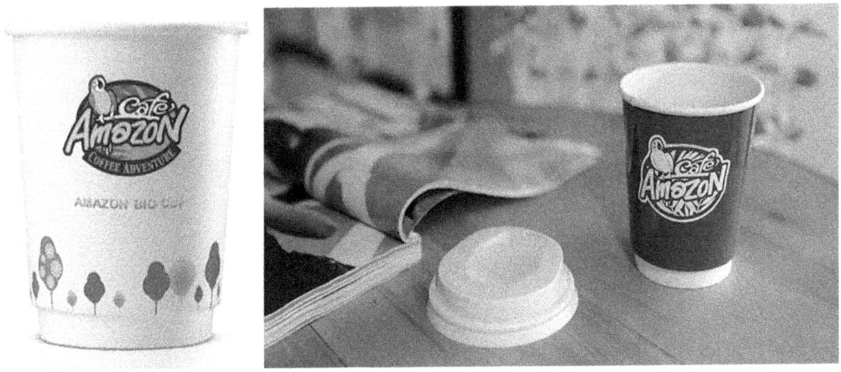

Figure 7.22 PBS coated coffee cup.

PLA/PBS to replace PVC.[148] EcolGreen uses a blend of 30–50% PBS, 30–40% PLA, and 10–20% nanocomposites to serve as a biodegradable finish material for artificial turfs in soccer fields, indoor golf ranges, manufacturing mats for landscape architecture, and gymnasium protection mats.[149]

In the food packaging industry, excellent heat sealability and flexibility make PBS an excellent material for packaging. One specific commercialized example is a PBS-coated paper coffee cup in Thailand's Café Amazon coffee shop chain (see Figure 7.22) in which a PBS coating replaces a conventional LDPE coating in the paper cup. A high heat deflection temperature makes it possible for the PBS coating to survive the heat from hot coffee while PLA cannot. Once dumped in a landfill, these PBS-coated paper cups will gradually turn into compost within 180 days at room temperature while a conventional LDPE-coated paper cup will not fully disintegrate. An independent study done by Western Michigan University also showed that over 96% of pulp fiber can be recovered from paper cups coated on both sides with PTTMCC's BioPBS™. This is a significantly higher fiber recovery rate than that from two-sided LDPE-coated paper cups.

Other than cup coatings, there are also various commercial PBS packaging products such as multi-layer snack bags, utensils, cling films, containers, and straws. SUCCIPACK, a European consortium dedicated to investigating PBS's potential in food packaging, conducted comprehensive PBS feasibility tests in material, processing, and food safety assessment protocols for a variety of foods including cheese, vegetables, seafood, and ready-to-eat meals.[123] In North America, BioAmber also formed a joint venture with NatureWorks on its PLA/PBS formulation specifically for food serviceware applications.

In general packaging, trash bags are one of the most popular PBS products. Companies including Chukoh Chemical Industries, Office Media, Masuyama Shoko, Kitamura Chemicals, KIRACS, and Showa Denko all sell PBS trash bags under their own brands. Wrapping films are another big category for PBS applications. One major focus in film manufacturing is to compound PBS with other biopolymers to achieve bigger design space and enhance biodegradability. For example, Metabolix compounds PHA with branched PBS to achieve greater tear/puncture resistance and tensile strength.[150] Mitsubishi Plastics used PBS compounds to modify properties of PLA films to achieve better shrink finish, adherence, and mechanical strength along both directions upon stretch.[151]

In medical devices, PBS is used in scaffolds and stents as a coating to shield the brittle parts from cracking due to its biocompatibility and good impact resistance. For example, Abbott applies PBS as an elastic primer layer on a surface of an implantable medical device such as vascular stents.[152] Donghua University developed a process to use PBS/PCL compounds to serve as porous scaffolds for tissue engineering.[153]

In textile applications, PBS's good water retention and responsiveness is of additional value, making it feasible for disposable hygiene products. Kimberly-Clark has patented grafted PBS formulations for use in personal care products such as diapers, sanitary napkins, and adult incontinence garments.[154] OJI Paper also has patented biodegradable absorbent sheets that contain PBS fibers.[155]

Other PBS applications can be found in stationery, toys, *etc.* Examples include Henkel using PBS/PLA as a key component in hot melt adhesives.[156] Minima Technology uses PBS/PLA in bullets for toy guns.

7.8.8 Conclusion

In this section, we reviewed PBS synthesis, properties, processing, as well as its applications in various industries. Overall, PBS has attracted much interest from both academia and industry due to its unique combination of room-temperature biodegradability, high heat resistance, balanced mechanical properties, good compounding compatibility, and wide processing window. While many of those properties already find initial adoption in applications ranging from agriculture and food packaging to textiles, PBS's commercialization potential is far from being fully utilized. Further formulation of PBS compounds and composites, as well as more application developments, are needed to allow a broader market adoption.

Abbreviations

ABS Acrylonitrile butadiene styrene
BDO Butanediol
DP Direct polymerization
ENR Epoxidized natural rubber
GPPS General purpose polystyrene
HC Homocrystal
HDT Heat deflection temperature
HIPS High impact polystyrene
HV Hydroxyvalerate
LDPE Low-density polyethylene

MW	Molecular weight
OTR	Oxygen transmission rate
PBAT	Polybutylene adipate-*co*-terephthalate
PBS	Polybutylene succinate
PBSA	Polybutylene succinate-*co*-adipate
PC	Polycarbonate
PCL	Polycaprolactone
PDLA	Poly D-lactic acid
PE	Polyethylene
PET	Polyethylene terephthalate
PHA	Polyhydroxyalkanoates
PHB	Polyhydroxybutyrate
PLA	Polylactic acid
PLLA	Poly L-lactic acid
PMMA	Polymethylmethacrylate
PP	Polypropylene
PS	Polystyrene
PHBV	Polyhydroxyalkanoate-type polymer
ROP	Ring opening polymerization
SC	Stereocrystal
SIC	Stress induced crystallization
scl-PHAs	Short-chain-length PHAs
WAXS	Wide angle X-ray scattering
WVTR	Water vapor transmission rate

References

1. W. H. Carothers, G. L. Dorough and F. J. van Natta, Studies of Polymerization and Ring Formation. X. The Reversible Polymerization of Six-Membered Cyclic Esters, *J. Am. Chem. Soc.*, 1932, **54**(2), 761–772.
2. Global Plastic Production 1950–2020|Statista. Accessed July 25, 2021, https://www.statista.com/statistics/282732/global-production-of-plastics-since-1950/.
3. EUBIO_Admin, Market, *European Bioplastics e.V.* (blog). Accessed July 25, 2021, https://www.european-bioplastics.org/market/.
4. D. R. Witzke, *Introduction to Properties, Engineering, and Prospects of Polylactide Polymers*, Michigan State University, 1997, https://www.proquest.com/openview/ea8449ca6c845317de14128f50b3dbfe/.
5. D. Garlotta, A Literature Review of Poly(Lactic Acid), *J. Polym. Environ.*, 2001, **9**(2), 63–84.
6. M. H. Hartmann, High Molecular Weight Polylactic Acid Polymers, in *Biopolymers from Renewable Resources*, ed. D. L. Kaplan, Macromolecular Systems—Materials Approach, Springer, Berlin, Heidelberg, 1998, pp. 367–411.

7. R. E. Drumright, P. R. Gruber and D. E. Henton, Polylactic Acid Technology, *Adv. Mater.*, 2000, **12**(23), 1841–1846.
8. D. E. Henton, P. Gruber, J. Lunt and J. Randall, Polylactic Acid Technology, in *Natural Fibers, Biopolymers, and Biocomposites*, ed. A. K. Mohanty, M. Misra and L. T. Drzal, CRC Press, 2005.
9. R. A. Auras, L.-T. Lim, S. E. M. Selke and H. Tsuji, *Poly(Lactic Acid): Synthesis, Structures, Properties, Processing, and Applications*, John Wiley & Sons, 2011.
10. L. T. Sin and B. S. Tueen, *Polylactic Acid: A Practical Guide for the Processing, Manufacturing, and Applications of PLA*, 2019.
11. Circular Economy – UK, USA, Europe, Asia & South America – The Ellen MacArthur Foundation. Accessed July 30, 2021, https://www.ellenmacarthurfoundation.org/.
12. E. T. H. Vink and S. Davies, Life Cycle Inventory and Impact Assessment Data for 2014 Ingeo™ Polylactide Production, *Ind. Biotechnol.*, 2015, **11**(3), 167–180.
13. J. Randall, N. Whiteman and P. Vollmann, Innovation Takes Root Conference, Pre-Conference Workshop, 2016.
14. M. Verbruggen, Closing plenary session, Innovation Takes Root Conference, 2016.
15. B. Bian, S. Bajracharya, J. Xu, D. Pant and P. E. Saikaly, Microbial Electrosynthesis from CO_2: Challenges, Opportunities and Perspectives in the Context of Circular Bioeconomy, *Bioresour. Technol.*, 2020, **302**, 122863.
16. K. Enomoto, M. Ajioka and A. Yamaguchi, Polyhydroxycarboxylic Acid and Preparation Process Thereof, *US Pat.*, US5310865A, filed December 18, 1992, and issued May 10, 1994, https://patents.google.com/patent/US5310865A/en?oq=+US+5310865+.
17. T. Kashima, T. Kameoka, C. Higuchi, M. Ajioka and A. Yamaguchi, Aliphatic Polyester and Preparation Process Thereof, *US Pat.*, US5428126A, filed January 5, 1995, https://patents.google.com/patent/US5428126A/en?oq=US+5428126.
18. F. Ichikawa, M. Kobayashi, M. Ohta, Y. Yoshida, S. Obuchi and H. Itoh, Process for Preparing Polyhydroxycarboxylic Acid, *US Pat.*, US5440008A, filed May 19, 1994, and issued August 8, 1995, https://patents.google.com/patent/US5440008A/en?oq=US+5440008+.
19. M. Ohta, Y. Yoshida, S. Obuchi and Y. Yoshida, *US Pat.*, US5440008A, 1994.
20. T. Nagano, T. Nasuno, H. Ome, Y. He, B. Yu, K. Zhu, X. Guo and P. Zuo, Method for Producing Aliphatic Polyester Resin, and an Aliphatic Polyester Resin Composition, *US Pat.*, US8173753B2, filed May 19, 2009, and issued May 8, 2012, https://patents.google.com/patent/US8173753B2/en?oq=US+8173753.
21. M. Hamaguchi, T. Naotsuka, Y. Takahashi, T. Takahashi and H. Ome, Method for Producing Crystallized Polyester, *US Pat.*, US8901275B2, filed June 17, 2010, and issued December 2, 2014, https://patents.google.com/patent/US8901275B2/en?oq=US+8901275.

22. T. Nagano, T. Tanaka and H. Ome, Process for Production of Poly(lactic Acid)-Type Resin, and Poly(lactic Acid)-Type Prepolymer, *US Pat.*, US9023953B2, filed June 27, 2011, and issued 2015, https://patents.google.com/patent/US9023953B2/en?oq=US+9023953.
23. T. Tanaka, S. Kumazawa and H. Ome, Polylactic Acid Resin and Method for Producing Same, *US Pat.*, US9080008B2, filed July 18, 2012, and issued July 14, 2015, https://patents.google.com/patent/US9080008B2/en?oq=US+9080008.
24. Y. He, P. Zuo, Z. Ye, M. Aoyama, X. Zeng, X. Fan, X. Dai and J. Lu, Solid-Phase Polymerization Method for Preparing High-Molecular-Weight Aliphatic Polyester, *US Pat.*, US9290616B2, filed May 28, 2012, and issued March 22, 2016, https://patents.google.com/patent/US9290616B2/en?oq=US+9290616.
25. C. A. Miller, B. Rush Fosmer, T. McMullin, D. Beacom and P. Suominen, Industrial Production of Lactic Acid, in *Comprehensive Biotechnology*, ed. M. Moo-Young, Elsevier, 2nd edn, 2011.
26. K. A. M. Thakur, R. T. Kean, E. S. Hall, J. J. Kolstad and E. J. Munson, Stereochemical Aspects of Lactide Stereo-Copolymerization Investigated by 1H NMR: A Case of Changing Stereospecificity, *Macromolecules*, 1998, **31**(5), 1487–1494.
27. T. M. Ovitt and G. W. Coates, Stereoselective Ring-Opening Polymerization of Meso-Lactide: Synthesis of Syndiotactic Poly(Lactic Acid), *J. Am. Chem. Soc.*, 1999, **121**(16), 4072–4073.
28. B. Tadeusz, A. Kowalski, J. Libiszowski, A. Duda and S. Penczek, Progress in Polymerization of Cyclic Esters: Mechanisms and Synthetic Applications, *Macromol. Symp.*, 2006, **240**(1), 47–55.
29. X. Yang, S. Kang, Y. Yang, K. Aou and S. L. Hsu, Raman Spectroscopic Study of Conformational Changes in the Amorphous Phase of Poly(Lactic Acid) During Deformation, *Polymer*, 2004, **45**(12), 4241–4248.
30. Y. Wang, S. M. Chiao, T.-F. Hung and S.-Y. Yang, Improvement in Toughness and Heat Resistance of Poly(Lactic Acid)/Polycarbonate Blend Through Twin-Screw Blending: Influence of Compatibilizer Type, *J. Appl. Polym. Sci.*, 2012, **125**(S2), E402–E412.
31. G. Zhang, J. Zhang, S. Wang and D. Shen, Miscibility and Phase Structure of Binary Blends of Polylactide and Poly(Methyl Methacrylate), *J. Polym. Sci., Part B: Polym. Phys.*, 2003, **41**(1), 23–30.
32. Y. Li and H. Shimizu, Improvement in Toughness of Poly(l-Lactide) (Plla) Through Reactive Blending with Acrylonitrile–Butadiene–Styrene Copolymer (Abs): Morphology and Properties, *Eur. Polym. J.*, 2009, **45**(3), 738–746.
33. J. Huang, M. S. Lisowski, J. Runt, E. S. Hall, R. T. Kean, N. Buehler and J. S. Lin, Crystallization and Microstructure of Poly(l-Lactide-Co-Meso-Lactide) Copolymers, *Macromolecules*, 1998, **31**(8), 2593–2599.
34. P. D. Santis and A. J. Kovacs, Molecular Conformation of Poly(s-Lactic Acid), *Biopolymers*, 1968, **6**(3), 299–306.

35. M. T. Takemori, Towards an Understanding of the Heat Distortion Temperature of Thermoplastics, *Polym. Eng. Sci.*, 1979, **19**(15), 1104–1109.
36. J. D. Schroeder, Method for Making Copolymers of Lactide and a 1:4-3:6 Dianhydrohexitol, *US Pat.*, US8268954B2, filed August 7, 2008, and issued September 18, 2012, https://patents.google.com/patent/US8268954B2/en?oq=US8268954+.
37. S. Talibuddin, J. Runt, L.-Z. Liu and B. Chu, Microstructure Development and Crystallization of Poly(Ethylene Oxide) and Melt-Miscible Peo Blends, *Macromolecules*, 1998, **31**(5), 1627–1634.
38. X. Ou and M. Cakmak, Influence of Biaxial Stretching Mode on the Crystalline Texture in Polylactic Acid Films, *Polymer*, 2008, **49**(24), 5344–5352.
39. J. A. Cicero, J. R. Dorgan, J. Janzen, J. Garrett, J. Runt and J. S. Lin, Supramolecular Morphology of Two-Step, Melt-Spun Poly(Lactic Acid) Fibers, *J. Appl. Polym. Sci.*, 2002, **86**(11), 2828–2838.
40. J. R. Randall, C. M. Ryan, J. Lunt and M. H. Hartmann, Impact Modified Melt-Stable Lactide Polymer Compositions and Processes for Manufacture Thereof, *US Pat.*, US5714573A, filed January 19, 1995, and issued February 3, 1998, https://patents.google.com/patent/US5714573A/en?oq=US+5714573+.
41. S. P. McCarthy, R. A. Gross and W. Ma, Polyactic Acid-Based Blends, *US Pat.*, US5883199A, filed April 3, 1997, and issued March 16, 1999, https://patents.google.com/patent/US5883199A/en?oq=US+5883199+.
42. O. Owusu and D. Venkatraman, Innovation Takes Root Conference, Pre-Conference Workshop, 2016.
43. Metabolix, Technical Brief: Modification of PLA with Mirel® a-PHA Copolymer, 2014.
44. J. Valentine, Applications of Lactic Acid Equivalents in Oil Field Services (OFS), Hydraulic Fracturing Production Chemicals Conference, 2019.
45. J. R. Dorgan, H. Lehermeier and M. Mang, Thermal and Rheological Properties of Commercial-Grade Poly(Lactic Acid)s, *J. Polym. Environ.*, 2000, **8**(1), 1–9.
46. J. J. Cooper-White and M. E. Mackay, Rheological Properties of Poly-(Lactides). Effect of Molecular Weight and Temperature on the Viscoelasticity of Poly(l-Lactic Acid), *J. Polym. Sci., Part B: Polym. Phys.*, 1999, **37**(15), 1803–1814.
47. L.-I. Palade, J. L. Hans and J. R. Dorgan, Melt Rheology of High L-Content Poly(Lactic Acid), *Macromolecules*, 2001, **34**(5), 1384–1390.
48. J. R. Dorgan, J. S. Williams and D. N. Lewis, Melt Rheology of Poly-(Lactic Acid): Entanglement and Chain Architecture Effects, *J. Rheol.*, 1999, **43**(5), 1141–1155.
49. R. E. Drumright, M. Hartmann and R. Wolf, Copolymers of monocyclic esters and carbonates and methods for making same, World Intellectual Property Organization WO2002100921A1, filed June 7, 2002,

and issued December 19, 2002, https://patents.google.com/patent/WO2002100921A1/en?oq=2002100921.

50. P. R. Gruber, J. J. Kolstad, D. R. Witzke, M. H. Hartmann and A. L. Brosch, Viscosity-Modified Lactide Polymer Composition and Process for Manufacture Thereof, *US Pat.*, US5594095A, filed July 27, 1994, and issued January 14, 1997, https://patents.google.com/patent/US5594095A/en?oq=US+5594095+.
51. J. R. Randall, K. Cink, and J. C. Smith, Branching Polylactide by Reacting Oh or Cooh Polylactide with Epoxide Acrylate (co)polymer. United States US7566753B2, filed June 23, 2005, and issued July 28, 2009, https://patents.google.com/patent/US7566753B2/en?oq=US+7566753+.
52. J. A. Cicero, J. R. Dorgan, S. F. Dec and D. M. Knauss, Phosphite Stabilization Effects on Two-Step Melt-Spun Fibers of Polylactide, *Polym. Degrad. Stab.*, 2002, **78**(1), 95–105.
53. L. L. Madison and G. W. Huisman, Metabolic Engineering of Poly(3-Hydroxyalkanoates): From DNA to Plastic, *Microbiol. Mol. Biol. Rev.*, 1999, **63**(1), 21–53.
54. A. J. Anderson and E. A. Dawes, Occurrence, Metabolism, Metabolic Role, and Industrial Uses of Bacterial Polyhydroxyalkanoates, *Microbiol. Rev.*, 1990, **54**(4), 450–472.
55. R. Geyer, J. R. Jambeck and K. L. Law, Production, Use, and Fate of All Plastics Ever Made, *Sci. Adv.*, 2017, **3**(7), e1700782.
56. R. Narayan, Bio-based and Biodegradable Polymer Materials: Rationale, Drivers, and Technology Exemplars, in *Degradable Polymers and Materials*, ACS Symposium Series 939, American Chemical Society, 2006, vol. 939, pp. 282–306.
57. F. Aeschelmann and M. Carus, Bio-based Building Blocks and Polymers in the World: Capacities, Production, and Applications–Status Quo and Trends Towards 2020, *Ind. Biotechnol.*, 2015, **11**(3), 154–159.
58. G.-Y. A. Tan, C.-L. Chen, L. Li, L. Ge, L. Wang, I. Razaad, Y. Li, L. Zhao, Y. Mo and J.-Y. Wang, Start a Research on Biopolymer Polyhydroxyalkanoate (PHA): A Review, *Polymers*, 2014, **6**(3), 706–754.
59. A. Steinbüchel and H. E. Valentin, Diversity of Bacterial Polyhydroxyalkanoic Acids, *FEMS Microbiol. Lett.*, 1995, **128**(3), 219–228.
60. A. Anjum, M. Zuber, K. M. Zia, A. Noreen, M. N. Anjum and S. Tabasum, Microbial Production of Polyhydroxyalkanoates (Phas) and Its Copolymers: A Review of Recent Advancements, *Int. J. Biol. Macromol.*, 2016, **89**, 161–174.
61. Z. Lasemi, G. N. Darzi and M. S. Baei, Media Optimization for Poly(β-Hydroxybutyrate) Production Using Azotobacter Beijerinckii, *Int. J. Polym. Mater. Polym. Biomater.*, 2013, **62**(5), 265–269.
62. A. Ishizaki and K. Tanaka, Production of Poly-B-Hydroxybutyric Acid from Carbon Dioxide by Alcaligenes Eutrophus Atcc 17697t, *J. Ferment. Bioeng.*, 1991, **71**(4), 254–257.
63. B. S. Kim, S. C. Lee, S. Y. Lee, H. N. Chang, Y. K. Chang and S. I. Woo, Production of Poly(3-Hydroxybutyric Acid) by Fed-Batch Culture of

Alcaligenes Eutrophus with Glucose Concentration Control, *Biotechnol. Bioeng.*, 1994, **43**(9), 892–898.
64. T. Fukui and Y. Doi, Efficient Production of Polyhydroxyalkanoates from Plant Oils by Alcaligenes Eutrophus and Its Recombinant Strain, *Appl. Microbiol. Biotechnol.*, 1998, **49**(3), 333–336.
65. P. Chakraborty, W. Gibbons and K. Muthukumarappan, Conversion of Volatile Fatty Acids into Polyhydroxyalkanoate by Ralstonia Eutropha, *J. Appl. Microbiol.*, 2009, **106**(6), 1996–2005.
66. R. Haas, B. Jin and F. T. Zepf, Production of Poly(3-Hydroxybutyrate) from Waste Potato Starch, *Biosci., Biotechnol., Biochem.*, 2008, **72**(1), 253–256.
67. M. Koller, P. Hesse, R. Bona, C. Kutschera, A. Atlić and G. Braunegg, Potential of Various Archae- and Eubacterial Strains as Industrial Polyhydroxyalkanoate Producers from Whey, *Macromol. Biosci.*, 2007, **7**(2), 218–226.
68. D. Bourque, Y. Pomerleau and D. Groleau, High-Cell-Density Production of Poly-B-Hydroxybutyrate (Phb) from Methanol by Methylobacterium Extorquens: Production of High-Molecular-Mass Phb, *Appl. Microbiol. Biotechnol.*, 1995, **44**(3), 367–376.
69. K.-D. Wendlandt, M. Jechorek, J. Helm and U. Stottmeister, Production of Phb with a High Molecular Mass from Methane, *Polym. Degrad. Stab.*, 1998, **59**(1), 191–194.
70. T. Yamane, X. Chen and S. Ueda, Growth-Associated Production of Poly(3-Hydroxyvalerate) from N-Pentanol by a Methylotrophic Bacterium, Paracoccus Denitrificans, *Appl. Environ. Microbiol.*, 1996, **62**(2), 380–384.
71. A. D. Tripathi, A. Yadav, A. Jha and S. K. Srivastava, Utilizing of Sugar Refinery Waste (Cane Molasses) for Production of Bio-Plastic Under Submerged Fermentation Process, *J. Polym. Environ.*, 2012, **20**(2), 446–453.
72. S. T. Kenny, J. N. Runic, W. Kaminsky, T. Woods, R. P. Babu, C. M. Keely, W. Blau and K. E. O'Connor, Up-Cycling of PET (Polyethylene Terephthalate) to the Biodegradable Plastic PHA (Polyhydroxyalkanoate), *Environ. Sci. Technol.*, 2008, **42**(20), 7696–7701.
73. E. Y. Lee, D. Jendrossek, A. Schirmer, C. Y. Choi and A. Steinbüchel, Biosynthesis of Copolyesters Consisting of 3-Hydroxybutyric Acid and Medium-Chain-Length 3-Hydroxyalkanoic Acids from 1,3-Butanediol or from 3-Hydroxybutyrate by Pseudomonas Sp. A33, *Appl. Microbiol. Biotechnol.*, 1995, **42**(6), 901–909.
74. A. A. Pantazaki, C. P. Papaneophytou, A. G. Pritsa, M. Liakopoulou-Kyriakides and D. A. Kyriakidis, Production of Polyhydroxyalkanoates from Whey by Thermus Thermophilus Hb8, *Process Biochem.*, 2009, **44**(8), 847–853.
75. S. Shahid, R. Mosrati, J. Ledauphin, C. Amiel, P. Fontaine, J.-L. Gaillard and D. Corroler, Impact of Carbon Source and Variable Nitrogen Conditions on Bacterial Biosynthesis of Polyhydroxyalkanoates: Evidence of an Atypical Metabolism in Bacillus Megaterium Dsm 509, *J. Biosci. Bioeng.*, 2013, **116**(3), 302–308.

76. G. W. Haywood, A. J. Anderson, D. R. Williams, E. A. Dawes and D. F. Ewing, Accumulation of a Poly(Hydroxyalkanoate) Copolymer Containing Primarily 3-Hydroxyvalerate from Simple Carbohydrate Substrates by Rhodococcus Sp. Ncimb 40126, *Int. J. Biol. Macromol.*, 1991, **13**(2), 83–88.
77. A. Bhattacharyya, A. Pramanik, S. K. Maji, S. Haldar, U. K. Mukhopadhyay and J. Mukherjee, Utilization of Vinasse for Production of Poly-3-(Hydroxybutyrate-Co-Hydroxyvalerate) by Haloferax Mediterranei, *AMB Express*, 2012, **2**(1), 34.
78. C. Hermann-Krauss, M. Koller, A. Muhr, H. Fasl, F. Stelzer and G. Braunegg, Archaeal Production of Polyhydroxyalkanoate (PHA) Co- and Terpolyesters from Biodiesel Industry-Derived by-Products, *Archaea*, 2013, **2013**, e129268.
79. J. V. Walsem, L. Zhong and S. S. Shih, Polymer Extraction Methods, *US Pat.*, US7252980B2, filed July 23, 2003, and issued August 7, 2007, https://patents.google.com/patent/US7252980B2/en?oq=US+7252980.
80. Y. Doi, *Microbial Polyesters*, VCH, 1990.
81. N. Grassie, E. J. Murray and P. A. Holmes, The Thermal Degradation of Poly(-(d)-Β-Hydroxybutyric Acid): Part 2—Changes in Molecular Weight, *Polym. Degrad. Stab.*, 1984, **6**(2), 95–103.
82. N. C. Loureiro, J. L. Esteves, J. C. Viana and S. Ghosh, Mechanical Characterization of Polyhydroxyalkanoate and Poly(Lactic Acid) Blends, *J. Thermoplast. Compos. Mater.*, 2015, **28**(2), 195–213.
83. P. A. Holmes, Biologically Produced (R)-3-Hydroxy-Alkanoate Polymers and Copolymers, in *Developments in Crystalline Polymers*, ed. D. C. Bassett, Dordrecht, Springer Netherlands, 1988, pp. 1–65.
84. T. L. Bluhm, G. K. Hamer, R. H. Marchessault, C. A. Fyfe and R. P. Veregin, Isodimorphism in Bacterial Poly(β-Hydroxybutyrate-Co-Β-Hydroxyvalerate), *Macromolecules*, 1986, **19**(11), 2871–2876.
85. R. K. Krishnaswamy, J. Baird and J. P. Christiano, Single Screw Extrusion of Bio-based and Biodegradable Poly(Hydroxy Butanoic Acid) Copolymers, *Society of Plastics Engineers*, 2009, **1**, 161–166.
86. G. Ronca, Heterogeneous Nucleation in Polymer Crystallization, PhD Thesis, Brunel University School of Engineering and Design PhD Theses, 1985, http://bura.brunel.ac.uk/handle/2438/7285.
87. N. Jacquel, K. Tajima, N. Nakamura, H. Kawachi, P. Pan and Y. Inoue, Nucleation Mechanism of Polyhydroxybutyrate and Poly(Hydroxybutyrate-Co-Hydroxyhexanoate) Crystallized by Orotic Acid as a Nucleating Agent, *J. Appl. Polym. Sci.*, 2010, **115**(2), 709–715.
88. P. Pan, G. Shan, Y. Bao and Z. Weng, Crystallization Kinetics of Bacterial Poly(3-Hydroxylbutyrate) Copolyesters with Cyanuric Acid as a Nucleating Agent, *J. Appl. Polym. Sci.*, 2013, **129**(3), 1374–1382.
89. A. R. Padwa, Y. Kann, R. S. Whitehouse and K. Sheehan, Branched Pha Compositions, Methods for Their Production, and Use in Applications, *US Pat.*, US9034989B2, filed June 19, 2009, and issued May 19, 2015, https://patents.google.com/patent/US9034989B2/en?oq=US+9034989.

90. R. K. Krishnaswamy and A. R. Padwa, Impact Modification of PLA Using Bio-based, Biodegradable Mirel™ PHB Copolymers, in *Annual Technical Conference – ANTEC, Conference Proceedings*, 2013, vol. 3, pp. 1998–2001.
91. L. Wei, A. G. McDonald and N. M. Stark, Grafting of Bacterial Polyhydroxybutyrate (PHB) Onto Cellulose Via in Situ Reactive Extrusion with Dicumyl Peroxide, *Biomacromolecules*, 2015, **16**(3), 1040–1049.
92. N. Altaee, G. A. El-Hiti, A. Fahdil, K. Sudesh and E. Yousif, Biodegradation of Different Formulations of Polyhydroxybutyrate Films in Soil, *SpringerPlus*, 2016, **5**(1), 762.
93. T. Ohura, Y. Aoyagi, K.-I. Takagi, Y. Yoshida, K.-I. Kasuya and Y. Doi, Biodegradation of Poly(3-Hydroxyalkanoic Acids) Fibers and Isolation of Poly(3-Hydroxybutyric Acid)-Degrading Microorganisms Under Aquatic Environments, *Polym. Degrad. Stab.*, 1999, **63**(1), 23–29.
94. A. P. Bonartsev, V. L. Myshkina, D. A. Nikolaeva, E. K. Furina, T. A. Makhina, A. P. Boskhomdzhiev, E. A. Ivanov, A. L. Iordanskii and G. A. Bonartseva, Biosynthesis, Biodegradation, and Application of Poly(3-Hydroxybutyrate) and Its Copolymers – Natural Polyesters Produced by Diazotrophic Bacteria, *Commun. Curr. Res. Educ. Top. Trends Appl. Microbiol.*, 2007, **1**, 295–307.
95. ASTM D6400-12, Specification for Labeling of Plastics Designed to Be Aerobically Composted in Municipal or Industrial Facilities, ASTM International, 2012, DOI: 10.1520/D6400-19.
96. S. M. Emadian, T. T. Onay and B. Demirel, Biodegradation of Bioplastics in Natural Environments, *Waste Manage.*, 2017, **59**, 526–536.
97. M. C. Morse, Anaerobic Biodegradation of Biocomposites for the Building Industry, Stanford University, 2009, http://proquest.umi.com/pqdweb?did=1835708881&Fmt=7&clientId=41143&RQT=309&VName=PQD.
98. A. E. Schauer, The Performance Evaluation of a Temperature Phased Anaerobic Digestion System at a Dairy Farm in Wrightstown, Wisconsin, University of Wisconsin-Green Bay, 2002.
99. J. Mergaert, C. Anderson, A. Wouters, J. Swings and K. Kersters, Biodegradation of Polyhydroxyalkanoates, *FEMS Microbiol. Lett.*, 1992, **103**(2–4), 317–321.
100. J. Mergaert, A. Webb, C. Anderson, A. Wouters and J. Swings, Microbial Degradation of Poly(3-Hydroxybutyrate) and Poly(3-Hydroxybutyrate-Co-3-Hydroxyvalerate) in Soils, *Appl. Environ. Microbiol.*, 1993, **59**(10), 3233–3238.
101. S. S. Miyazaki, A. R. Yep, F. Kolton, É. B. Hermida, F. Povolo, E. G. Fernandes and E. Chiellini, Biodegradable Polymeric Films Based on Microbial Poly(3-Hydroxybutyrate). Effect of Gamma-Radiation on Mechanical Properties and Biodegradability, *Macromol. Symp.*, 2003, **197**(1), 57–64.
102. D. Jendrossek, M. Backhaus and M. Andermann, Characterization of the Extracellular Poly(3-Hydroxybutyrate) Depolymerase of Comamonas Sp. and of Its Structural Gene, *Can. J. Microbiol.*, 1995, **41**(13), 160–169.

103. B. H. Briese, B. Schmidt and D. Jendrossek, Pseudomonas Lemoignei Has Five Poly(Hydroxyalkanoic Acid) (Pha) Depolymerase Genes: A Comparative Study of Bacterial and Eukaryotic Pha Depolymerases, *J. Environ. Polym. Degrad.*, 1994, **2**(2), 75–87.
104. C. A. Woolnough, T. Charlton, L. H. Yee, M. Sarris and L. J. R. Foster, Surface Changes in Polyhydroxyalkanoate Films During Biodegradation and Biofouling, *Polym. Int.*, 2008, **57**(9), 1042–1051.
105. D. M. Abou-Zeid, R.-J. Müller and W.-D. Deckwer, Degradation of Natural and Synthetic Polyesters Under Anaerobic Conditions, *J. Biotechnol.*, 2001, **86**(2), 113–126.
106. K. Budwill, P. M. Fedorak and W. J. Page, Methanogenic Degradation of Poly(3-Hydroxyalkanoates), *Appl. Environ. Microbiol.*, 1992, **58**(4), 1398–1401.
107. M. T. Gutierrez-Wing, B. E. Stevens, C. S. Theegala, I. I. Negulescu and K. A. Rusch, Anaerobic Biodegradation of Polyhydroxybutyrate in Municipal Sewage Sludge, *J. Environ. Eng.*, 2010, **136**(7), 709–718.
108. T. Volova, A. Boyandin, M. Gladyshev, I. Gitelson and S. Prudnikova, Biodegradation of Polyhydroxyalkanoates in Natural Water Environments, *J. Sib. Fed. Univ., Biol.*, 2015, **8**, 168–186.
109. C. Thellen, M. Coyne, D. Froio, M. Auerbach, C. Wirsen and J. A. Ratto, A Processing, Characterization and Marine Biodegradation Study of Melt-Extruded Polyhydroxyalkanoate (PHA) Films, *J. Polym. Environ.*, 2008, **16**(1), 1–11.
110. J. Mergaert, A. Wouters, J. Swings and C. Anderson, In Situ Biodegradation of Poly(3-Hydroxybutyrate) and Poly(3-Hydroxybutyrate-Co-3-Hydroxyvalerate) in Natural Waters, *Can. J. Microbiol.*, 1995, **41**(13), 154–159.
111. European Bioplastics, Bioplastics Market Development Update 2019, European Bioplastics, 2019, https://www.european-bioplastics.org/wp-content/uploads/2019/11/Report_Bioplastics-Market-Data_2019_short_version.pdf.
112. I. H. S. Markit, World Analysis – Polypropylene, Houston, Texas, 2017.
113. J. H. Schut, Waiting for Bio PBS, *Plastics Engineering Blog*, 2014.
114. A. A. Shah, F. Hasan, A. Hameed and S. Ahmed, Biological Degradation of Plastics: A Comprehensive Review, *Biotechnol. Adv.*, 2008, **26**(3), 246–265.
115. W. H. Carothers and J. A. Arvin, Studies on Polymerization and Ring Formation. II. Poly-Esters, *J. Am. Chem. Soc.*, 1929, **51**(8), 2560–2570.
116. V. Tserki, P. Matzinos, E. Pavlidou, D. Vachliotis and C. Panayiotou, Biodegradable Aliphatic Polyesters. Part I. Properties and Biodegradation of Poly(Butylene Succinate-Co-Butylene Adipate), *Polym. Degrad. Stab.*, 2006, **91**(2), 367–376.
117. H. Azim, A. Dekhterman, Z. Jiang and R. A. Gross, *Candida Antarctica* Lipase B-Catalyzed Synthesis of Poly(Butylene Succinate): Shorter Chain Building Blocks Also Work, *Biomacromolecules*, 2006, **7**(11), 3093–3097.
118. S. Velmathi, R. Nagahata, J.-I. Sugiyama and K. Takeuchi, A Rapid Eco-Friendly Synthesis of Poly(Butylene Succinate) by a Direct Polyesterification

Under Microwave Irradiation, *Macromol. Rapid Commun.*, 2005, **26**(14), 1163–1167.
119. R. Ishioka, E. Kitakuni and Y. Ichikawa, Aliphatic Polyesters: 'Bionolle', in *Biopolymers, Biology, Chemistry, Biotechnology, Applications, Volume 4, Polyesters III – Applications and Commercial Products*, ed. Y. Doi and A. Steinbüchel, Wiley-VCH, New York, 2002, pp. 275–297.
120. *Optimization of Polymer Nanocomposite Properties*, ed. V. Mittal, Wiley-VCH, Weinheim, 2010.
121. A. S. Meraldo Antonio, *et al.*, Comparing the Performance and Addressable Markets for Biopolymers, Lux Research, 2015.
122. J. Xu and B.-H. Guo, Microbial Succinic Acid, Its Polymer Poly(Butylene Succinate), and Applications, in *Plastics from Bacteria: Natural Functions and Applications*, ed. G. G.-Q. Chen, Microbiology Monographs, Springer, Berlin, Heidelberg, 2010, pp. 347–388.
123. Final report from SUCCIPACK Development of active, intelligent and sustainable food PACKaging using Polybutylenesuccinate, https://cordis.europa.eu/project/id/289196/reporting/es.
124. PTT MCC BioPBSTM technical presentation, 2017 (http://www.mcpp-global.com/en/asia/products/product/biopbstm-general-properties/).
125. N. Ramaswamy, Bio-based, Biodegradable Polymers and Coatings for Food Packaging, UMass Lowell course presentation.
126. P. Boonprasith, J. Wootthikanokkhan and N. Nimitsiriwat, Mechanical, Thermal, and Barrier Properties of Nanocomposites Based on Poly-(Butylene Succinate)/Thermoplastic Starch Blends Containing Different Types of Clay, *J. Appl. Polym. Sci.*, 2013, **130**(2), 1114–1123.
127. R. K. Basha, K. Konno, H. Kani and T. Kimura, Water Vapor Transmission Rate of Biomass Based Film Materials, *Eng. Agric. Environ. Food*, 2011, **4**(2), 37–42.
128. A. Bhatia, R. K. Gupta, S. N. Bhattacharya and H. J. Choi, Analysis of Gas Permeability Characteristics of Poly(Lactic Acid)/Poly(Butylene Succinate) Nanocomposites, *J. Nanomater.*, 2012, **2012**, 1–11.
129. J. Xu and B.-H. Guo, Poly(Butylene Succinate) and Its Copolymers: Research, Development and Industrialization, *Biotechnol. J.*, 2010, **5**(11), 1149–1163.
130. T. Fujimaki, Processability and Properties of Aliphatic Polyesters, 'Bionolle', Synthesized by Polycondensation Reaction, *Polym. Degrad. Stab.*, 1998, **59**(1), 209–214.
131. Y. Tokiwa, B. P. Calabia, C. U. Ugwu and S. Aiba, Biodegradability of Plastics, *Int. J. Mol. Sci.*, 2009, **10**(9), 3722–3742.
132. H. Maeda, Y. Yamagata, K. Abe, F. Hasegawa, M. Machida, R. Ishioka, K. Gomi and T. Nakajima, Purification and Characterization of a Biodegradable Plastic-Degrading Enzyme from *Aspergillus Oryzae*, *Appl. Microbiol. Biotechnol.*, 2005, **67**(6), 778–788.
133. Y.-F. Shih and T.-M. Wu, Enzymatic Degradation Kinetics of Poly(Butylene Succinate) Nanocomposites, *J. Polym. Res.*, 2009, **16**(2), 109–115.

134. K. Masaki, N. R. Kamini, H. Ikeda and H. Iefuji, Cutinase-Like Enzyme from the Yeast Cryptococcus Sp. Strain S-2 Hydrolyzes Polylactic Acid and Other Biodegradable Plastics, *Appl. Environ. Microbiol.*, 2005, **71**(11), 7548–7550.
135. I. Taniguchi, S. Nakano, T. Nakamura, A. El-Salmawy, M. Miyamoto and Y. Kimura, Mechanism of Enzymatic Hydrolysis of Poly(Butylene Succinate) and Poly(Butylene Succinate-Co-L-Lactate) with a Lipase from *Pseudomonas Cepacia*, *Macromol. Biosci.*, 2002, **2**(9), 447–455.
136. A. Bhatia, R. K. Gupta, S. N. Bhattacharya and H. J. Choi, Compatibility of Biodegradable Poly (Lactic Acid) (PLA) and Poly (Butylene Succinate) (PBS) Blends for Packaging Application, *Korea Aust. Rheol. J.*, 2007, **19**(3), 125–131.
137. A. Soroudi and I. Jakubowicz, Recycling of Bioplastics, Their Blends and Biocomposites: A Review, *Eur. Polym. J.*, 2013, **49**(10), 2839–2858.
138. J. W. Park and S. S. Im, Phase Behavior and Morphology in Blends of Poly(l-Lactic Acid) and Poly(Butylene Succinate), *J. Appl. Polym. Sci.*, 2002, **86**(3), 647–655.
139. M. Harada, T. Ohya, K. Iida, H. Hayashi, K. Hirano and H. Fukuda, Increased Impact Strength of Biodegradable Poly(Lactic Acid)/Poly-(Butylene Succinate) Blend Composites by Using Isocyanate as a Reactive Processing Agent, *J. Appl. Polym. Sci.*, 2007, **106**(3), 1813–1820.
140. R. Homklin and N. Hongsriphan, Mechanical and Thermal Properties of PLA/PBS Co-Continuous Blends Adding Nucleating Agent, *Energy Procedia*, 2013, **34**, 871–879.
141. J. W. Park and S. S. Im, Phase Behavior and Morphology in Blends of Poly(l-Lactic Acid) and Poly(Butylene Succinate), *J. Appl. Polym. Sci.*, 2002, **86**(3), 647–655.
142. SK Chemicals, R&D Center, SKYGREEN sheet extrusion molding guide.
143. SK Chemicals, R&D Center, SKYGREEN sheet injection molding guide.
144. G. Kannan, L. C. Ahmed, B. Duckworth, P. W. Bell and S. K. Subramanian, Coated Granular Fertilizers, Methods of Manufacture Thereof, and Uses, *US Pat.*, US20170088480A1, filed April 27, 2015, and issued March 30, 2017, https://patents.google.com/patent/US20170088480A1/en?oq=US+20170088480.
145. Y. Lee, *et al.*, Method for Manufacturing a Biodegradable Agricultural Mulching Film for Directly Seeding Rice, and Method for Directly Seeding Rice Using the Biodegradable Agricultural Mulching Film Manufactured by the Method. World Intellectual Property Organization WO2011034262A1, filed January 5, 2010, and issued March 24, 2011, https://patents.google.com/patent/WO2011034262A1/en?oq=2011034262.
146. M. Izydorczyk, K. Sulak, T. Mik, D. Ciechanska, I. Krucinka, M. Puchalski and P. Siwek, Biodegradable seedling pots and method of their manufacture. World Intellectual Property Organization WO2016105217A1, filed December 22, 2014, and issued June 30, 2016, https://patents.google.com/patent/WO2016105217A1/en?oq=2016105217.

147. N. Jacquel, R. Saint-Loup, F. Fenouillot-Rimlinger, J.-P. Pascault and A. Rousseau, Materiales Compuestos a Base De Pbs Y De Sílice, ES2576964T3, filed October 16, 2012, and issued July 12, 2016, https://patents.google.com/patent/ES2576964T3/en?oq=ES2576964T3.
148. C. Z. Huang, S. Y. Lee, and P. S. Kim, Wallpaper and Method for Manufacturing Same, *US Pat.*, US9309628B2, filed October 27, 2009, and issued April 12, 2016, https://patents.google.com/patent/US9309628B2/en?oq=US+9309628.
149. J. H. Park, J. W. Kim, J. H. Choi, H. C. Nam, J. S. Lee and I. H. Kim, Biodegradable finish material, KR100833584B1, 2008.
150. R. K. Krishnaswamy and X. Sun, PHA Compositions Comprising PBS and PBSA and Methods for their Production, *US Pat.*, US20160053111A1, filed May 1, 2015, and issued February 25, 2016, https://patents.google.com/patent/US20160053111A1/en?oq=US+20160053111.
151. J. Terada, M. Terada and J. Takagi, Polylactic acid-based shrinkable sheet material and packaging material or shrinkabale labelling material by using the same, JP2000281818A, filed April 1, 1999, and issued October 10, 2000, https://patents.google.com/patent/JP2000281818A/en?oq=JP+2000281818.
152. L. W. Kleiner, Y. Tang, M. O. Trollsas, S. F. A. Hossainy and S. Prabhu, Coatings Having an Elastic Primer Layer, *US Pat.*, US9056155B1, filed May 29, 2007, and issued June 16, 2015, https://patents.google.com/patent/US9056155B1/en?oq=US+9056155+B1.
153. Y. Qing, C. Sishi, T. Zhiqing, S. Xinyuan and P. Lanlan, Three-dimensional stephanoporate organization engineering bracket material, fibre cementing method preparing same and applications thereof, *China Pat.*, CN101249277A, filed April 11, 2008, and issued August 27, 2008, https://patents.google.com/patent/CN101249277A/en?oq=CN101249277+A.
154. J. Wang and D. Schertz, Water-Responsive Biodegradable Polymer Compositions and Method of Making Same, *US Pat.*, US20020193517A1, filed March 12, 2001, and issued December 19, 2002, https://patents.google.com/patent/US20020193517A1/en?oq=20020193517.
155. T. Yaesawa and T. Miura, Absorbent sheet and method for manufacturing the same, JP2016176176A, filed May 20, 2016, and issued October 6, 2016, https://patents.google.com/patent/JP2016176176A/en?oq=JP+2016176176A.
156. M. Takenaka, A. Takamori and T. Hayakawa, Hot Melt Adhesive, *US Pat.*, US20160075924A1, filed November 24, 2015, and issued March 17, 2016, https://patents.google.com/patent/US20160075924A1/en?oq=US+20160075924+A1.

CHAPTER 8

Protein-based Plastics

D. GREWELL,*[a] J. SCHRADER[b] AND G. SRINIVASAN[c]

[a] College of Engineering and Engineering Technology, Northern Illinois University, 590 Garden Rd, Dekalb, IL 60115, USA; [b] Department of Horticulture, Iowa State University, Ames, IA, USA; [c] Center for Industrial Research & Service, Ames, IA, USA
*Email: Dgrewell@niu.edu

8.1 Chemistry

A material's behavior is defined by its fundamental building blocks. Therefore, it is critical to fully understand the chemical composition and structure of materials used for the manufacture of specific products.[1] The fundamental building blocks of protein-based plastics are polymerized amino acids, and approximately 21 standard amino acids serve as building blocks (monomers) for proteins.[2,3] They are polymerized into short and long chains by plants and animals. Short chains, with a molecular weight of 200–1000 g mol^{-1}, are called peptides. Longer chains form proteins which have a typical molecular weight ranging between 100 000 and 500 000 g mol^{-1} and function as enzymes, energy-storage structures, or structural components, such as keratin found in hair and fingernails. In addition to their composition, such as the type and number of monomers, the functionality of proteins is also determined by their conformation. Long chains can form a wide range of structures, creating helix or sheet conformations when folding on each other (see Figure 8.1). Secondary bonds, such as hydrogen bonds, or primary bonds, such as di-sulfide bonds, maintain these structures. Reconfiguration (or denaturing) of these structures alters their functionality.[4,5]

To use soy proteins as building blocks for plastic materials requires a change in their structure: the conformations have to be unfolded and the

Green Chemistry Series No. 79
Bioplastics and Biocomposites: A Practical Introduction
Edited by David Grewell

Published by the Royal Society of Chemistry, www.rsc.org

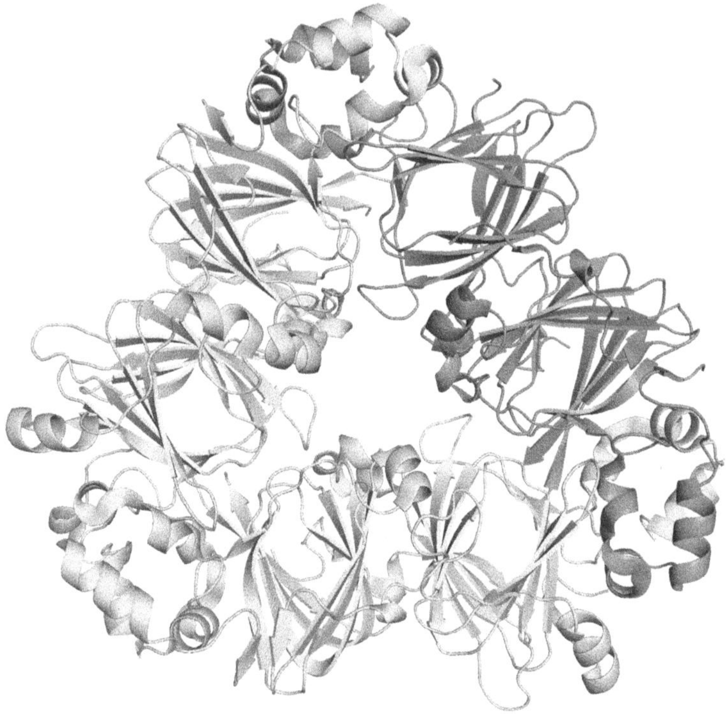

Figure 8.1 Protein structure of a typical 11S soy protein. Reproduced from ref. 4 with permission from American Chemical Society, Copyright 2014.

molecules linearized to make them similar to traditional plastic molecules. For linearization to occur, the proteins must be dissolved in a solvent and chemically altered to break the di-sulfide bonds.[5,6] Using an extrusion process that heats the protein structure accomplishes further denaturing, and shearing results in a material that behaves like a traditional petrochemical thermoplastic, see Figure 8.2.[4]

Compounding soy-protein plastics with other plastics is possible and useful in specific applications. These compounds are often used to enhance mechanical properties or water stability. While chemical modifications can reduce water absorption of soy-protein plastics from 200% to less than 20%, even 20% greatly reduces the strength of soy-protein plastics. Compounding soy-protein plastics with other plastics, *e.g.*, PLA, results in a relatively water-resistant blend. Blending with PLA and other polymers also enhances the tensile strength compared to non-compounded soy plastic, which typically have a tensile strength of 10 MPa. Soy/PLA polymers blends (50/50 by weight) exhibit tensile strengths over 5 MPa.

Here, we will provide information on only a few of the many possible formulations of soy-protein plastics. Table 8.1 shows the compositions, recommended processing conditions, and tensile strengths of two

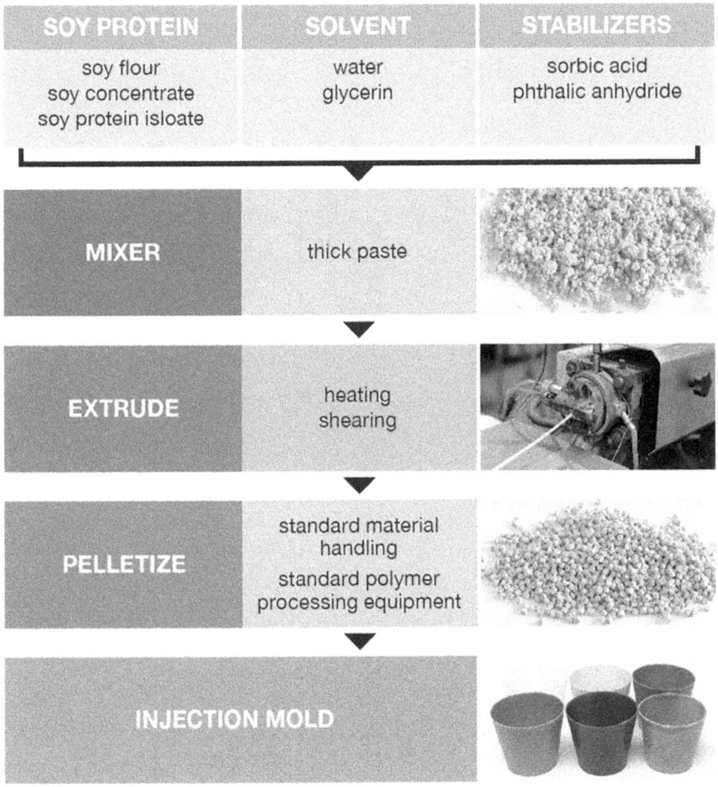

Figure 8.2 Illustration of general steps required in producing plastic products from soy protein. Reproduced from ref. 4 with permission from American Chemical Society, Copyright 2014.

promising soy plastics. Other formulations include soy-flour, polyethylene, and/or polypropylene composites.[4,7–11]

8.2 Material Science

To meet design requirements, material properties can be tailored through chemical pathways. Critical properties include stiffness, strength, solvent resistance, temperature deflection, and fatigue, among many others. Problem-solving techniques common in material science and informatics are applied to achieve defined properties.[12] First, the base material is characterized and compared to the design requirements. Depending on application demands, additional chemical treatment or tailoring by adding fillers or reinforcing fibers may be required. Alternatively, other resins may be blended in. Heat treatment can also be used to alter chemical or mechanical properties. Other critical engineering aspects, such as machinability, formability, and shape stability, must be considered during formulation development. For example, if long reinforcing fibers are required to provide

Table 8.1 General soy-protein plastic formulations. Reproduced from ref. 4 with permission from American Chemical Society, Copyright 2014.

	Soy (SPA) formulation	PLA/PEG formulation	PLA/PEG/SPA formulation	Water stable soy
Extrusion temperature	90–125 °C	145–210 °C	110–145 °C	Mixed at 60 °C
Injection molding temperature	130 °C	145 °C	145 °C	N/A
Tensile strength (ULT. STR.)	1.11 MPa	13.67 MPa	8.40 MPa	N/A
Stabilizer	75 g phthalic anhydride + 75 g adipic acid			10 mL BERSET 2700 (2.5% concentration) crosslinker
Solvent	650 g water + 150 g glycerol	10% PEG 8000		355 mL water
Anti-microbial	10 g potassium sorbate			
Di-sulfide bond opener	10 g sodium sulfite			
Polymer	500 g soy flour + 500 g soy protein isolate	90% PLA	60% PLA/PEG + 40% SPA	

sufficient strength, processing by injection molding will no longer be an option and formulations with a very narrow range of thermal stability will be nearly impossible to thermoform. The engineering aspects of material, formability, and final product are greatly affected by formulation and chemistry.

8.3 Engineering

Design, manufacturing, and costs of a product are closely interrelated with material properties, product characteristics, and final requirements. Chemistry and formulation of the material will determine available manufacturing processes. In addition, the properties of the formulation determine material-specific design features needed to meet product requirements. For example, weaker materials will require the incorporation of more rib-stiffeners and/or thicker walls to achieve the same performance accomplished by a stronger material. Thus, weaker materials will require more raw material and will often be more difficult to fabricate. These engineering aspects are critical throughout the stages of the design process, starting with the chemical modification of the proteins, to the final design, and the choice of processing technology. Today's protein-based plastics were developed for mass production, and therefore it was critical that the material be suitable for injection molding. Consequently, the chemistry and formulations were tailored to result in a thermoplastic material that can be plasticized in an extruder.

8.4 Product Design

The success of the final product design is one of the most difficult aspects of product development to be measured. The product must meet its defined functionality requirements, be sufficiently robust (engineered) to ensure it will not fail, and it must be accepted by the end user, whose perception is often hard to predict. To reduce the risk of product rejection by the end user, designers have to consider qualitative aspects such as aesthetics, feel, ergonomics, and smell. A case study of a soy-protein-based plastics application demonstrates this critical aspect of design. In a trial of disposable cutlery made of soy-based protein plastics, the prototypes failed because of water solvency issues (an engineering failure), but they were also rejected by the end user because of residual odors. If the end product is not properly designed to meet the expectations of the end user, it will never successfully enter the market.

8.5 Economics

Protein-based plastics will successfully establish themselves on the market only if they make a profit, thus the entire product cost must be considered. These costs include design, engineering, testing, manufacturing, distribution, marketing, and opportunity. The costs of the raw materials, such as proteins and solvents, as well as the costs to process, design, *etc.*, for a final product will define the final costs of any product. If the final cost of a product is equal to or below the market price of an existing product, it will be able to capture market share. However, if the market is willing to pay a premium, special considerations come into play. The public is willing to pay a premium for organic foods, and it is anticipated that a similar trend may be seen with bio-renewable, biodegradable plastics such as protein-based plastics. Even though the raw materials of bioplastics, such as soy-proteins, seem cost prohibitive, the growing awareness of their ecological advantage may promote economic success. In addition, the energy consumption during production of protein plastics is lower than for comparable petroleum-based products, and the price of energy will greatly affect the relative costs. From a holistic, long-term perspective, biodegradable plastics cause less social-economic damage compared to nondegradable plastics. Given the importance and cost of CO_2 reduction, final waste treatment, and damage caused to the environment by nondegradable, petroleum-based plastics, biodegradable plastics are more economical for certain applications. The imbalance of CO_2-fixation and CO_2-release of petroleum-based plastics (which is 10^5 years for petroleum compared one to ten years for bio-based carbon) can be reduced by using biodegradable and bio-renewable plastics which are derived annually from renewable feedstock.[13] Under the assumption that CO_2 emissions accelerate climate changes that result in more severe (and costly) natural disasters, such as extreme droughts and flooding, the reduced contributions to these environmental costs by bio-based plastics makes them cost-effective over the long term. While petroleum plastics can be incinerated and their energy recovered, their use still results in the release of fossil-based CO_2. Landfilling of these materials can also result in CO_2 and/or methane release from microbial decomposition depending on the plastic and the landfill conditions. In contrast, biodegradable plastics can be composted in a relatively short time.

At the time of writing, the cost of feedstocks for a typical soy-protein-based plastic material was less than $1 per pound, making it relatively competitive with petrochemical plastics. Composites that incorporate other low-cost fillers, such as dried distillers grains with solubles (DDGS) or corn stover, are even more cost competitive. The use of flours instead of protein isolates lowers the cost of the base resins of soy-protein plastics to approximately $0.75 per pound, which is competitive with commodity plastics such as polyethylene and polypropylene, as seen in Table 8.2.[4]

8.6 Applications

Soy-protein plastics have found a wide range of applications. The most interesting examples utilize the intrinsic advantages of the soy material: (1) golf tees that can biodegrade after use and provide a fertilizing effect by making the organic nitrogen released by the degrading protein available to

Table 8.2 Standard soy-flour plastic formulation and costs estimates (phthalic anhydride (PA), soy protein isolate (SPI)). Reproduced from ref. 4 with permission from American Chemical Society, Copyright 2014.

	Parts	Fraction	%	$/lbs	Price
SPI	50	0.26	26.1	2.36	$0.62
Soy flour/meal	50	0.26	26.1	0.28	$0.07
Glycerin	22.5	0.12	11.7	0.1	$0.01
PA	7.5	0.04	3.9	1.1	$0.04
Water	60	0.31	31.3	0	$0.00
Sodium sulfite	1	0.01	0.5	0.9	$0.00
Potassium sorbate	0.5	0.00	0.3	0.31	$0.00
	191.5		100		$0.75

Figure 8.3 Proposed applications for soy-protein plastics (Center for Crops Utilization Research). Reproduced from ref. 4 with permission from American Chemical Society, Copyright 2014.

the green; (2) using soy-protein plastics in dog toys that can supplement protein intake as the pet chews the toy; (3) the use of soy-protein-based materials for horticulture pots (containers); (4) disposable utensils and composite boards; (5) expendable dry lubricant applicator products for heavy duty machinery. Figure 8.3 shows applications, some of which have already been commercialized.[4]

Despite their environmental problems, single-use petroleum-based plant containers are ubiquitous in the container-crops horticulture industry because they are cheap and meet the functional requirements of the application. Increasingly, both horticulturists and their customers are less tolerant of non-sustainable growing practices. Lowering the environmental impact of growing systems has become a major priority. Protein-based plastics fulfill all requirements met by petroleum plastics while having a much lower negative environmental impact. In addition, decomposing proteins are able to supply some or most of the fertilizer required for the growing plant.[14] In early efforts to utilize protein-based plastics for plant containers, researchers incorporated an underused byproduct of the poultry industry: using keratin from poultry feathers.[15] Evaluations of the material were promising but the use of keratin polymers for plant containers has not gained momentum, mainly due to the high cost of development and the low profit margins for the application. In addition, the pricing model is highly dependent on the poultry industry and the amount of feathers generated as waste product.[16] Another potential protein for use in bio-based plant containers is zein protein from corn. Prototypes of zein-based plant containers showed very high nitrogen-fertilizer release that was detrimental to healthy plant growth. The use of zein plastics for this application will require either blending with a high-carbon polymer to lower the concentration of zein and maintain a suitable carbon–nitrogen ratio or the development of some other mechanism to slow the release of nutrients.[17,18]

Early growing trials with high-content soy-protein plastics showed an excessive fertilizer effect similar to that of zein. This effect was attributed to the resulting imbalance in carbon to nitrogen ratio in the soil. To mitigate this effect and to lower material cost, a soy-based polymer was formulated that included soy flour (lower nitrogen content, higher carbohydrate, and lower cost than soy-protein isolate) and blended with carbon-rich biopolymers such as polylactic acid and polyamide.[19] Blends of PLA/soy and PLA/soy/DDGS provided a balanced fertilizer effect that significantly reduced the amount of synthetic fertilizer needed for the culture of short- and medium-cycle greenhouse crops.[20] Blending these materials resolved many of the issues that made the materials unsuitable for this application when used alone. The soy component in the blend provided the plant-available nitrogen (fertilizer) and improved the biodegradability of the used container. The PLA provided strength, durability, carbon–nitrogen balance, and effectively controlled the rate of nitrogen release from the container.[21] The DDGS incorporated in some of the formulations helped reduce cost and balanced the nutrient ratios.

The composite blends were tested for physical properties, biodegradation rates, performance under application conditions (greenhouse trials), cost of materials, effectiveness of the added function (fertilizer effect), improvements in sustainability, aesthetic appeal, ease of processing, and consumer perception. The PLA/soy composite containers reduced the amount of synthetic fertilizer required to grow an ornamental greenhouse crop, such as marigolds, by 87% compared to plants grown in a standard petroleum-based container (see Figure 8.4). Fertilizer demand by tomato transplants was reduced by 44%. The added function of nutrient release resulted in (1) reduced cost for fertilizer, (2) reduced cost for labor to apply the fertilizer, and (3) reduced negative environmental impact by replacing synthetic fertilizer with natural fertilizer

Figure 8.4 Photograph of common plant in traditional plastic pot (left) with no fertilizer and protein-based pot (right) with no fertilizer. Reproduced from ref. 4 with permission from American Chemical Society, Copyright 2014.

nutrients from soy. While synthetic nitrogen fertilizer is energy intensive (produced by the Haber–Bosch process), the nitrogen available from soy-based plastics results from natural, biological nitrogen fixation performed in the roots of the soybean plant, a process fueled by natural photosynthesis.[22] In addition, the improvement in sustainability afforded by the plant-derived, soy-based nitrogen fertilizer is significant. Compared to synthetic fertilizers, plant-derived nitrogen fertilizer, such as that from soybean, uses 7.9 times less energy and causes 6.4 times less global-warming impact, 935 times less ozone depletion, and 1.8 times less acidifying emissions.[4,23,24]

The added function of nutrient provision by soy-based plastic containers can easily be seen with marigold plants grown using 87% less synthetic fertilizer than the standard protocol. Plants grown in the control treatment (left, petroleum-based polypropylene container) and in the PLA/soy composite container (right) were provided with the same amount of fertilizer, 100 parts-per-million nitrogen, during the first two weeks only. The fertilizer effect of the PLA/soy container is readily apparent, and the plant grown in this container is of better quality than those produced by standard protocols.

8.7 Summary

Most product development strategies are interdisciplinary in nature, and because of the relatively limited experience with protein-based materials in industry, they prove extraordinarily challenging. Despite the challenges, industry recognizes the potential of protein-based materials and is strongly interested in adopting them for single-use products such as horticulture pots. With the success of these applications, protein-based materials will be considered for many other uses. Work is progressing on soy protein-based fibers for textile products, an application that represents a large potential market. In addition, soy-flour composites are a growing market that demonstrates the increasing importance of these materials. Protein plastics will be most competitive in applications where they provide additional functions that petroleum plastics cannot.

Abbreviations

DDGS	Dried distillers grains with solubles
PA	Phthalic anhydride
PET	Polyethylene terephthalate
PHA	Polyhydroxyalkanoates
PLA	Polylactic acid
SPI	Soy protein isolate

References

1. K. G. Harding, J. S. Dennis, H. Von Blottnitz and S. T. L. Harrison, Environmental analysis of plastic production processes: Comparing petroleum-based polypropylene and polyethylene with biologically-based poly-β-hydroxybutyric acid using life cycle analysis, *J. Biotechnol.*, 2007, **130**(1), 57–66.
2. J. Locks, Properties and applications of compostable starch-based plastic material, *Polym. Degrad. Stab.*, 1998, **59**(1–3), 245–249.
3. (a) K. A. Rosentrater and A. W. Otieno, Considerations for manufacturing bio-based plastic products, *J. Polym. Environ.*, 2006, **14**(4), 335–346; (b) L. Jurgen, Properties and applications of compostable starch-based plastic material, *Polym. Degrad. Stab.*, 1998, **59**(1–3), 245–249.
4. R. Brentin, *Soy-Based Chemicals and Materials*, American Chemical Society, 2014, DOI: 10.1021/bk-2014-1178.
5. M. K. Campbell and S. O. Farrell, *Biochemistry*, 2011.
6. A. Achouri, W. Zhang and S. Xu, Enzymatic hydrolysis of soy protein isolate and effect of succinylation on the functional properties of resulting protein hydrolysates, *Food Res. Int.*, 1999, **31**(9), 617–623.
7. U. Kalapathy, N. S. Hettiarachchy and K. C. Rhee, Effect of drying methods on molecular properties and functionalities of disulfide bond-cleaved soy proteins, *J. Am. Oil Chem. Soc.*, 1997, **74**(3), 195–199.
8. B. Liu, L. Jiang, H. Liu and J. Zhang, Synergetic Effect of Dual Compatibilizers on in Situ Formed Poly(Lactic Acid)/Soy Protein Composites, *Ind. Eng. Chem. Res.*, 2010, **49**, 6399–6406.
9. J.-F. Su, X.-Y. Yuan, Z. Huang and W.-L. Xia, Properties stability and biodegradation behaviors of soy protein isolate/poly(vinylalcohol) blend films, *Polym. Degrad. Stab.*, 2010, **95**, 1226–1237.

10. L. Song, J. Zhi, P. Zhang, Q. Zhao, N. Li, M. Qiao and J. Liu, Synthesis and characterization of a new soy protein isolate/polyamic acid salt blend films, *J. Food Sci. Technol.*, 2015, **52**(5), 3072–3078.
11. O. Ozdemir, S. Lukubira, A. Ogale and P. Dawson, Fabrication and Analysis of Soy Flour Filled Polyethylene Fibers, Graduate Research and Discovery Symposium (GRADS), 2014, Paper 100.
12. S. Domenek, M.-H. Morel, A. Redl and S. Guilbert, Thermosetting of wheat protein based bioplastics: modeling of mechanism and material properties, *Macromol. Symp.*, 2003, **197**, 181–191.
13. N. Ramani, *Drivers for Biodegradable/Compostable Plastics & Role of Composting in Waste Management & Sustainable Agriculture.*
14. J. A. Schrader, G. Srinivasan, D. Grewell, K. G. McCabe and W. R. Graves, Fertilizer effects of soy-plastic containers during crop production and transplant establishment, *HortScience*, 2013, **48**, 724–731.
15. M. R. Evans and D. L. Hensley, Plant growth in plastic, peat, and processed poultry feather fiber growing containers, *HortScience*, 2004, **39**, 1012–1014.
16. D. Kuack, From coop to container to consumer, *HRI News*, 2012, www.hriresearch.org/index.cfm?page=Content&categoryID=157&ID=894.
17. M. S. Helgeson, W. R. Graves, D. Grewell and G. Srinivasan, Degradation and nitrogen release of zein-based bioplastic containers, *J. Environ. Hortic.*, 2009, **27**, 123–127.
18. M. S. Helgeson, W. R. Graves, D. Grewell and G. Srinivasan, Zein-based bioplastic containers alter root-zone chemistry and growth of geranium, *J. Environ. Hortic.*, 2010, **27**, 123–127.
19. D. Grewell, G. Srinivasan, J. Schrader, W. Graves and M. Kessler, Sustainable materials for a horticultural application, *Plast. Eng.*, 2014, **70**(3), 44–52.
20. C. Currey, J. Schrader, K. McCabe, W. Graves, D. Grewell, G. Srinivasan and S. Madbouly, Bioplastics for greenhouses – Soy what?, *GrowerTalks*, 2014, **77**(9), 70–74.
21. C. Currey, J. Schrader, K. McCabe, W. Graves, D. Grewell, G. Srinivasan and S. Madbouly, Soy containers: Growing promise, growing plants, *GrowerTalks*, 2014, **77**(10), 60–65.
22. N. Pelletier, E. Audsley, S. Brodt, T. Garnett, P. Henriksson, A. Kendall, K. Jan Kramer, D. Murphy, T. Nemecek and M. Troell, Energy Intensity of Agriculture and Food Systems, *Annu. Rev. Environ. Res.*, 2011, **36**, 223–246.
23. N. Pelletier, N. Arsenault and P. Tyedmers, Scenario modeling potential eco-efficiency gains from a transition to organic agriculture: life cycle perspectives on Canadian canola, corn, soy, and wheat production, *Environ. Manage.*, 2008, **42**(6), 989–1001.
24. W. Groot and T. Borén, Life cycle assessment of the manufacture of lactide and PLA biopolymers from sugarcane in Thailand, *Int. J. Life Cycle Assess.*, 2010, **15**(9), 970–984.

CHAPTER 9

Starch-based Biopolymer Films

A. TRAVALINI,[a,b] I. DEMIATE[b] AND B. LAMSAL*[a]

[a] Department of Food Science and Human Nutrition, Iowa State University, Ames, IA 50011, USA; [b] Department of Food Engineering, Ponta Grossa State University, Ponta Grossa, PR, 84030900, Brazil
*Email: lamsal@iastate.edu

9.1 Starch and Its Molecular Structure

Starch, the principal source of dietary calories to the human population, is the major reserve polysaccharide (polymerized sugars) present in photosynthetic tissues in many types of plant storage organs such as seeds, tubers, roots, stems, and fruits.[1,2] Starch occurs in nature as water-insoluble microscopic granules (Figure 9.1) which are essentially composed of two main polysaccharides: amylose and amylopectin. Some minor components such as lipids and proteins are also present which can interact with the polysaccharides. Starches from different botanical origins have differences in their biosynthesis mechanisms and may exhibit distinct molecular structure and properties as well as diversity in shape, size, composition, and other macro constituents.[3]

Intrinsic characteristics of starch, for example, amylose content, amylopectin branch chain length, X-ray diffraction pattern, granule size and shape, and semi-crystalline structure of granules, can be modified by physical, chemical, and enzymatic methods, or their combinations.[4] These modifications contribute to overcoming possible limitations of the native polysaccharide. For example, water holding capacity, syneresis, retrogradation, and poor functional properties of starch-based films can be modified and improved upon thermal and mechanical stress during processing.[5] A vast portfolio of modified starches is commercially available

Green Chemistry Series No. 79
Bioplastics and Biocomposites: A Practical Introduction
Edited by David Grewell

Published by the Royal Society of Chemistry, www.rsc.org

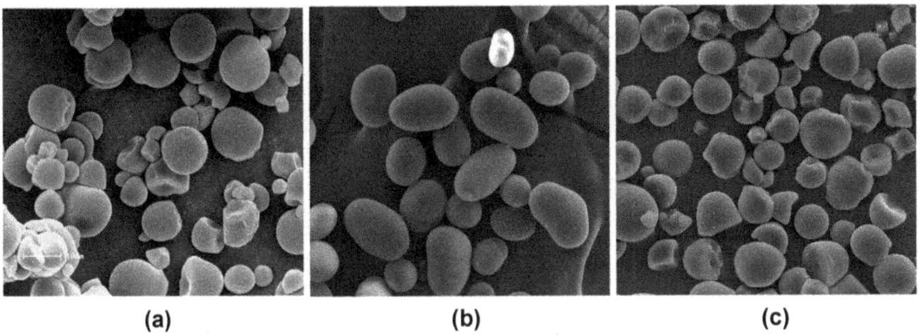

Figure 9.1 Different types of starch granules: (a) corn starch (1000×), (b) bean starch (1000×), and (c) cassava starch (1500×).

worldwide; however, there are still many cost effective possibilities that can be developed with a wider range of thermal–mechanical properties than conventionally modified starches.[6]

Considering starch macromolecular structure and its physicochemical properties after hydrothermal processing, a variety of applications, including the production of biodegradable films, are possible. Petroleum-derived polymers have been widely used over the course of many decades. However, many studies have investigated the possibility of replacing synthetic plastics with biodegradable plastics for lesser impact on the environment (Table 9.1). In this perspective, starch is considered as a potential raw material for biodegradable materials with varied applications such as packaging. Corn starch is the most frequently employed to produce thermoplastic starch (TPS) by industries in several countries.[7]

9.1.1 Starch Molecular Structure

9.1.1.1 Amylopectin and Amylose

Structurally, starch granules are made of amylose and amylopectin, the ratio of which differs based on the source and type. Amylopectin consists of a large, branched molecule composed of α-(1 → 4) linear bonds and α-(1 → 6) branched bonds of D-glucopyranose (Figure 9.2a). The molecular weight of amylopectin ranges approximately between 10^7–10^9 Da. Amylose, the minor component of normal starches, consists of α-(1 → 4) bonds of D-glucopyranose residues. The molecular weight of amylose varies between 10^5 and 10^6 Da.[8] The amylose content of most normal starches is between 20 and 30%; however, the ratio of amylose to amylopectin varies with the botanical source which confers different characteristics to different types of starch. Amylopectin external side chains are arranged as parallel-oriented double helices forming the crystalline domains, as shown in Figure 9.2B.[9,10] Those external side chains of amylopectin have shorter (DP 23–29) to longer (DP 30–44) average lengths; maize, waxy maize, and cassava starches

Table 9.1 Different formulations of starch films and their properties.

Starch source	Reinforcement	Concentration	Main results	Ref.
Pea, potato, cassava	Rice bran	10%	More resistant to fracture and less extensible films, with lower oxygen permeability, but without changes in water vapor permeability	Cano *et al.*, 2014[20]
Corn	Nanocrystal cellulose from rice straw	2.5, 5, 10, and 15%	Tensile strength and modulus significantly increased with increasing CNC	Agustin *et al.*, 2014[30]
Cassava	Nanocrystal cellulose from cassava bagasse	5, 10, and 20%	Decreased hydrophilic character and capacity of water uptake	Teixeira *et al.*, 2009[25]
Cassava	Nanocrystal cellulose from sweet potato	0.2, 0.3, 0.4, and 0.5%	WVP and moisture absorption decreased	Ma *et al.*, 2017[27]
White potato	Starch nanoparticles	3, 5, and 7%	Mechanical properties were improved	Martinez *et al.*, 2016[22]
Potato	Microcrystalline cellulose	1, 3, 5, 7, 10, and 15%	Improved mechanical properties and increased glass transition; decreased water absorption and solubility	Wilpiszewska and Czech, 2014[21]
Corn	Starch nanocrystals	0, 0.5, 1.0, 1.5, and 2.0%	Increased the tensile stress and elongation at break, and maintained Young's modulus values	Ren *et al.*, 2017[33]
Potato	Cellulose fiber	20%	Increased the water solubility and reduced WVP	Halal *et al.*, 2018[24]
Not informed	Nanocrystal cellulose from *Cladophora rupestris*	1, 3, and 5%	Reinforcement of 1% NCC to starch-based film improved its tensile strength by 78%.	Sucaldito and Camacho, 2017[111]

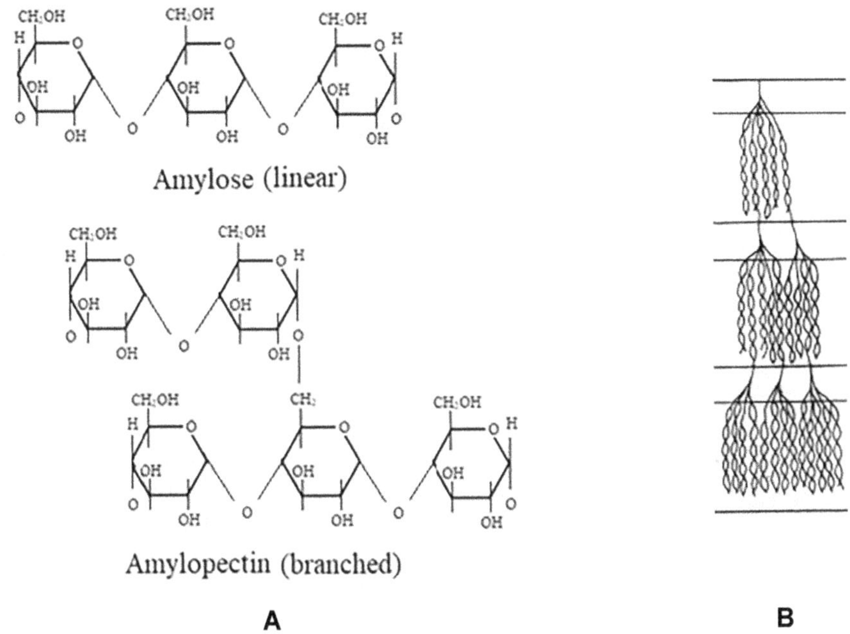

Figure 9.2 Structure of amylose and amylopectin fragments (A) and molecular modelling of the branching point of amylopectin (B).[11] Reproduced from ref. 11 with permission from Elsevier, Copyright 1997.

are examples of the former, whereas potato and high amylose maize starches, examples of the latter.

Starches with short and long amylopectin branches display the A-type and B-type X-ray diffraction patterns, respectively.[11] Amylose is a predominantly linear molecule distributed in amorphous regions of the starch granules.

9.1.1.2 *Crystalline Structure of Starch*

The X-ray diffraction scans of starch granules show peak intensity at different diffraction angles, classifying them as A-, B-, and C-type crystallinity. A-type granules present peaks at 2-theta angles of 15, 17, 18, and 23°, while B-type granules have peak intensities at 5.6, 15, 17, 22, and 23°.[2] The A-type crystallinity is characteristic of cereal starch, and B-type is characteristic of tubers (root-based starches) and some high-amylose cereal starches.[2] The C-type (legume, root, and some fruit starches) consists of a mixture of A and B types. The double helical structures in A-type and B-type are the same but the helices in A-type are more compact compared to B-type.[12] The relative content of amylose and amylopectin determines the crystallinity of the starch granules, such as amylopectin confers denser crystalline regions. Side chains of the highly branched amylopectin molecules are folded in a densely packed double helical crystalline structure while predominantly linear amylose molecules are distributed in amorphous regions of the starch granules.[13] Table 9.2 shows the X-ray diffraction pattern, relative crystallinity, and amylose content of starches from select sources, for example, pea, banana, potato, cassava, corn, and rice. As seen in Table 9.2, pea starch presents a C-type structure, corn and rice starches are of A-type. Starches from banana and from roots and tubers (cassava and potato, respectively) are variously reported as A-, B-, and C-types. The amylose contents are also variable when reported by different authors for the same sources; these same discrepancies were also observed for the relative granule crystallinity.

The different polymorphic forms of starch crystalline structures affect several properties of importance. Starch granule digestibility by porcine pancreatic α-amylase (responsible for starch digestion), for instance, is considerably favoured in A-type when compared to B-type polymorphs. The B-type starch may develop a superior crystalline structure with granules displaying a smooth, nonporous surface.[11] Films made from high amylose corn starch (B-type) possessed more stable mechanical properties compared to others that had starches with normal levels of amylose (20–30%). Those films made with high amylose starch were also more stable over time at different relative humidity levels.[14] In another study, maize starches with different amylose contents (4.3, 29.0, 61.5, and 77.4%) were also tested for edible film production and improved mechanical properties were found for those with the highest amylose levels.[15]

9.1.2 Conventional Sources of Starch

Roots and tubers, legumes (pulses), and fruits have generally high starch contents (40–80%, w/w, dry basis). Roots and tuber crops are plants yielding starchy roots, rhizomes, stems, and tubers. They contain 70–80% moisture and 16–24% starch. A great number of roots and tubers are grown in the tropics for edible purposes including yams, sweet potato, cassava, cocoyam, canna, and ginger.[16]

Compared to the roots and tubers or fruits, grains and cereals (wheat, rice, corn, oats, barley, millet, sorghum, and quinoa) contain high starch levels aside from higher contents of protein, lipids, and minerals. Among these different resources, starches obtained from some cereals (corn and wheat) or from some tubers and roots (particularly potato and cassava) are the main commercial products and largely dominate the global starch market.[17] By the year 2015, world starch production reached 85 million metric tons per year with a predominance of corn starch at approximately 80% of the global starch market.[18] Tubers and root starches (particularly potato and cassava) represented 4% and 12% of the global starch market, respectively.[1]

Several studies have reported the use of conventional starches in the development of biodegradable films such as potato,[19–24] cassava,[25–28] and corn.[29–33] Films produced from starches with high amylose content were shown to result in better mechanical properties in films. In addition, during

Table 9.2 X-ray diffraction (XRD) pattern, relative crystallinity, and amylose content of starch from different sources.

Starch	XRD pattern	Relative crystallinity/%	Amylose content/%	Type	Shape	Size/μm	Ref.
Pea (*Pisum sativum*)	C	—	24.9	Legume (pulse)	Rentiform	5–10	Tester *et al.*, 2004;[10] Cano *et al.*, 2014;[20] Qin *et al.*, 2016[112]
	C	38.9	40				
Potato (*Solanum tuberosum*)	C	37.1	17.9	Tuber	Lenticular	5–100	Tester *et al.*, 2004;[10] Cano *et al.*, 2014;[20] Qin *et al.*, 2016[112]
	B	—	20.46				
	A	36.9	28				
Corn (*Zea mays*)	A	19.58	19.70	Cereal	Spherical	2–30	Tester *et al.*, 2004;[10] Romero-Bastida *et al.*, 2015;[113] Qin *et al.*, 2016;[112] Yang *et al.*, 2017;[114] Ghanbari *et al.*, 2018[57]
	A	27.6	—				
	A	21.8	—				
	—	15	30				
	A	34.5	26.5				
Banana (*Musa* spp.)	C	57.3	16.36	Fruit	—	—	Agama-Acevedo *et al.*, 2015[115]
	B	28–30	23.8				
Cassava (*Manihot esculenta*)	C	38	—	Root	Spherical	5–45	Tester *et al.*, 2004;[10] Mei *et al.*, 2015;[116] Qin *et al.*, 2016[112]
	A	37.64	—				
	A	36.7	18.9				
Rice (*Oryza sativa*)	A	41.0	14.6	Cereal	Polyhedral	3–8	Tester *et al.*, 2004;[10] Kong *et al.*, 2015;[117] Van Hung *et al.*, 2016[118]
	A	38.0	21.6				
	A	32.3	28.1				
	A	21.7	33.4				
	A	26.7	31.7				

the drying, films rich in amylose presented greater stiffness and resistance to fracture, however, with less extensibility and permeability to oxygen.[34]

9.1.3 Non-conventional Sources of Starch

There are several studies reporting the physicochemical characterization and tentative applications of non-conventional starches for bioplastics development, some of them include pea starch,[35] mango kernel starch,[36] mung bean starch,[37] sorghum starch,[38] banana starch,[39] sago starch,[40,41] quinoa starch,[42] and Chinese yam starch.[43] Ríos-Soberanis *et al.*[7] have produced TPS from Ramón seeds (*Brosimum alicastrum* Swarts) intended as a substitute to corn starch, which is the predominant TPS in use. It is mentioned that this unconventional starch source is not used as food, which would be an advantage compared to using corn for producing TPS.

When compared to the conventional starch processing, the production of non-conventional starches is challenging due to the higher costs and need for developing or adapting industrial processing. Some sources of starch may have higher protein contents requiring additional steps of purification, while others may need higher volumes of water during processing and generate higher effluent volumes with potentially severe environmental impact. Size and shape of the starch granules from non-conventional sources vary considerably as do their degree of crystallinity and amylose contents.[44] A big variation relative to physicochemical, thermal, and technological properties of non-conventional root and tuber starches from the Ivory Coast has been reported.[45] Starches from different sources can result in films with distinct properties. The differences in amylose–amylopectin ratio, as well as in amylopectin branch chain length, will definitely influence the mechanical and barrier properties of starch-based films.[46]

9.2 Properties of Starch Films and Their Characterization

Semi-crystalline starch granules with complex structures must be dissociated to become amorphous TPS, a prerequisite to forming films. Hydrothermal treatment, shearing, and the addition of plasticizers are very important for starch gelatinization and conversion into TPS.[47]

The development of starch-based raw materials has been widely studied in the last few decades; chief among them are packaging materials that play an important role in containing and preserving foods throughout the supply chain. The most frequently used packaging materials are based on polymers from non-renewable sources which are associated with environmental concerns.[48] Several researchers have reported the development of new food packaging materials based on natural polymers such as starch, cellulose, gelatin, and alginates among others.[49]

9.2.1 Film Physical Properties

New bioplastic films must be evaluated for relevant properties/behaviours to compare their commercially available properties/behaviours. Examples of such relevant physical properties include thickness, surface hydrophobicity, water solubility of the film, water content of the film, and water absorption by the film; such properties indicate the application potentials for the films (Table 9.3). The film thickness is measured with a digital micrometer, and surface hydrophobicity can be assessed by measuring the contact angle of the film surfaces to water drops. The water solubility of the film indicates the amount of film dissolved when soaked in water for a given time. The water content of the film is confirmed in order to evaluate the water uptake of the film by gravimetric analysis after the film is dried in an oven at 105 °C for 24 hours[42] and can determine the film application. Normally, absorption of water by the film is determined by placing it in a controlled temperature environment saturated with humidity, and after 24 hours the increase in film weight is measured and expressed as the incorporation of moisture.

9.2.2 Mechanical Properties

The most commonly studied mechanical properties of films are tensile stress, elongation at break, and tensile modulus or Young's modulus.[42,50] Tensile stress is the maximum strength offered by the material when subjected to a tensile force expressed in MPa or $N\,m^{-2}$. This parameter consists of a ratio between the measured force (F_{max}) and the initial cross-sectional area (A_{min}) of the specimen [eqn (9.1)].

$$\alpha_{max} = \frac{F_{max}}{A_{min}} \tag{9.1}$$

Elongation at break is the percentage ratio (%) between the elongation of the specimen at break and its initial length. This parameter evaluates the flexibility of the material because a higher percentage of elongation until rupture indicates a higher elasticity of the material. The tensile modulus (E) is generally expressed as MPa or GPa, where the rigidity of the material is evaluated. Tensile modulus is the relationship between tensile stress and deformation in the elastic region, and rigidity is generally proportional to the tensile modulus value.

The mechanical properties of starch films are directly related to their components and the thickness, in addition to the processing techniques. However, the plasticizer in the formulation is of extreme importance considering that it directly influences the mechanical properties of resulting films making them more flexible and deformable. For example, in the development of food packaging, knowing the mechanical properties of the materials becomes essential; however, the integrity of the packaging directly influences the preservation of the food during the storage, transportation,

Table 9.3 Physical properties of reinforced films prepared from different starch sources.

Film	Solubility/%	Water content/%	Density/g cm^{-3}	Opacity	Ref.
Corn starch + NCC[a]	18.5	33	1.36	—	Slavutsky and Bertuzzi, 2014[31]
Potato starch	24	19	—	—	Zhang *et al.*, 2011[19]
Sugar palm	19	8	1.5	—	Sanyang *et al.*, 2015[66]
Corn starch + nanoclay 3%	24.03	15	1.48	222AU nm (400–800 nm)	Slavutsky *et al.*, 2012[29]
Corn starch + nanoclay 10%	21.35	14.5	1.62	249AU nm (400–800 nm)	
Corn starch	13.48	9.1	—	138AU nm (400–800 nm)	Garcia *et al.*, 2006[52]
Corn starch + glycerol	30.3	25.2	—	109.6AU nm (400–800 nm)	
Potato starch	28.08	19.6	—	0.32 (A_{600} mm^{-1})	Oleyaei *et al.*, 2016[23]
Potato starch + TiO_2 1%	26.64	15.03		3.61 (A_{600} mm^{-1})	
Cassava starch + chitosan 25%	30	15.47	—	—	Zhao *et al.*, 2018[28]
Cassava starch + chitosan 25%	28	12.31	—	—	
Banana starch	20	11.3	1.04	62.1%	Sartori and Menegalli, 2016[119]
Banana starch + ascorbic acid	18	12.2	1.12	65.6%	
Banana starch + lauric acid/oleic acid	12	17.3	1.11	75.8%	
Cassava starch	2	12	—	—	Kim *et al.*, 2015[26]
Cassava starch + gumarabic	15	13	—	—	
Cassava starch	—	12	2.41	—	Müller *et al.*, 2009[120]
Cassava starch + fibers 10%	—	11.5	1.47	—	
Cassava starch + fibers 50%	—	9	1.31	—	
Wheat starch	30.16	44.5	—	—	Basiak *et al.*, 2017[90]
Corn starch	44.76	36.7	—	—	
Potato starch	14.52	31.6	—	—	
Yam starch	32	52	—	—	Gutiérrez *et al.*, 2015[121]
Phosphated yam starch	47	37	—	—	
Cassava starch	24	34	—	—	
Phosphated cassava starch	36	32	—	—	
Corn starch	24	7.9	—	—	Ortega-Toro *et al.*, 2015[122]
Corn starch + 20% polycaprolactone	13.7	6.1	—	—	
Corn starch + 40% polycaprolactone	13	4.6	—	—	

[a]NCC: nanocrystal cellulose.

and retail operations. Starches from different sources (maize, rice, wheat, and potato) and with distinct particle sizes (small-particle maize starch and small and large wheat starch granules) were reported as renewable and biodegradable fillers in low-density polyethylene films.[51] There was a direct effect of the granule size on the film thickness and on the light transmission through the film. The tensile strength of the films was negatively correlated with starch granule size (diameter).

The mechanical properties of starch films (Table 9.4) can be measured through various mechanical tests, such as perforation, traction, tensile, elongation, and relaxation tests. The tensile properties evaluate the strength of the material to the elongation and the rupture when subjected to a tensile force.

As seen in Table 9.4, when a reinforcement in formulation is added, for example, cassava starch with cellulose nanocrystals, pea starch with starch nanocrystals, tensile stress generally increases while the elongation at break decreases. The plasticizer (glycerol) increases the elasticity of the film, increasing the elongation at break while decreasing the tensile stress, as reported for corn starch.[52] Comparing the data from bean and pea starches, tensile stress and Young's modulus showed similar values. The tuber and cereal starches produced different values for all the tests suggesting that the sources of the starch can influence the mechanical properties of resulting films.

9.2.3 Thermal Properties

The thermal properties of materials provide information about the stability, degradation, gelatinization, and melting and glass transition when subjected to thermal conditions. The glass transition temperature (T_g) is a gradual transition from rigid to soft materials in amorphous materials. Materials that are below T_g in the glassy state and have low molecular mobility resulting in increased stiffness; while above T_g, the materials are more flexible because of the greater mobility of the molecules. Glass transition temperature is an important parameter to be evaluated depending on the application of the polymer material, mainly due to the processes of recrystallization and storage.[53] Some analytical methods employed to evaluate thermal characteristics of biopolymer films are mentioned below. In contrast, crystalline materials have a melt temperature (T_m), above which they go from a solid to liquid or liquid-like state.

9.2.3.1 Differential Scanning Calorimetry

Differential scanning calorimetry (DSC) is a widely employed technique to study the gelatinization of starch, which is an irreversible phase change in the crystalline structure as a result of exposure to heat and/or moisture. This technique has been widely used in recent decades in order to measure the temperature and the heat flow in samples as the temperature is increased. Physical and chemical changes are measured quantitatively and qualitatively through endothermic and exothermic processes.[54] Normally, the sample is

Table 9.4 Mechanical properties of reinforced films prepared from different starch sources.

Film	Tensile stress/MPa	Elongation at break/%	Young modulus/MPa	Ref.
Bean starch	7.14	2.46	16.29	Rompothi *et al.*, 2017[37]
Corn starch + CNC[a]	17.4	9.1	520	Slavutsky and Bertuzzi, 2014[31]
Cassava starch	35	6	—	Kim *et al.*, 2015[26]
Wheat starch	3.4	116	40.3	Lu *et al.*, 2005[123]
TPS + chitosan 5	12.4	1.69	1136.7	Lopez *et al.*, 2014[102]
Corn starch + glycerol	9.2	8	800	Jiménez *et al.*, 2012[124]
Corn starch + glycerol	15.7	4.1	587	Ortega-Toro *et al.*, 2016[32]
Cassava starch + glycerol	6	80	350	Dang and Yoksan, 2015[107]
TPS[b]	7.9	29.2	0.30	Lendvai *et al.*, 2016[99]
Potato	10.61	—	0.08	Santacruz *et al.*, 2015[125]
Cassava	9.27	—	0.12	
Cassava starch + CNC 0.2	10	19	—	Ma *et al.*, 2017[27]
Cassava starch + CNC 0.5	22.5	8.5	—	
Cassava starch	47.05	329	315.78	Teodoro *et al.*, 2015[126]
Pea starch	5.76	29.23	21.15	Li *et al.*, 2015[127]
Pea starch + SNC[c] 9%	6.68	26.7	27.56	
Corn starch	47.4	3.6	—	Garcia *et al.*, 2006[52]
Corn starch + glycerol	7.1	22.5	—	
Corn starch	15.7	4.1	278	Ortega-Toro *et al.*, 2015[122]
Corn starch + 20% polycaprolactone	9.9	1.7	405	
Corn starch + 40% polycaprolactone	7.1	2.0	390	
Wheat starch	3.29	15.21	0.12	Basiak *et al.*, 2017[90]
Corn starch	3.72	19.13	0.10	
Potato starch	6.56	5.67	5.33	

[a] CNC: cellulose nanocrystal.
[b] TPS: thermoplastic starch.
[c] SNC: starch nanocrystal.

placed into a sealed crucible (starch suspension with water in a ratio of approximately 1 : 3 w/w) with an empty crucible used as a reference. The sample and reference are heated and cooled by a set user-defined condition. DSC can measure glass transition temperature (T_g), onset (T_o), peak (T_p), and conclusion (T_c) temperatures for gelatinization, melting temperature (peak, T_m), and gelatinization enthalpy (ΔH). Thermal properties of different starch films and their differences for different types of starch are shown in Table 9.5.

It can be seen in Table 9.5 that the addition of plasticizer results in increased onset temperature of gelatinization as well as a reduction in melting temperature. Cassava starch (root) showed the lowest onset temperature of gelatinization (154 °C), followed by corn (cereal) at 172 °C, and pulse (187 °C). Cassava starch (5–45 μm) and corn (2–30 μm) have a spherical shape. Again, there is evidence that the type of starch, shape, and size influence the thermal properties.

Corn starch films produced by casting were tested for thermal stability using DSC. Citric acid was used to cross link corn starch and the films from native and citric acid cross-linked starch were compared. The films produced from cross-linked starch had a broad melting peak (80–130 °C[55]) and were considerably more heat stable.[56] Corn starch films reinforced with cellulose nanofibers were produced by extrusion and their thermal properties were evaluated; the interaction of cellulose nanofibers and the starch matrix increased the temperature of degradation compared with thermoplastic starch film without reinforcement.[57] DSC thermal analysis of banana starch resulted in higher enthalpy values (17.5 J g^{-1}), when compared with corn and potato starches. Banana starch gel had fast retrogradation kinetics and largest thermal stability (high temperature of dissociation of retrograded starch), which was associated with its amylopectin chain length.[58]

9.2.3.2 *Thermogravimetric Analysis*

Thermogravimetric analysis (TGA) provides information on the thermal stability of materials. In general, the temperatures correspond to dehydration (removal of water molecules) and decomposition of the material components. The loss of mass of a sample as a function of temperature is observed at a constant heating rate.[59] Some parameters that may affect the results are the heating rate (°C min^{-1}), the furnace atmosphere (inert gas or oxidant), and the type of pan used.

Amylose content of corn starch-based films was positively related with thermal stability. This behaviour was attributed to the compact crystallinity of amylose that reduced depolymerization and breakage of chemical bonds compared with amorphous aggregation.[60]

9.2.3.3 *Dynamic Mechanical Testing*

Dynamic mechanical testing consists of a technique used to determine the glass transition temperature and the linear coefficient of thermal expansion.

Table 9.5 Thermal properties (DSC) and water vapor permeability of different starch film formulations.

Film	T_o[a]/°C	T_m[b]/°C	ΔH_m[c]/J g^{-1}	WVP/g m^{-1} s^{-1} Pa^{-1} × 10^{-11}	Ref.
TPS[d] + CNF[e] 0.5	—	107.0	0.3	—	Ghanbari *et al.*, 2018[63]
TPS + chitosan 5	135.4	154.8	105.2	—	Lopez *et al.*, 2014[102]
Sago starch	123.6	157.2	—	—	Abdorreza *et al.*, 2011[128]
Pea starch	186.6	188.5	23.5	—	Li *et al.*, 2015[127]
Pea starch + SNC[f] 9%	187.5	190.1	22.2		
Corn starch	171.7	210.2	45.5	7.6	Dai *et al.*, 2015[129]
Corn starch + 2% Taro SNP	187.4	223.2	50.8	5.1	
Corn starch + 5% Taro SNP	184.1	218.1	50.0	4.1	
Corn starch + 10% Taro SNP	187.3	222.3	55.3	3.3	
Corn starch + glycerol	—	—	—	9.14	Jiménez *et al.*, 2012[124]
Potato starch	—	132.3	—	27.5	Oleyaei *et al.*, 2016[23]
Potato starch + TiO_2 1%	—	142.6	—	23.3	
Banana starch	—	—	—	7	Sartori and Menegalli, 2016[119]
Banana starch + ascorbic acid	—	—	—	6.5	
Potato starch	—	—	—	21.1	Noshirvani *et al.*, 2018[130]
Potato starch + PVA[g]	—	210	—	20.6	
Corn starch	—	—	—	7.90	Basiak *et al.*, 2017[90]
Corn starch	—	—	—	—	Ortega-Toro *et al.*, 2015[122]
Corn starch + 20% polycaprolactone	—	63.3	82.3	—	
Corn starch + 40% polycaprolactone	—	63.3	81.9	—	
Cassava starch	154	157.9	172.4	—	Vu and Lumdubwong, 2016[131]
Cassava starch + glycerol	160.5	164.6	157.2	—	
Cassava starch + sorbitol	176.3	180.6	138.2	—	
Cassava starch + mungbean	146.2	154.1	165.3	—	
Cassava starch	—	—	—	4.10	Bersaneti *et al.*, 2016[132]
Cassava starch + fructooligosaccharides 1%	—	—	—	2.86	
Cassava starch + fructooligosaccharides 5%	—	—	—	2.71	

[a] T_o: onset temperature.
[b] T_m: melting temperature (peak).
[c] ΔH_m: melting enthalpy.
[d] TPS: thermoplastic starch.
[e] CNF: cellulose nanofibrils.
[f] SNC: starch nanocrystal.
[g] PVA: polyvinyl alcohol.

A specimen is subjected to a constant load as a function of time and temperature, heating, and cooling. As per the deformations occurring in the test specimen, the angular coefficient is obtained.[61] This test is widely used in the characterization of amorphous or semi-crystalline polymers, for example, starch, where it is possible to determine the temperature at which the polymer passes from a solid state to a state of greater mobility without structural changes (improved E', E'', and tan δ dynamic properties to thermoplastic state[62,63]).

9.2.4 Transmission and Barrier Properties

Water vapor and gas molecules can be transported through a polymer or film by diffusion, solubility, and/or permeability. Diffusion occurs when a molecule permeates through a polymer matrix and depends on the morphology of the plastic. Solubility depends on hydrophilicity, the surface of the material, and the medium into which the system is inserted. The permeability is a combination of diffusion and solubility.[64] Polymers tend to be relatively permeable to gas and liquid molecules compared to other materials such as a metal film. Typical penetrating molecules through polymers are gases such as oxygen (O_2), nitrogen (N_2), or carbon dioxide (CO_2) or vapors such as water (H_2O).

The permeability of polymer films to gases and vapors becomes an important property to evaluate for a range of applications. Among the possible applications are packaging, protection of substrates, and encapsulation, among others.[64] The water vapor permeability (see Table 9.5) is the barrier property frequently studied in starch bioplastics.

9.2.4.1 Water Vapor Permeability

The water vapor permeability (WVP) is quantified by measuring the water vapor transmission rate for various film thicknesses (mm or μm) and atmospheric pressures (Pa). Generally, the test is conducted by sealing a test cup with the polymer film containing a desiccant agent and the test cup is inserted into a humidity-controlled chamber. WVP is defined as the time rate of water vapor transmission through a certain area with a certain thickness induced by the unit vapor pressure difference between two surfaces.[42,65] Typically, the test is performed at room temperature with relative humidity inside the cup being 0% and 50% in the chamber (outside).

Starch films exhibit higher water vapor permeabilities compared to traditional plastics because of their hydrophilic nature; however, additives/reinforcements can reduce their permeability.[66] Low water vapor permeability may be necessary in the packaging of low and intermediate-moisture foods, avoiding moisture absorption, and microbial growth, as well as maintaining crispness.[67] In one study, physically modified maize starch processed through a high-speed homogenizer at different shear rates was used for producing starch film. Water resistance and water vapor permeability improved considerably in the film which would be suitable for food packaging.[68] Acetylated maize starch with 1.5% w/w glycerol was used for producing films that showed 87% reduction on WVP and 34% increase on flexibility, compared with the film made from native maize starch.[69]

WVP values for different starch film formulations are shown in Table 9.5. The results for neat corn starch films are similar (between 7.6 and 9.14 $g\,m^{-1}\,s^{-1}\,Pa^{-1} \times 10^{-11}$) which are reduced by the addition of starch nanoparticles. Cassava (4.10 $g\,m^{-1}\,s^{-1}\,Pa^{-1} \times 10^{-11}$) and potato starch films (21.1 $g\,m^{-1}\,s^{-1}\,Pa^{-1} \times 10^{-11}$) presented lower and higher values, respectively, than corn starch films. The water vapor permeability value of the film can be influenced by the size of the starch granules, particularly the largest granule size found in potato starch (up to 100 μm). This size corresponds to the highest water vapor permeability value. The WVP values found for corn and cassava were similar because of the similarity in granule size (2–30 μm for corn starch and 5–45 μm for cassava starch).

9.2.4.2 *Oxygen Permeability*

Oxygen permeability is defined as the amount of oxygen that is able to pass through a unit area of a film per unit time under certain test conditions. Permeability to oxygen varies in polymers used for packages, for example, oxidation can limit the preservation of food or oxidation can reduce nutritional and sensory properties of some foods. However, a high permeability can be a positive aspect for short-term packaging of fresh red meat, allowing the formation of oxymyoglobin which is responsible for the desirable bright red color of the meat.[70] In contrast, low permeability is required in packages that involve foods rich in fats in order to avoid oxidation and free radical formation.

9.3 Formulation of Starch-based Plastics

Food grade polymers have received considerable attention as alternative materials and are often used to produce films, edible coatings, and improve the shelf life of foods.[71] Several studies have reported the use of starch from various sources to prepare films by melt-processing or by casting and demonstrated for food packaging. These sources include starches extracted from cassava,[72–77] potato,[78–81] corn,[77,82–84] and wheat,[85–87] among others. The differences in the properties of each type of starch, for example, rheological properties, solubility in water, and the size and the morphology of the granules, can influence the properties of the resulting films. In addition, several studies have reported the effect of starch types on films properties.[20,22,25,27,30]

9.3.1 Plasticizers

Plasticizers are typically low molecular weight, small molecules that are compatible with a polymer. They can occupy the spaces between the polymer chains, reducing the secondary forces and overall stiffness of the resulting plastic. The molecular organization of the polymers is modified, and the energy required to promote molecular mobility is reduced, resulting in the formation of hydrogen bonds between the chains, making the films softer and more flexible.[88] The addition of many plasticizers to starch also results in similar effects.[89] For example, Cano *et al.*[20] developed films with pea, potato, and cassava starches, and glycerol as the plasticizer. The water vapor and oxygen permeability were lower for films developed with cassava starch, as well as lower tensile stress and elongation at break for the potato starch films. Basiak *et al.*[90] developed films of wheat, corn, and potato starch with glycerol. The water vapor permeability and elongation at break were lower for films with potato starch, while tensile stress was lower for films developed with wheat starch.

Thus, it can be seen that the use of plasticizers in biopolymer films serves to reduce the intermolecular forces in the biopolymer chains, increasing the flexibility and mobility of the material.[91] The choice of a plasticizer is important for starch-based films so they can take on plastic-like properties and can allow for easier removal of the film from casting supports.[36,83] Generally, the plasticizers are hygroscopic and the water content in the film/polymer is dependent on the ambient conditions to which the material is exposed.[92] There are several types of commercial plasticizers; however, those most used in the development of starch-based bioplastics are water, glycerol, and sorbitol. Glycerol has a relatively lower molecular weight and a higher hydrophilicity compared to other polyols such as sorbitol. Usually, the addition of glycerol to starch increases moisture content, absorption, and solubility in water. There is also an increase in the water vapor and oxygen permeability. Table 9.6 details many examples of starch film formulations reported, plasticizers used, gelatinization conditions, and drying conditions for films.

Plasticizers, as mentioned above, decrease intermolecular forces and increase the mobility of starch chains, reducing brittle zones, resulting in lower glass transition temperatures. The use of water as a plasticizer results in increased crystallinity and moisture. However, when glycerol or sorbitol is applied, the rate of recrystallization is reduced because of the greater interaction between glycerol/sorbitol and starch.[83,93] In recent years, more interest arose in developing bioplastic films with natural plasticizers because of their low toxicity and migration.[94,95] Among these groups of plasticizers are monosaccharides (glucose, mannose, fructose, sucrose), vegetable oil triglycerides, lecithin, waxes, amino acids, surfactants, water, and fatty acid esters.

9.4 Thermoplastic Starch and Its Applications

Most plastics used by industry are derived from petroleum. In view of environmental conservation, it is necessary to reduce and/or replace these synthetic polymers with bio-based, and preferably biodegradable, materials.[20] Starch is one such abundant bio-based and biodegradable feedstock.

Starch granules in the presence of water and heat can be swelled and transformed through gelatinization. Heating of aqueous starch suspension

Table 9.6 Plasticizer, heating (gelatinization), and drying conditions for some starch film formulations.

Type of starch	Plasticizer	Gelatinization conditions	Drying conditions of films	Ref.
Sugar palm	Glycerol Sorbitol Glycerol—sorbitol	95 °C, 15 min	40 °C, 24 h	Sanyang *et al.*, 2016[89]
Corn	Glycerol	160 °C, 30 min	160 °C, 2 min (compression molding)	Ortega-Toro *et al.*, 2016[32]
Cassava	Glycerol	Room temperature, 48 h	65 °C, 18 h	Dang and Yoksan, 2015[107]
Potato Cassava	Glycerol Glucose	90 °C, 5 min	30 °C, 24 h	Santacruz *et al.*, 2015[125]
Corn	Glycerol	75 °C	25 °C	Alves *et al.*, 2015
Cassava	Glycerol	80 °C, 10 min	60 °C, 2 h	Ma *et al.*, 2017[27]
Cassava	Glycerol	70 °C, 30 min	60 °C, 24 h	Teodoro *et al.*, 2015[126]
Pea	Glycerol	Boiling temperature, 30 min	40 °C, overnight	Li *et al.*, 2015[127]
Banana	Glycerol	81 °C, 15 min	20 °C	Sartori and Menegalli, 2016[119]
Cassava	Glycerol	Reactor (75–150 °C), 10 min	40 °C, 48 h	Zhao *et al.*, 2018[28]
Pea	Glycerol	95 °C, 30 min	40 °C, 48 h	Cano *et al.*, 2015[35]
Cassava	Glycerol	80 °C	40 °C, 16 h	Müller *et al.*, 2009[120]
Potato	Glycerol	90 °C, 30 min	60 °C, 15 h	Oleyaei *et al.*, 2016[23]
Potato	Glycerol	80 °C, 10 min	55 °C, 16 h	Noshirvani *et al.*, 2018[130]
Sugar palm	Glycerol Sorbitol	90 °C, 30 min	50 °C, overnight	Poeloengasih *et al.*, 2016[133]
Cassava starch Mungbean starch	Glycerol Sorbitol	80 °C, 20 min	60 °C, 6 h	Vu and Lumdubwong, 2016[131]
Wheat starch Corn starch Potato starch	Glycerol	85 °C, 30 min	25 °C, 48 h	Basiak *et al.*, 2017[90]

results in breaking of the hydrogen bonds and hydration of the hydroxyl groups of the glucoses, resulting in an irreversible swelling of the granules. The ability of swelling and solubility are variable based on many factors such as type of starch, ratio of amylose and amylopectin, molecular weight, degree, and length of amylopectin branching and conformation. Each type of starch has a typical gelatinization temperature; this temperature is dependent on the presence of water, molecular structure of amylopectin, ratio of crystalline and amorphous structures, and amylose and amylopectin ratio in starch. Above the gelatinization temperature, the granules lose their natural birefringence and cause a reduction in the viscosity of the starch.[12] This process is typically a prerequisite for a starch to be amenable for bioplastic applications and has been extensively studied for decades. Bioplastics are formed from the gelatinization/melting of starch granules with the addition of small amounts of plasticizer. Because starch polymers melt similarly to synthetic polymers, it is often referred to in the literature as thermoplastic starch.[96]

When starch is heated above its gelatinization temperature and stirred together with plasticizers such as water, glycerol, sorbitol, and others, it undergoes a transformation into a thermoplastic material. TPS has been extensively studied in the last few decades because of its low cost, renewability, easy availability, and easy chemical and mechanical modification.[7,97] However, films made from TPS typically have relatively lower mechanical strength. Several approaches have been investigated to improve it, for example, addition of fibers and nanoparticles and modification of the starch itself.[25,57,98–100] In addition, because of their higher specific surface area, nanoparticles have been reported to improve characteristics of TPS films/composites.[101]

9.4.1 Reinforced Starch Films and Composites

Because starch films inherently have structural and barrier limitations, it is often necessary to improve the mechanical, thermal, and barrier properties through additives/reinforcements. As starch is a biodegradable material, the addition of compatible agroindustry wastes results in a low-priced and biodegradable plastic. Studies show that incorporation of different reinforcements in TPS such as fibers,[19,98] pulp,[19] nanofibers,[57] and chitin/chitosan[102] improved mechanical properties. Cellulose nanoparticles can be applied as reinforcement in starch-based matrices leading to improvement in structural and mechanical properties in packages and a decrease in the crack growth rate of nanocomposites. There are many studies on the application of reinforcements in non-biodegradable polymer matrices.[103,104] The development of thermoplastic starch composites with reinforcements has also been studied,[99,100] resulting in a biodegradable composite.

9.5 Processing Conditions for Thermoplastic Starch

Plastic films can be produced in different ways, the main ones being extrusion and casting.[57,92] The processed thermoplastic starch results in a biodegradable material that can replace synthetic polymers in various areas, such as packaging, and disposable materials, among many others.[102,105] Often pre-drying is important prior to extrusion.[57,106] Generally, films are subjected to a series of standard tests, such as thickness, density, permeability to water vapor and oxygen, water absorption, water content, as well as mechanical, thermal, and structural tests.[35,40,107]

9.5.1 Casting of Starch Films

Thermoplastic starch can be cast into films with a solution casting process. This process consists of placing the heated polymeric solution in a mold and subjecting the assembly to drying under controlled temperature, humidity, pressure, and time, upon which the solvent will be evaporated.[108] Starch films can be obtained with a dispersion, followed by solubilization and gelatinization of the starch in solvent, usually water, and formation of a film-forming solution that is poured on a base and dried.[107] The thickness of the formed film will depend on the amount of suspension added over the base and starch concentration in the formulation. After complete evaporation of the solvent, the support with the film-forming solution should be stored in a place with a higher relative humidity in order to facilitate the removal of the films without cracking and damaging the format.[108]

9.5.2 Extrusion Systems

Typically, the thermoplastic starch can be processed using single or dual-screw extruders; however, twin-screw extruders provide greater homogeneity of the blend because of the greater shear force.[109] Inside the equipment, a pressure in the range of 5–6 MPa is commonly employed, moisture between 15 and 23%, and variable flow and speed, depending on the material and equipment used.[15,110] The extrusion process has many advantages such as better homogeneity, higher starch throughput, higher temperatures, and a faster process compared to the solution-casting method.[102]

Abbreviations

CNC	Cellulose nanocrystal
CNF	Cellulose nanofibrils
DSC	Differential scanning calorimetry
E	Tensile modulus
NCC	Nanocrystal cellulose
PVA	Polyvinyl alcohol
T_c	Conclusion temperature
T_g	Glass transition temperature
TGA	Thermogravimetric analysis
T_m	Melting temperature (peak)
T_o	Onset temperature

T_p	Peak temperature
TPS	Thermoplastic starch
WVP	Water vapor permeability
ΔH	Gelatinization enthalpy
ΔH_m	Melting enthalpy

References

1. E. Bertoft, *Agronomy*, 2017, **7**, 56.
2. W. He and C. Wei, *Food Hydrocoll.*, 2017, **73**, 162–175.
3. P. Halley and L. Avérous, *Starch polymers: from genetic engineering to green applications*, San Diego, USA, 2014.
4. Z. Din, H. Xiong and P. Fei, *Crit. Rev. Food Sci. Nutr.*, 2017, **57**, 2691–2705.
5. M. Pratiwi, D. N. Faridah and H. N. Lioe, *Starch - Stärke*, 2018, **70**, 1700028.
6. B. Kaur, F. Ariffin, R. Bhat and A. A. Karim, *Food Hydrocoll.*, 2012, **26**, 398–404.
7. C. R. Ríos-Soberanis, R. Javier Estrada-León, V. Manuel Moo-Huchin, M. José Cabrera-Sierra, J. Manuel Cervantes-Uc, L. Arturo Bello-Pérez and E. Pérez-Pacheco, *J. Appl. Polym. Sci.*, 2016, **133**, 44235.
8. R. Hoover, T. Hughes, H. J. Chung and Q. Liu, *Food Res. Int.*, 2010, **43**, 399–413.
9. F. Corzana, M. S. Motawia, C. du Penhoat, F. van den Berg, A. Blennow, S. Perez and S. B. Engelsen, *J. Am. Chem. Soc.*, 2004, **126**, 13144–13155.
10. R. F. Tester, J. Karkalas and X. Qi, *J. Cereal Sci.*, 2004, **39**, 151–165.
11. J. Jane, K. Wong and A. E. McPherson, *Carbohydr. Res.*, 1997, **300**, 219–227.
12. L. Copeland, J. Blazek, H. Salman and M. C. Tang, *Food Hydrocoll.*, 2009, **23**, 1527–1534.
13. J. Jane, Z. Ao, S. A. Duvick, M. Wiklund, S.-H. Yoo, K.-S. Wong and C. Gardner, *J. Appl. Glycosci.*, 2003, **50**, 167–172.
14. J. W. Lawton, *Carbohydr. Polym.*, 1996, **29**, 203–208.
15. M. Li, P. Liu, W. Zou, L. Yu, F. Xie, H. Pu, H. Liu and L. Chen, *J. Food Eng.*, 2011, **106**, 95–101.
16. P. Halley and L. Avérous, *Starch polymers: from genetic engineering to green applications*, Newnes, 2014.
17. V. K. Khlestkin, S. E. Peltek and N. A. Kolchanov, *Carbohydr. Polym.*, 2018, **181**, 460–476.
18. J. Waterschoot, S. V. Gomand, E. Fierens and J. A. Delcour, *Starch-Stärke*, 2015, **67**, 14–29.
19. Y. Zhang, M. Thompson and Q. Liu, *Carbohydr. Polym.*, 2011, **86**, 700–707.
20. A. Cano, A. Jiménez, M. Cháfer, C. Gónzalez and A. Chiralt, *Carbohydr. Polym.*, 2014, **111**, 543–555.

21. K. Wilpiszewska and Z. Czech, *Starch - Stärke*, 2014, **66**, 660–667.
22. S. Martinez, C. Rivon, O. P. Troncoso and F. G. Torres, *Starch - Stärke*, 2016, **68**, 935–942.
23. S. A. Oleyaei, Y. Zahedi, B. Ghanbarzadeh and A. A. Moayedi, *Int. J. Biol. Macromol.*, 2016, **89**, 256–264.
24. S. L. M. El Halal, G. P. Bruni, J. A. do Evangelho, B. Biduski, F. T. Silva, A. R. G. Dias, E. da Rosa Zavareze and M. de Mello Luvielmo, *Starch - Stärke*, 2018, **70**, 1–10.
25. E. de M. Teixeira, D. Pasquini, A. A. S. Curvelo, E. Corradini, M. N. Belgacem and A. Dufresne, *Carbohydr. Polym.*, 2009, **78**, 422–431.
26. S. R. B. Kim, Y.-G. Choi, J.-Y. Kim and S.-T. Lim, *LWT - Food Sci. Technol.*, 2015, **64**, 475–482.
27. X. Ma, Y. Cheng, X. Qin, T. Guo, J. Deng and X. Liu, *LWT - Food Sci. Technol.*, 2017, **86**, 318–326.
28. Y. Zhao, J. S. Teixeira, M. M. Gänzle and M. D. A. Saldaña, *J. Supercrit. Fluids*, 2018, **137**, 101–110.
29. A. M. Slavutsky, M. A. Bertuzzi and M. Armada, *Braz. J. Food Technol.*, 2012, **15**, 208–218.
30. M. B. Agustin, B. Ahmmad, S. M. M. Alonzo and F. M. Patriana, *J. Reinf. Plast. Compos.*, 2014, **33**, 2205–2213.
31. A. M. Slavutsky and M. A. Bertuzzi, *Carbohydr. Polym.*, 2014, **110**, 53–61.
32. R. Ortega-Toro, A. Muñoz, P. Talens and A. Chiralt, *Food Hydrocoll.*, 2016, **56**, 9–19.
33. L. Ren, X. Yan, J. Zhou, J. Tong and X. Su, *Int. J. Biol. Macromol.*, 2017, **105**, 1636–1643.
34. R. Colussi, V. Z. Pinto, S. L. M. El Halal, B. Biduski, L. Prietto, D. D. Castilhos, E. da Rosa Zavareze and A. R. G. Dias, *Food Chem.*, 2017, **221**, 1614–1620.
35. A. I. Cano, M. Cháfer, A. Chiralt and C. González-Martínez, *J. Food Eng.*, 2015, **167**, 59–64.
36. A. Nawab, F. Alam, M. A. Haq, Z. Lutfi and A. Hasnain, *Int. J. Biol. Macromol.*, 2017, **98**, 869–876.
37. O. Rompothi, P. Pradipasena, K. Tananuwong, A. Somwangthanaroj and T. Janjarasskul, *Carbohydr. Polym.*, 2017, **157**, 748–756.
38. B. Biduski, F. T. da Silva, W. M. da Silva, S. L. de M. El Halal, V. Z. Pinto, A. R. G. Dias and E. da Rosa Zavareze, *Food Chem.*, 2017, **214**, 53–60.
39. F. M. Pelissari, M. M. Andrade-Mahecha, P. J. do A. Sobral and F. C. Menegalli, *J. Colloid Interface Sci.*, 2017, **505**, 154–167.
40. L. Nouri and A. M. Nafchi, *Int. J. Biol. Macromol.*, 2014, **66**, 254–259.
41. T. J. Gutiérrez and V. A. Alvarez, *J. Polym. Environ.*, 2017, **25**, 701–716.
42. C. H. Pagno, T. M. H. Costa, E. W. de Menezes, E. V. Benvenutti, P. F. Hertz, C. R. Matte, J. V. Tosati, A. R. Monteiro, A. O. Rios and S. H. Flôres, *Food Chem.*, 2015, **173**, 755–762.
43. L. Wang, X. Liu and J. Wang, *Int. J. Food Prop.*, 2017, **20**, 1239–1250.
44. S. Sukhija, S. Singh and C. S. Riar, *Starch - Stärke*, 2016, **68**, 160–168.

45. F. A. Tetchi, A. Rolland-Sabaté, G. N. Amani and P. Colonna, *J. Sci. Food Agric.*, 2007, **87**, 1906–1916.
46. N. L. García, L. Famá, A. Dufresne, M. Aranguren and S. Goyanes, *Food Res. Int.*, 2009, **42**, 976–982.
47. A. Dufresne and J. Castaño, *Starch - Stärke*, 2017, **69**, 1500307.
48. Y. Zhu, C. Romain and C. K. Williams, *Nature*, 2016, **540**, 354.
49. A. López-Córdoba, C. Medina-Jaramillo, D. Piñeros-Hernandez and S. Goyanes, *Food Hydrocoll.*, 2017, **71**, 26–34.
50. S. L. Valença, S. Griza, V. G. de Oliveira, E. M. Sussuchi and F. G. C. de Cunha, *Composites, Part B*, 2015, **70**, 1–8.
51. S.-T. Lim, J.-L. Jane, S. Rajagopalan and P. A. Seib, *Biotechnol. Prog.*, 1992, **8**, 51–57.
52. M. A. Garcia, A. Pinotti and N. E. Zaritzky, *Starch - Stärke*, 2006, **58**, 453–463.
53. R. Xie, Y. Lee, M. P. Aplan, N. J. Caggiano, C. Müller, R. H. Colby and E. D. Gomez, *Macromolecules*, 2017, **50**, 5146–5154.
54. V. Vamadevan and E. Bertoft, *Starch - Stärke*, 2015, **67**, 55–68.
55. J. Delville, C. Joly, P. Dole and C. Bliard, *Carbohydr. Polym.*, 2003, **53**, 373–381.
56. N. Reddy and Y. Yang, *Food Chem.*, 2010, **118**, 702–711.
57. A. Ghanbari, T. Tabarsa, A. Ashori, A. Shakeri and M. Mashkour, *Int. J. Biol. Macromol.*, 2018, **112**, 442–447.
58. P. Zhang and B. R. Hamaker, *Carbohydr. Polym.*, 2012, **87**, 1552–1558.
59. V. Tsanaktsis, E. Vouvoudi, G. Z. Papageorgiou, D. G. Papageorgiou, K. Chrissafis and D. N. Bikiaris, *J. Anal. Appl. Pyrolysis*, 2015, **112**, 369–378.
60. L. Fu, J. Zhu, S. Zhang, X. Li, B. Zhang, H. Pu, L. Li and Q. Wang, *Carbohydr. Polym.*, 2018, **181**, 528–535.
61. M. TabkhPaz, D.-Y. Park, P. C. Lee, R. Hugo and S. S. Park, *J. Compos. Mater.*, 2017, **52**, 1045–1060.
62. Y. Fourati, Q. Tarrés, P. Mutjé and S. Boufi, *Carbohydr. Polym.*, 2018, **199**, 51–57.
63. A. Ghanbari, T. Tabarsa, A. Ashori, A. Shakeri and M. Mashkour, *Carbohydr. Polym.*, 2018, **197**, 305–311.
64. Y. Cui, S. I. Kundalwal and S. Kumar, *Carbon*, 2016, **98**, 313–333.
65. ASTM, in *Annual Book of ASTM Standards*, Philadelphia, 2000.
66. M. L. Sanyang, S. M. Sapuan, M. Jawaid, M. R. Ishak and J. Sahari, *Polymers*, 2015, **7**, 1106–1124.
67. F. Versino, O. V. Lopez, M. A. Garcia and N. E. Zaritzky, *Starch - Stärke*, 2016, **68**, 1026–1037.
68. M. Shahbazi, M. Majzoobi and A. Farahnaky, *J. Food Eng.*, 2018, **223**, 10–21.
69. O. V. López, M. A. García and N. E. Zaritzky, *Carbohydr. Polym.*, 2008, **73**, 573–581.
70. K. W. McMillin, *Meat Sci.*, 2017, **132**, 153–162.
71. S. Jiang, C. Liu, X. Wang, L. Xiong and Q. Sun, *LWT-Food Sci. Technol.*, 2016, **69**, 251–257.

72. S. Mali, L. S. Sakanaka, F. Yamashita and M. V. E. Grossmann, *Carbohydr. Polym.*, 2005, **60**, 283–289.
73. J. P. Maran, V. Sivakumar, R. Sridhar and V. P. Immanuel, *Ind. Crops Prod.*, 2013, **42**, 159–168.
74. J. Colivet and R. A. Carvalho, *Ind. Crops Prod.*, 2017, **95**, 599–607.
75. C. M. Léis, A. R. Nogueira, L. Kulay and C. C. Tadini, *J. Clean. Prod.*, 2017, **143**, 76–89.
76. D. Piñeros-Hernandez, C. Medina-Jaramillo, A. López-Córdoba and S. Goyanes, *Food Hydrocoll.*, 2017, **63**, 488–495.
77. C. L. Luchese, T. Garrido, J. C. Spada, I. C. Tessaro and K. de la Caba, *Int. J. Biol. Macromol.*, 2018, **106**, 834–839.
78. O. Moreno, L. Atarés and A. Chiralt, *Carbohydr. Polym.*, 2015, **133**, 353–364.
79. P. Liu, S. Sun, H. Hou and H. Dong, *Food Hydrocoll.*, 2016, **61**, 351–357.
80. M. Davoodi, G. Kavoosi and R. Shakeri, *Int. J. Biol. Macromol.*, 2017, **104**, 173–179.
81. A. Podshivalov, M. Zakharova, E. Glazacheva and M. Uspenskaya, *Carbohydr. Polym.*, 2017, **157**, 1162–1172.
82. X. Chen, X.-W. He, B. Zhang, X. Fu, J. Jane and Q. Huang, *Int. J. Biol. Macromol.*, 2017, **104**, 481–486.
83. H.-Y. Kim, J. Jane and B. Lamsal, *Ind. Crops Prod.*, 2017, **95**, 175–183.
84. L. Ren, Y. Fu, Y. Chang, M. Jiang, J. Tong and J. Zhou, *Starch - Stärke*, 2017, **69**, 1600025.
85. J. Bonilla, L. Atarés, M. Vargas and A. Chiralt, *J. Food Eng.*, 2013, **114**, 303–312.
86. C. Menzel, G. Seisenbaeva, P. Agback, M. Gällstedt, A. Boldizar and K. Koch, *Carbohydr. Polym.*, 2017, **172**, 365–373.
87. X. Song, G. Zuo and F. Chen, *Int. J. Biol. Macromol.*, 2018, **107**, 1302–1309.
88. C. M. Jaramillo, T. J. Gutiérrez, S. Goyanes, C. Bernal and L. Famá, *Carbohydr. Polym.*, 2016, **151**, 150–159.
89. M. L. Sanyang, S. M. Sapuan, M. Jawaid, M. R. Ishak and J. Sahari, *J. Food Sci. Technol.*, 2016, **53**, 326–336.
90. E. Basiak, A. Lenart and F. Debeaufort, *Int. J. Biol. Macromol.*, 2017, **98**, 348–356.
91. K. Bin Liew, Y. T. F. Tan and K.-K. Peh, *Drug Dev. Ind. Pharm.*, 2014, **40**, 110–119.
92. X. Feng, A. Vo, H. Patil, R. V. Tiwari, A. S. Alshetaili, M. B. Pimparade and M. A. Repka, *J. Pharm. Pharmacol.*, 2016, **68**, 692–704.
93. S. Mali, M. V. E. Grossmann, M. A. García, M. N. Martino and N. E. Zaritzky, *J. Food Eng.*, 2006, **75**, 453–460.
94. J. Chen, Z. Liu, J. Jiang, X. Nie, Y. Zhou and R. E. Murray, *RSC Adv.*, 2015, **5**, 56171–56180.
95. M. G. A. Vieira, M. A. da Silva, A. C. G. Maçumoto, L. O. dos Santos and M. M. Beppu, *Mater. Res.*, 2014, **17**, 386–391.
96. F. Xie, P. J. Halley and L. Avérous, *Prog. Polym. Sci.*, 2012, **37**, 595–623.

97. N. Khanoonkon, R. Yoksan and A. A. Ogale, *Eur. Polym. J.*, 2016, **76**, 266–277.
98. F. X. Espinach, M. Delgado-Aguilar, J. Puig, F. Julian, S. Boufi and P. Mutjé, *Compos. Part B Eng.*, 2015, **81**, 98–106.
99. L. Lendvai, J. Karger-Kocsis, Á. Kmetty and S. X. Drakopoulos, *J. Appl. Polym. Sci.*, 2016, **133**, 42397.
100. N. Gray, Y. Hamzeh, A. Kaboorani and A. Abdulkhani, *Ind. Crops Prod.*, 2018, **115**, 298–305.
101. V. C. Correia, S. F. Santos, R. S. Teixeira and H. S. Junior, *Constr. Build. Mater.*, 2018, **160**, 376–384.
102. O. Lopez, M. A. Garcia, M. A. Villar, A. Gentili, M. S. Rodriguez and L. Albertengo, *LWT - Food Sci. Technol.*, 2014, **57**, 106–115.
103. F. M. Al-Oqla and S. M. Sapuan, *J. Clean. Prod.*, 2014, **66**, 347–354.
104. H.-M. Ng, L. T. Sin, T.-T. Tee, S.-T. Bee, D. Hui, C.-Y. Low and A. R. Rahmat, *Composites, Part B*, 2015, **75**, 176–200.
105. O. V. López, L. A. Castillo, M. A. García, M. A. Villar and S. E. Barbosa, *Food Hydrocoll.*, 2015, **43**, 18–24.
106. K. M. Dang and R. Yoksan, *Carbohydr. Polym.*, 2015, **115**, 575–581.
107. H. Y. Kim, J. Lin Jane and B. Lamsal, *Ind. Crops Prod.*, 2017, **95**, 175–183.
108. I. Martinez-Pardo, R. A. Shanks, B. Adhikari and R. Adhikari, *Carbohydr. Polym.*, 2017, **173**, 170–177.
109. P. González-Seligra, L. Guz, O. Ochoa-Yepes, S. Goyanes and L. Famá, *LWT-Food Sci. Technol.*, 2017, **84**, 520–528.
110. F. Xie, L. Yu, B. Su, P. Liu, J. Wang, H. Liu and L. Chen, *J. Cereal Sci.*, 2009, **49**, 371–377.
111. M. R. Sucaldito and D. H. Camacho, *Carbohydr. Polym.*, 2017, **169**, 315–323.
112. Y. Qin, C. Liu, S. Jiang, L. Xiong and Q. Sun, *Ind. Crops Prod.*, 2016, **87**, 182–190.
113. C. A. Romero-Bastida, L. A. Bello-Perez, G. Velazquez and J. Alvarez-Ramirez, *Carbohydr. Polym.*, 2015, **127**, 195–201.
114. N. Yang, X. W. Xu, R. R. Wang, W. L. Peng, L. Cai, J. Song, W. Li, X. Luo, L. Niu, Y. Wang, M. Jin, L. Chen, J. Luo, M. Deng, L. Wang, Q. Pan, F. Liu, D. Jackson, X. Yang, L. Chen and J. Yan, *Nat. Commun.*, 2017, **8**, 1874.
115. E. Agama-Acevedo, M. C. Nuñez-Santiago, J. Alvarez-Ramirez and L. A. Bello-Pérez, *Carbohydr. Polym.*, 2015, **124**, 17–24.
116. J. Q. Mei, D. N. Zhou, Z. Y. Jin, X. M. Xu and H. Q. Chen, *Food Chem.*, 2015, **187**, 378–384.
117. X. Kong, P. Zhu, Z. Sui and J. Bao, *Food Chem.*, 2015, **172**, 433–440.
118. P. Van Hung, N. L. Vien and N. T. L. Phi, *Food Chem.*, 2016, **191**, 67–73.
119. T. Sartori and F. C. Menegalli, *Food Hydrocolloids*, 2016, **55**, 210–219.
120. C. M. Müller, J. B. Laurindo and F. Yamashita, *Food Hydrocolloids*, 2009, **23**, 1328–1333.
121. T. J. Gutiérrez, M. S. Tapia, E. Pérez and L. Famá, *Food Hydrocolloids*, 2015, **45**, 211–217.

122. R. Ortega-Toro, J. Contreras, P. Talens and A. Chiralt, *Food Packag. Shelf Life*, 2015, **5**, 10–20.
123. Y. Lu, L. Tighzert, F. Berzin and S. Rondot, *Carbohydr. Polym.*, 2005, **61**, 174–182.
124. A. Jiménez, M. J. Fabra, P. Talens and A. Chiralt, *Food Hydrocolloids*, 2012, **29**, 265–271.
125. S. Santacruz, C. Rivadeneira and M. Castro, *Food Hydrocolloids*, 2015, **49**, 89–94.
126. A. P. Teodoro, S. Mali, N. Romero and G. M. Carvalho, *Carbohydr. Polym.*, 2015, **126**, 9–16.
127. X. Li, C. Qiu, N. Ji, C. Sun, L. Xiong and Q. Sun, *Carbohydr. Polym.*, 2015, **121**, 155–162.
128. M. N. Abdorreza, L. H. Cheng and A. A. Karim, *Food Hydrocolloids*, 2011, **25**, 56–60.
129. L. Dai, C. Qiu, L. Xiong and Q. Sun, *Food Chem.*, 2015, **174**, 82–88.
130. N. Noshirvani, W. Hong, B. Ghanbarzadeh, H. Fasihi and R. Montazami, *Int. J. Biol. Macromol.*, 2018, **107**, 2065–2074.
131. H. P. N. Vu and N. Lumdubwong, *Carbohydr. Polym.*, 2016, **154**, 112–120.
132. G. T. Bersaneti, J. Mantovan, A. Magri, S. Mali and M. A. P. C. Celligoi, *Carbohydr. Polym.*, 2016, **151**, 1132–1138.
133. C. D. Poeloengasih, Y. Pranoto, S. N. Hayati, H. Hernawan, V. T. Rosyida, D. J. Prasetyo, T. H. Jatmiko, W. Apriyana and A. Suwanto, *AIP Conf. Proc.*, 2016, **1711**, 080003.

CHAPTER 10

Bio-based Polymers and Resins in Paints and Coatings

DEAN C. WEBSTER

Department of Coatings and Polymeric Materials, North Dakota State University, Fargo, ND 58108, USA
Email: Dean.Webster@ndsu.edu

10.1 Introduction

Paints and coatings are used extensively around the world for their protective and decorative properties. Coatings serve to protect the underlying material from degradation – increasing the material's lifetime – and also provide a good appearance. Paints are comprised of a binder – usually an organic resin or polymer – plus pigments to give color, solvents to regulate the viscosity, and numerous additives to further modify the properties. During the curing or drying process, paints convert from a liquid to a dry film. This process involves evaporation of the solvent, and, in many cases, a curing or crosslinking reaction. Historically, organic solvents were used; however, evaporation of organic solvents create VOCs – volatile organic compounds – which are now regulated in most parts of the world. Thus, a shift to paints using water as the carrier has occurred. While paints are applied as liquids, coatings can also be made from solids, with powder coatings being a prime example.

As observed by Lichtenthaler and Peters, prior to the development of the petrochemical industry in the early decades of the 20th century, most of the materials used by society were derived from renewable feedstocks – agricultural materials (see Figure 10.1).[1] The use of coal and then natural gas and oil resulted in a rapid introduction of many new chemicals that were quickly adopted for use by many industries, supplanting or generally

Green Chemistry Series No. 79
Bioplastics and Biocomposites: A Practical Introduction
Edited by David Grewell

Published by the Royal Society of Chemistry, www.rsc.org

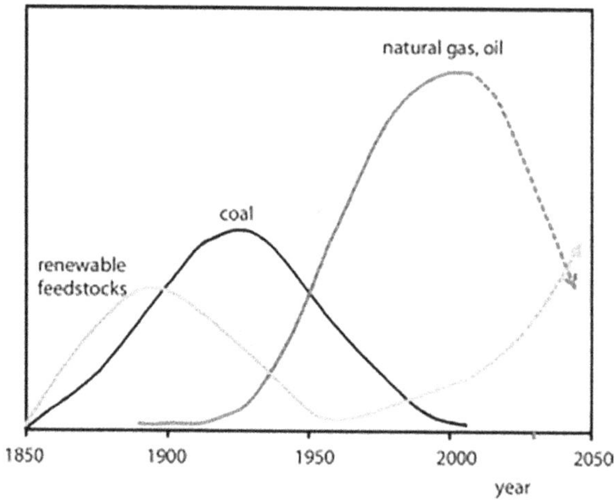

Figure 10.1 Use of raw materials in the chemical industry. Reproduced from ref. 1 with permission from Elsevier, Copyright 2004.

ignoring the use of renewable feedstocks for the production of chemicals of commerce.

The paint and coatings industry was no different. As far back as the Renaissance, paints were made using naturally occurring materials such as seed oils, milk, egg whites, rosin, and turpentine along with various inorganic pigments. As painting of houses and other buildings became popular in the 1800s and the paint industry grew to meet the demand, paints were still made using materials derived from nature such as linseed oil and turpentine. Paints based on nitrocellulose were essential to the success of the early automobile industry. And, as more petrochemical raw materials became available, resins that combined seed oils and petrochemical monomers – alkyd resins – came to the fore in the industry. However, as Lichtenthaler and Peters note, eventually polymers and resins derived from petrochemicals began to dominate the industry. Epoxy resins, acrylics, and polyurethanes demonstrated significant performance advantages in coatings over the resins derived from renewable resources.

However, today there is renewed interest in using renewable resources in paint and coatings systems. While initially driven by the potential of limited petrochemicals and rising prices, the current interest is from increasing awareness of concepts of sustainability and corporate sustainability goals, as well as the potential to have new materials with new property profiles. A challenge is that coatings derived from petrochemical materials have exceptional properties of durability and corrosion protection. New materials derived from sustainable raw materials have to be able to be competitive with the current petrochemical materials in terms of performance and price in order to be successful.

This chapter will first discuss some of the historical uses of materials derived from renewable resources – many of which are still in use to some extent today – and transition to a discussion of some new approaches to sustainable materials that are being introduced today. Some of the ongoing research in materials derived from renewable resources will then be surveyed.

10.2 Drying Oils

Vegetable or seed oils are extracted from various plant sources. Many of these naturally occurring oils have the ability to "dry" – that is, form a solid film – on their own and are thus called drying oils. Linseed oil, extracted from flax seeds, and tung oil, also known as China wood oil, extracted from the nut of the tung tree, are two important examples of drying oils.

Compositionally, these oils are triglycerides of fatty acids; three fatty acids are esterified to one molecule of glycerol (see Figure 10.2). Fatty acids are monofunctional carboxylic acids with a long linear aliphatic chain of 4 to 28 carbon atoms. Most of the commonly used oils have 14 to 22 carbon atoms in the fatty acid chains and these acids have different degrees of unsaturation. Different plants produce oils having differing fatty acid compositions. Drying occurs due to the reaction of the oil with atmospheric oxygen: a process known as autoxidation.[2,3] The level and type of unsaturation in the fatty acids dictates the type of drying behavior the oil will have. The more unsaturation, the more active is the drying behavior. Oils where the double bonds are conjugated usually dry faster than those oils where the double bonds are non-conjugated; however, most oils in use have non-conjugated fatty acids.

Linseed oil is obtained from the seeds of the flax plant, where we also get flax fibers and linen. Linseed oil has about 47% linolenic fatty acid (three double bonds), 24% linoleic acid (two double bonds), 19% oleic acid (one double bond) with the balance being saturated fatty acids. The high amount of unsaturated fatty acids is what lends linseed oil to drying behavior. The drying speed can be accelerated through the use of compounds called driers; essentially catalysts to increase the rate of autoxidation of the fatty acids.[2,4]

Linseed oil was used as a primary binder for paints used from the late 1800s through to the mid-20th century. Linseed oil was valued as a binder for its ability to flow and level and then dry to a hard and tack free paint film in a reasonable period of time. Paints containing white lead carbonate and zinc oxide along with a linseed oil binder were considered the best quality paints for houses and buildings in the early 20th century. As new types of resins and pigments were introduced, linseed oil-based paints fell out of favor.

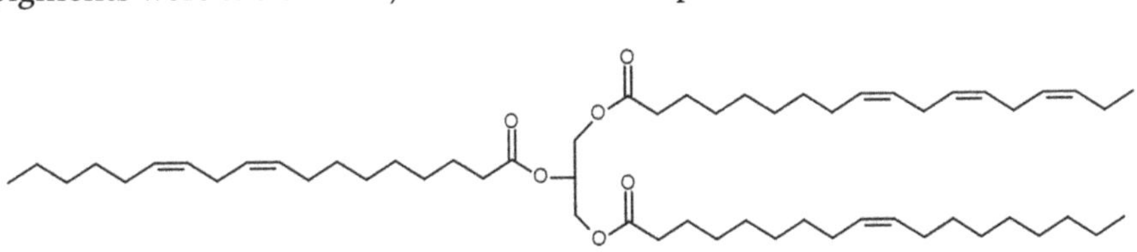

Figure 10.2 Example structure of a triglyceride oil.

10.3 Alkyd Resins

Alkyd resins were invented in the 1920s by Roy Kienle and coworkers at General Electric. Initially, Kienle studied reactions of glycerol and dibasic acids such as phthalic anhydride, and the term 'alkyd' was coined by combining alcohol and acid and hardening the "c" sound to a "k." Thus, originally the term alkyd would apply to any polyester resin. However, the discovery of a resin incorporating a fatty acid or oil, leading to an air-drying resin, led to the term alkyd resin being mainly used for those resins made from polyols, polyacids, and a fatty acid or oil.[5] Figure 10.3 shows a representative structure of an alkyd resin molecule.

Alkyd resins were found to be superior to drying oils or drying oil varnishes in paint formulations. They could dry faster and form harder films. This is due to the incorporation of other monomers into the resin such as the rigid aromatic monomer phthalic anhydride as well as increasing the degree of polymerization (*i.e.*, the molecular weight). Increasing the molecular weight results in the incorporation of numerous fatty acids into the polymer molecule and increases the effective functionality for curing by autoxidation. Increasing the number of functional groups per molecule results in faster drying because fewer chemical reactions are needed between chains to result in a crosslinked network. As a consequence, semi-drying oils such as soybean oil, dehydrated castor oil, and tall oil fatty acids could also be used in alkyd resins as well as drying oils such as linseed oils. Non-drying oils – for example, coconut oil – can also be used to form alkyd resins that cure through solvent evaporation or by reaction of hydroxyl groups with an amino resin.

Since alkyd resin technology is highly versatile, the technology developed rapidly in the 1930s and 40s and was the primary type of binder used in paints and coatings well into the 1980s. Many different compositional variations are possible with alkyd resins, enabling different compositions to be developed for different applications. Alkyd resins can be formulated with different amounts of oil/fatty acid content, known as the "oil length." Resins with higher amounts of oil (called "long oil" alkyds) are used for air drying coatings, while resins having lower amounts of oil (called "short oil" alkyds) are more commonly used for baking enamels by reacting their excess hydroxyl groups with curing agents such as amino resins. Different acid and polyol monomers can also be used to tailor the molecular structure of the alkyd resin and its properties. The molecular weight of the resin is varied by changing the ratio of polyol and acid monomers. A fairly sound

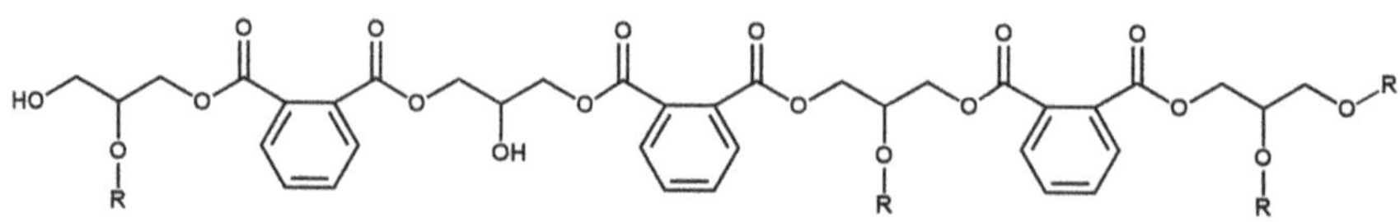

Figure 10.3 Representative structure of an alkyd resin where R represents a fatty acid chain.

mathematical basis for formulating alkyd resins (branching polymerization) was established initially by Carothers and elaborated by Flory and Patton.[6–8]

There are also additional variations of alkyd resins that broadened their utility. Alkyd resins can be grafted with monomers such as styrene to yield resins that have very fast drying times. Urethane alkyds are made using a diisocyanate in place of all or much of the diacid monomer. Urethane alkyds found use particularly in wood finishing where the hydrogen bonding of the polyurethane component results in fast drying and good abrasion resistance.

10.4 Cellulose-derived Resins

Cellulose is the most abundant biopolymer on earth and is a major component of woody plants, see Figure 10.4. A large volume of cellulose is obtained from the paper pulping process.[9] Cellulose is highly crystalline and difficult to dissolve in solvents; however, it can be functionalized, yielding materials having better solubility and film forming properties suitable for use in coatings.

Nitrocellulose is made by nitration of cellulose and can be used to make fast drying lacquers—coatings that dry by solvent evaporation. Nitrocellulose lacquers were used in the early part of the 20th century for automotive coatings and resulted in improved productivity due to their fast-drying nature. Nitrocellulose lacquers have also found use as wood coatings, especially for musical instruments, and special applications such as fingernail polish. While nitrocellulose lacquers produce a highly glossy finish, they oxidize rapidly when outdoors, requiring constant polishing. Another drawback of nitrocellulose in coatings is the large amount of solvent needed to obtain a suitable viscosity. Formulations made using nitrocellulose might only have 10–12 percent solids content. As environmental regulations led to reductions in the amount of solvent emitted from coatings, nitrocellulose lacquers fell out of favor. Another drawback of nitrocellulose is that it is highly flammable.

Cellulose esters can be made by esterifying the hydroxyl groups in cellulose with organic acids (or their anhydrides).[10] Cellulose acetate can be used for fiber and film applications. For use in coatings, cellulose acetate butyrates (CABs) or cellulose acetate propionates (CAPs) are used. While these can be used as primary film formers, it is more common for the CABs and CAPs to be used as additives. These help improve dry-to-touch times, film

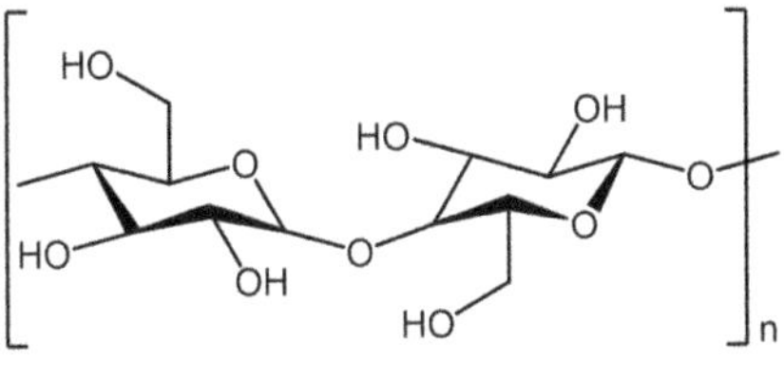

Figure 10.4 Structure of cellulose.

hardness, and UV stability. One particular use of CABs is to improve the orientation of metal flake in coating films.[11] Cellulose esters continue to be used in coatings systems.

Cellulose ethers, specifically hydroxyethyl cellulose (HEC), are made by reacting cellulose with ethylene oxide.[12] This renders HEC soluble in water; however, it forms highly viscous solutions. Thus, HEC is used as a thickener in aqueous paint formulations and as a protective colloid in emulsion polymerizations.[13] While newer thickener technologies have been introduced, HEC continues to be used in many aqueous paint systems.

10.5 Natural Resins

Many naturally occurring resins have been used through the ages as binders for paints and coatings. These include resins such as shellac, copal, and oriental (Japanese and Chinese) lacquers. Shellac is from the secretions of the female lac bug, found on trees in the forests of India and Thailand. The shellac is collected, filtered to remove impurities, and sold as either solid flakes or in a solvent, typically alcohol. Shellac was used as a wood varnish until it was superseded by nitrocellulose. Today, shellac is used as a stain blocking primer as well as for enteric and edible coatings. While oriental lacquers are still used in small quantities for specialized applications, there is some renewed interest in further understanding the chemistry behind the drying behavior as well as the good optical and physical properties of these lacquers.[3,14]

10.6 Recent Advances

With the current resurgence of interest in bio-based and renewable materials, significant research and commercialization activity is being carried out. Raw materials being used include seed/vegetable oils, compounds derived from cellulose and lignin, and chemicals derived from lignin. All of these areas have significant potential to transform materials from being based on petrochemicals to a renewable basis; however, significant "catching up" needs to be done to realize a fully bio-based materials economy. Fortunately, a lot is now known about material structure–property relationships that can be used to help guide the research toward commercial products. Plus, there are fairly firm economic targets in place that help to guide research and commercialization efforts.

The following provides an overview of some of the approaches being explored to derive materials for coatings binder systems derived from renewables.

10.7 Vegetable/Seed Oils in Coatings Cured by Autoxidation

While alkyd resins are still in wide use as a coatings binder, there is a resurgence in interest in alkyds and alkyd-like resins due to the fact that they are partially based on renewable resources.[15] The fact that unsaturated fatty acids can undergo autoxidative curing is a benefit for coatings systems because a one-component coating can be made that cures upon exposure to atmospheric oxygen.

Alkyds typically require the use of solvents to attain viscosity suitable for application. As environmental regulations are limiting the amount of solvent that can be used in paints and coatings, it has been challenging to redesign alkyds to have lower viscosity and still maintain suitable film properties. One way to overcome this problem is to use water as the main carrier. Alkyd emulsions have been developed and commercialized where the alkyd resin is dispersed in water using high shear mixing and a surfactant to stabilize the particles.[16–18] Alkyd emulsions form coatings films that can dry quickly and have high gloss, which is preferred in some regions over the medium gloss that can be achieved from latex paints.

There has been interest in incorporating unsaturated fatty acids into latex binder systems using free radical emulsion polymerization. The objective is to have a latex that can further cure through autoxidation after film formation and lead to coatings films having improved durability. Chen and Bufkin reported on the incorporation of unsaturated fatty acid monomers into a latex.[19,20] While autoxidative crosslinking was observed, it was found that the plasticizing effect of the fatty acid overcame any film property improvements gained through curing.

In a series of studies, the group of Thames explored the use of methacrylated and acrylated monomers derived from vegetable oils such as castor and soybean in latex polymers *via* either emulsion polymerization or miniemulsion polymerization.[21–32] Improvements in the properties of the coating films, such as the glass transition temperature, could be observed due to the oxidative crosslinking of the fatty acids. A challenge observed is that during the free radical polymerization, chain transfer to the unsaturated fatty acids can occur and reduce the rate of polymerization.[25]

The group of Voronov has extensively studied the incorporation of a novel vegetable oil based methacrylamide monomer in latexes using both conventional and miniemulsion polymerization methods.[33–41] Using a range of vegetable oils having differing degrees of unsaturation, the effect of chain transfer on the polymerization kinetics has been elucidated.[34–36] The use of methyl-β-cyclodextrin was explored as a method to improve the transport of the hydrophobic fatty acid monomer during emulsion polymerization, and it was also discovered that it could mitigate the effects of chain transfer, thus overcoming a long-observed issue in making latexes with unsaturated fatty acids.[33]

A different type of fatty acid functional molecule is sucrose ester resins. Sucrose ester resins are made from the esterification of fatty acids – for example from soybean oil – onto the hydroxyl groups of sucrose. This results in an alkyd-like molecule that has a low viscosity. While explored initially in the 1960s, these resins did not achieve commercial use at that time.[42–44] More recently, Procter and Gamble developed a process for the efficient manufacture of sucrose ester resins having a high degree of fatty acid substitution on the sucrose, around 7.7 of the eight hydroxyl groups.[45,46] Sucrose ester resins can undergo autoxidative curing and have been explored as reactive diluents for alkyd coatings, enabling the reduction of VOC from solvent.[47] As seen in Figure 10.5, incorporation of sucrose octaester into a coating made from a medium oil alkyd resin results in improvements in drying time, VOC content, solvent resistance, and impact resistance, with only a small reduction in film hardness.

The Chisholm group has synthesized novel fatty acid vinyl ether monomers from oils such as soybean oil by transesterification of the oil with a hydroxyalkyl vinyl ether. Using cationic polymerization, the vinyl ether monomer can be polymerized or copolymerized to generate a polymer having pendant fatty acid chains.[48–50] Figure 10.6 shows the reaction scheme for the synthesis of the poly(vinyl ether) polymers. An advantage of this approach is that the unsaturation in the fatty acids is not involved in the cationic polymerization and thus is preserved and can be used to generate coatings that cure through autoxidation.[39,51–54] Since relatively high molecular weight polymers can be made using cationic polymerization, the oxidative curing speed of the polymers is much more rapid than that of the oil alone or of a typical alkyd resin.

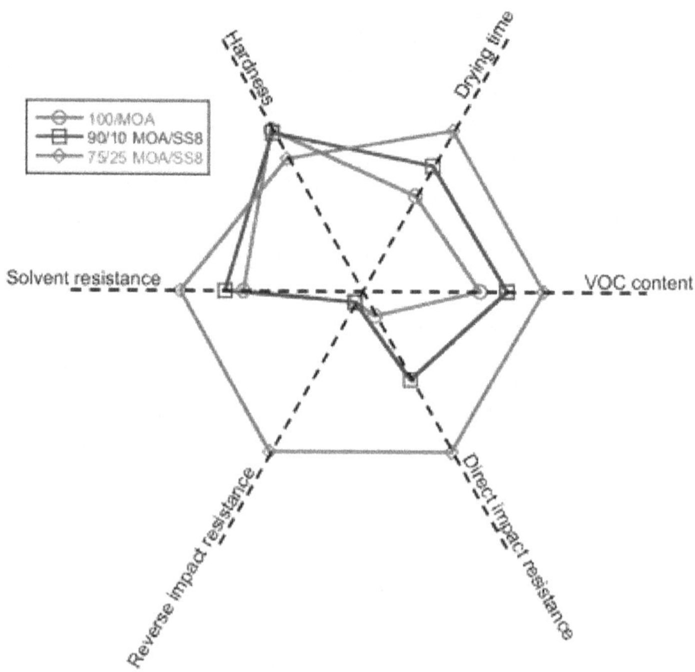

Figure 10.5 Comparison of properties of a medium oil alkyd (MOA) with blends of the MOA with sucrose octasoyate (SS8). Reproduced from ref. 47 with permission from Springer Nature, Copyright 2018.

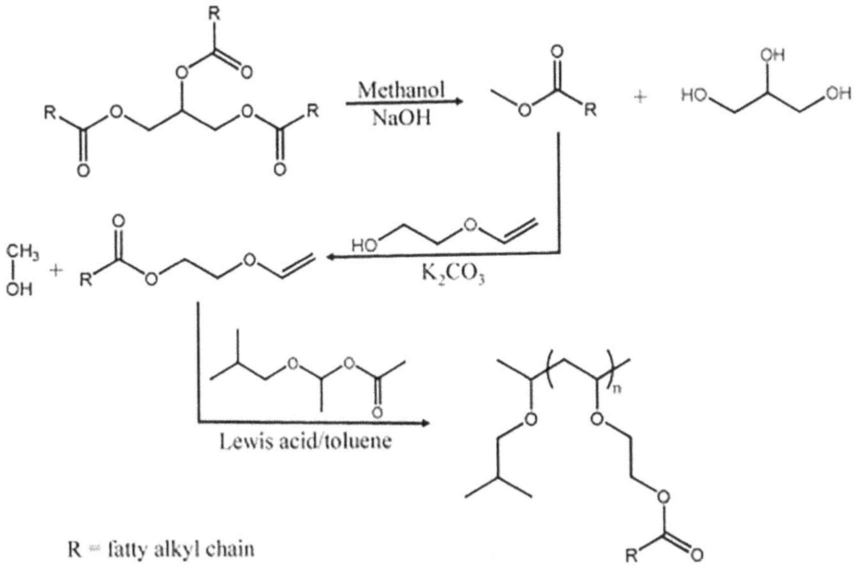

Figure 10.6 Reaction scheme for the synthesis of poly(vinylether) polymers from seed oils. Reproduced from ref. 52 with permission from Elsevier, Copyright 2018.

10.8 Derivatives of Vegetable/Seed Oils

Due to the presence of double bonds in the fatty acids of vegetable/seed oils, various chemical reactions can be employed to convert these into other reactive functional groups. One of the most popular reactions is epoxidation, which converts the double bonds into epoxy groups. Further derivatives can be made through further reaction of the epoxy groups.

While epoxidized oils have a multiplicity of epoxy groups, they cannot be used as drop-in replacements for conventional epoxy resins used in coatings such as the bisphenol-A based epoxy resins. This is due to the fact that the epoxy groups in an epoxidized oil are disubstituted and much less reactive than the terminal epoxy groups in conventional epoxy resins. However, epoxidized oils have been able to be cured into materials having reasonable properties using photoinitiated cationic polymerization as well as through catalytic cationic polymerization.[55–59]

The Webster group has explored the synthesis and use of derivatives of sucrose ester resins. Sucrose ester resins can be epoxidized to yield compact resins having a large number of reactive groups and moderate viscosity (see Figure 10.7).[60] The epoxy resins can be cured using cyclic anhydrides and through polymerization of the epoxy groups, either using photoinitiation or thermal initiation.[61–63] Due to the high number of functional groups and the rigid sucrose core, the thermosets have good hardness, strength, and solvent resistance. Completely bio-based thermosets can be made by curing the epoxidized sucrose ester resin with naturally occurring carboxylic

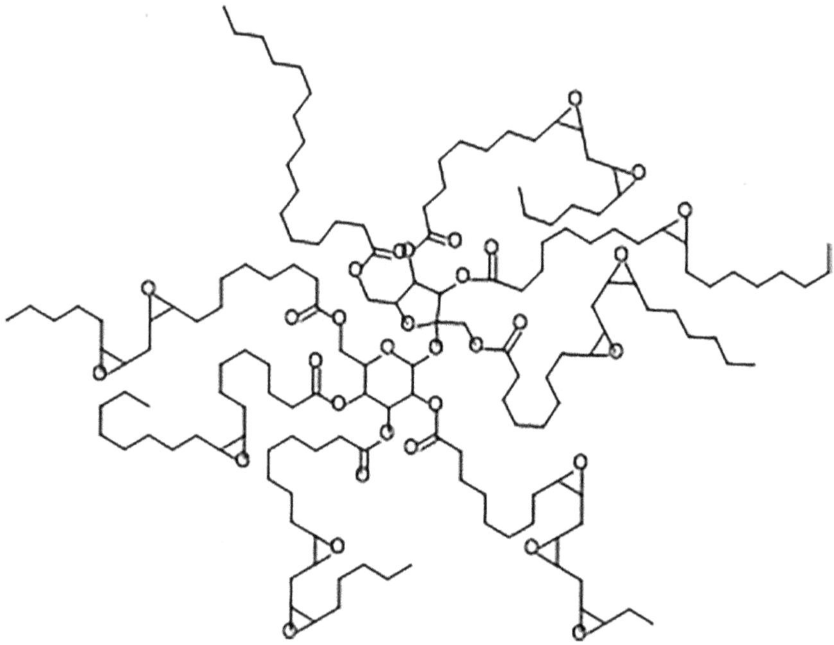

Figure 10.7 Representative structure of epoxidized sucrose soyate (ESS).

acids, such as citric or tartaric acids.[64,65] Water and polar organic solvents can be used to catalyze the curing reaction. The use of an acid functional oligomer based on isosorbide was also explored as a curing agent, and it was found that a coating with a good combination of high hardness while having excellent flexibility could be achieved.[66] Kalita *et al.* also showed that the double bonds in the fatty acids of the soy-based polyvinyl ether polymers can be epoxidized and crosslinked to yield coatings having high hardness and glass transition temperature.[54]

Epoxidized oils can be further reacted with unsaturated acids such as acrylic or methacrylic acid to yield resins that can be cured using free radical methods, such as photopolymerization.[67–69] Acrylated epoxidized soybean and linseed oils are available commercially. Similarly, acrylated and methacrylated epoxidized sucrose soyate have been synthesized and studied as photopolymerized coatings, see Figure 10.8.[70,71]

Since the majority of the fatty acids in castor oil are from ricinoleic acid, which has a double bond and a hydroxyl group, castor oil is a naturally occurring polyol. Castor oil can be used in the preparation of polyurethane coatings.[72] Similarly, polyols can also be made from vegetable oils through various routes, typically starting with epoxidized oils and then reacting with an alcohol.[73] However, while polyester and acrylic polyols are commonly used with curing agents such as melamine-formaldehyde resins and polyisocyanates in coatings, vegetable oil polyols have a somewhat limited use in coatings systems due to their lower reactivity (due to the secondary hydroxyl

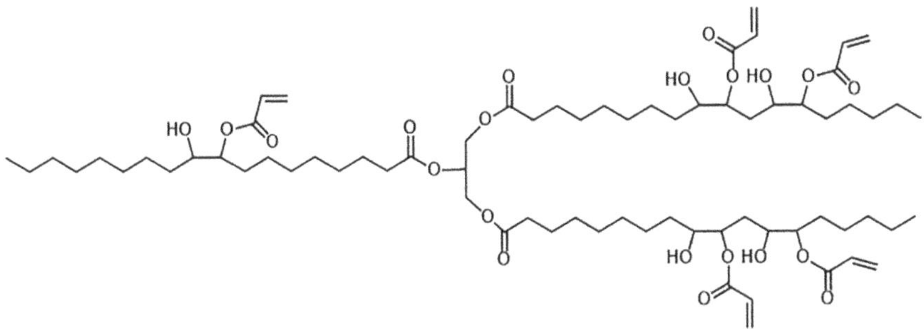

Figure 10.8 Representative structure of acrylated epoxidized soybean oil (AESO).

groups) and low film hardness due to their aliphatic character. Using a hydroformylation-reduction route to form vegetable oil polyols can result in primary alcohols to increase the reactivity.[74] Vegetable oil polyols have found their primary usage in polyurethane foams.

Researchers at Dow Chemical developed novel vegetable oil-based polyols from first synthesizing an AB monomer from the hydroformylation-reduction of methyl oleate, followed by oligomerization using a diol as an initiating molecule.[75] These polyols were evaluated in several different types of coatings formulations as well as in polyurethane dispersions (PUD). When blended with acrylic polyols, the use of the novel polyols led to VOC reduction. In PUDs, the novel polyols showed improved hydroxylic stability and acid etch resistance.

Polyols were also made by reacting epoxidized sucrose soyate with an alcohol or carboxylic acid. The polyols can be used in 2k and 1k polyurethane coatings as well as in melamine-formaldehyde cured coatings.[76–78] In stark contrast to coatings made using conventional soy-based polyols, the coatings made from the methoxylated epoxidized sucrose soyate polyols demonstrated excellent hardness, solvent resistance, and high glass transition temperature – properties on par with coatings made using petrochemical polyols. This is attributed to the higher crosslink density that can be obtained from these highly functional polyols.

Aqueous PUDs are typically made by reacting a flexible polyol and a diol containing a hydrophilic group with a diisocyanate to form a prepolymer, followed by dispersion in water and chain extension.[79–81] PUDs have become a versatile technology and have found application in paints and coatings, leather finishing, textile finishing, and many other applications. The use of vegetable oil-based polyols as replacements for the flexible polyol in PUDs has been studied extensively.[82,83] Larock *et al.*, in a series of studies, incorporated vegetable oils as the soft segments into polyurethane dispersions as well as hybrid polyurethane–acrylic waterborne polymers.[84–89] Further work by Kessler studied rheological properties of castor oil-based polyurethane dispersions.[90] Antibacterial coatings were also made from vegetable oil-based cationic polyurethane dispersions.[91,92]

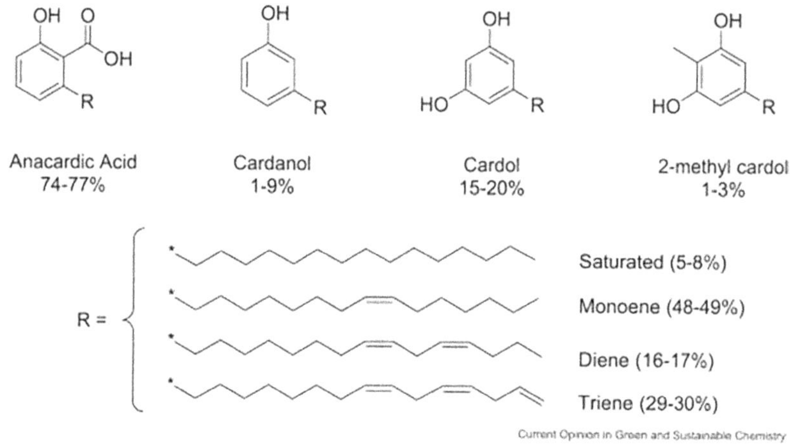

Figure 10.9 Constituents of cashew nutshell liquid (CNSL). Reproduced from ref. 93 with permission from Elsevier, Copyright 2018.

Cardanol is one of the constituents of cashew nutshell liquid (CNSL). It is a unique molecule that has a phenolic group at the head and an unsaturated hydrocarbon tail. Figure 10.9 shows the constituents of CNSL.[93] During purification, anacardic acid decarboxylates to cardanol, increasing the yield of cardanol from CNSL. Cardanol can be functionalized with epoxy groups to be used in coatings and other products. The Cardolite company has commercialized a monofunctional epoxy resin, a difunctional epoxy made from a cardanol dimer, and a multifunctional epoxy made by reacting cardanol with formaldehyde. A study of the difunctional cardanol epoxy resin compared to bisphenol-A epoxy resins in amine cured coatings showed that the cardanol-based coating had lower glass transition temperature and mechanical properties than the bisphenol-A based system due to the long aliphatic chains.[94] A number of other derivatives of cardanol continue to be explored.[93]

10.9 Monomers and Polymers Derived from Cellulose

A number of routes are being explored to derive chemicals from cellulose using various processes of fermentation and derivatization.[95] Some of the important monomers that are being explored for potential use in coatings are shown in Figure 10.10.

Furan dicarboxylic acid is viewed as a potential replacement for terephthalic acid. While the main emphasis is on its use in thermoplastics for packaging, some limited evaluation in polyester polyols used in coatings has been carried out.[96–99] While isosorbide is an attractive polymer building block due to its cycloaliphatic structure, the reactivity of the hydroxyl groups is poor, which makes it difficult to incorporate into a polyester. However, several papers have shown that isosorbide is a useful component of a polyester for use in coatings applications.[66,99–101]

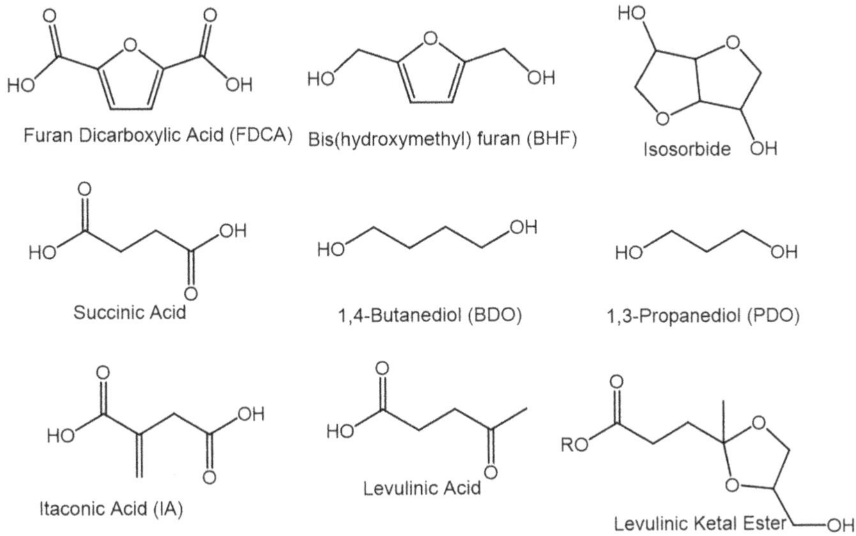

Figure 10.10 Structures of monomers derived from cellulose.

Succinic acid has been produced from petrochemicals by the reduction of maleic or fumaric acid. Several companies developed routes to make bio-based succinic acid from cellulose.[102] Succinic acid can be used to make polyester polyols and has some properties similar to adipic acid but may yield higher modulus and better abrasion resistance than an adipate in polyurethane dispersions.[103,104]

Succinic acid can be further reduced to make bio-based 1,4-butanediol, another known petrochemical. Bio-based 1,3-propanediol has been commercialized by DuPont Tate & Lyle.[105–107] These polyols can potentially be used in the preparation of polyester polyols for use in coatings applications or as soft segment polyols in polyurethane dispersions; however, their utility in coatings applications remains to be seen.

A bio-based route to the synthesis of itaconic acid has also been developed.[108] With its two acid groups, it can be incorporated into polyesters and its unsaturated groups can be polymerized *via* a free radical mechanism. One avenue of exploration is the use of itaconic acid in polyesters and polyurethanes that can be cured using UV light.[109–113]

Levulinic acid is a unique chemical having carboxylic acid and ketone groups in its structure that can be produced from cellulose.[114] One application of levulinic acid is to convert it to methyl tetrahydrofuran, which is considered as a green replacement for tetrahydrofuran. In polymer chemistry and coatings, several derivatives are being explored. Levulinic acid and its esters can be reacted with two moles of phenol to result in diphenolic acid/esters, which can then be reacted with epichlorohydrin to create novel epoxy resins.[115,116] These can be cured with amines to form coatings having properties comparable to those made from bisphenol-A epoxy resins.

Levulinic ketal esters can be synthesized from the reaction of levulinic ester with glycerol. This creates a novel A–B type monomer that can be reacted to form polyesters that are useful as plasticizers or for polyurethanes.[117] However, there do not appear to be significant studies of these applications in the open literature at this time.

Lignin is the earth's most abundant aromatic biopolymer and is isolated from woody plants during the extraction of cellulose, for example, in the paper pulping process. While there is interest in being able to extract aromatic chemicals from lignin, due to its complex structure, obtaining specific aromatic chemicals from lignin at high yields has proven challenging.[118–120] However, several chemicals that can be derived from lignin, such as vanillin and its derivatives, eugenol, guaiacol, and several others, have been explored as building blocks for resins that can be used in coatings systems.

Stanzione and coworkers synthesized a novel bisphenol from the reaction of vanillyl alcohol with guaiacol. This bisphenol was then reacted with epichlorohydrin to form a novel epoxy resin (see Figure 10.11).[121] When cured with *p*-aminocyclohexyl methane, the properties of the thermoset were similar to those of a bisphenol-A epoxy resin with higher elastic modulus and a bit lower glass transition temperature due to the plasticizing effects of the methoxy side chains.

The group of Caillol has carried out extensive work on the use of vanillin and derivatives of vanillin – vanillyl alcohol and vanillic acid – as functional components in thermosetting systems, including coatings applications.[122–128] A summary of the work is shown in Figure 10.12 and illustrates the synthesis of various compounds having epoxy, allyl, amine, thiol, cyclic carbonate, hydroxyl, and carboxylic acid groups that can be used for forming crosslinked networks. Due to the aromatic core derived from vanillin, many of the thermosets derived from these functional compounds have similar strength and thermal properties to their petrochemical counterparts and thus could have promising potential in material applications. However, it is not currently known if vanillin and other chemicals that can be derived from lignin will be able to be produced in sufficient quantities at an economical cost.

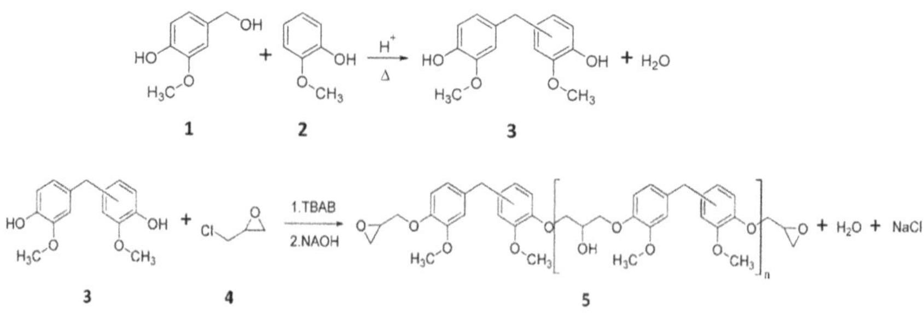

Figure 10.11 Synthesis of novel bio-based bisphenol (top) and epoxy resin (bottom). Reproduced from ref. 121 with permission from American Chemical Society, Copyright 2016

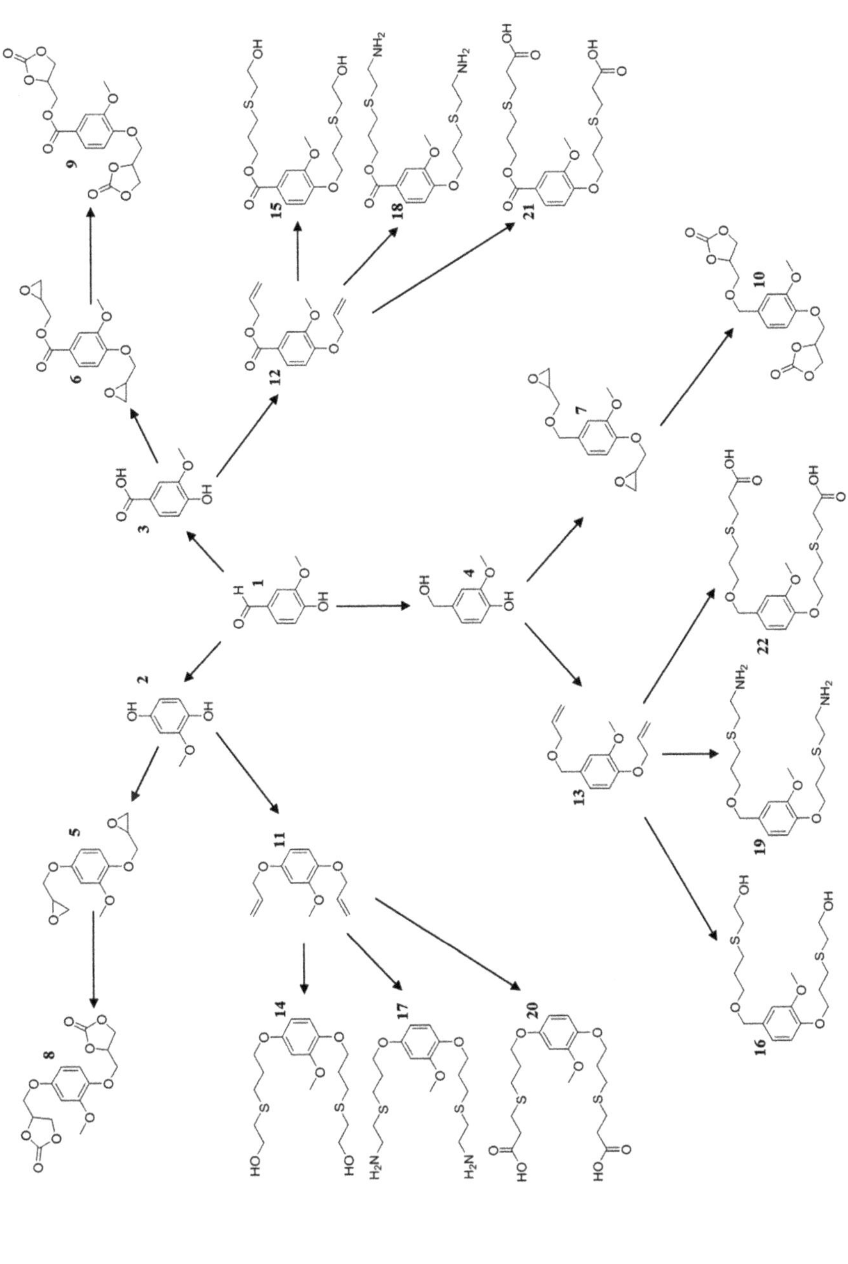

Figure 10.12 Derivatives of vanillin with functional groups for further curing and crosslinking. Reproduced from ref. 128 with permission from the Royal Society of Chemistry.

10.10 Summary and Outlook

The paint and coatings industry has historically relied on resins and polymers derived from naturally occurring or bio-based raw materials. Resins derived from vegetable and seed oils, cellulose, and other naturally occurring materials have been used in the far past, and many materials derived from seed oils and cellulose continue to be used today. Many petrochemical technologies are in use today and provide coatings with high levels of durability and corrosion protection coupled with excellent appearance. With increasing interest in sustainability, there has been a surge in research into bio-based chemicals that can be used in resin and binder systems for coatings. Much of this research has been exploratory in nature – that is, exploring the fundamental chemistry of synthesis of the resins and some preliminary studies of the performance of coating films. However, in order to move this work to commercial viability, there are a number of hurdles that must be overcome. First, many of the current petrochemical materials are highly optimized and well-understood. Further, the production facilities have been in existence and are also efficient, so that the costs of these resins are at a minimum and in line with demand and performance. Any new bio-based chemical or resin needs to be cost competitive with existing materials to be successful as a replacement for large volume resins. If these new chemicals or resins need new production facilities or have new processes, achieving cost parity can be very challenging.

Another consideration is material performance relative to current petrochemical materials. Today, introducing a bio-based material just because it is bio-based will probably not be commercially successful. To be commercially successful, either the cost must be lower at a comparable performance to current materials or there must be some performance advantage to the bio-based material. Thus, while carrying out fundamental exploratory research on bio-based materials may still have some value, it is important to consider what advantage the bio-based material might have over current materials. In many cases, chemicals and resins can be made that have compositions and structures with no petrochemical counterpart. What advantages might that bring to performance? And then, thinking about potential performance needs for coatings resin technology, what advantages and improvements might be brought by bio-based materials that cannot be realized with current petrochemical technology? Keeping these considerations in mind will result in successful development of new types of resins and binders for coatings based on renewable materials and increase materials sustainability for the future.

Abbreviations

CAB	Cellulose acetate butyrates
CAP	Cellulose acetate propionates
CNSL	Cashew nutshell liquid
HEC	Hydroxyethyl cellulose

PUD	Polyurethane dispersions
VOC	Volatile organic compound

References

1. F. W. Lichtenthaler and S. Peters, Carbohydrates as green raw materials for the chemical industry, *C. R. Chim.*, 2004, **7**, 65–90.
2. M. D. Soucek, T. Khattab and J. Wu, Review of autoxidation and driers, *Prog. Org. Coat.*, 2012, **73**, 435–454.
3. J. Honzíček, Curing of Air-Drying Paints: A Critical Review, *Ind. Eng. Chem. Res.*, 2019, **58**, 12485–12505.
4. J. Mallegol, J. Lemaire and J. L. Gardette, Drier influence on the curing of linseed oil, *Prog. Org. Coat.*, 2000, **39**, 107–113.
5. F. N. Jones, Alkyd Resins, *Ullmann's Encyclopedia of Industrial Chemistry*, 2003.
6. W. H. Carothers, Polymers and polyfunctionality, *Trans. Faraday Soc.*, 1936, **32**, 39–49.
7. P. J. Flory, *Principles of Polymer Chemistry*, Cornell University Press, Ithaca, 1953.
8. T. C. Patton, *Alkyd Resin Technology: Formulating Techniques and Allied Calculations*, Interscience Publishers, New York, 1962.
9. D. Klemm, B. Heublein, H.-P. Fink and A. Bohn, Cellulose: Fascinating biopolymer and sustainable raw material, *Angew. Chem., Int. Ed.*, 2005, **44**, 3358–3393.
10. K. J. Edgar, C. M. Buchanan, J. S. Debenham, P. A. Rundquist, B. D. Seiler, M. C. Shelton and D. Tindall, Advances in cellulose ester performance and application, *Prog. Polym. Sci.*, 2001, **26**, 1605–1688.
11. D. Bhattacharya, K. Seo, L. T. Germinario, M. D. Clark, K. W. McCreight and C. Williams, Novel techniques to investigate the impact of cellulose esters on the rheological properties and appearance in automotive basecoat systems, *J. Coat. Technol. Res.*, 2007, **4**, 139–150.
12. T. G. Majewicz, P. E. Erazo-Majewicz and T. J. Podlas, Cellulose Ethers, *Encyclopedia of Polymer Science and Technology*, 2002.
13. L. M. Zhang, Cellulosic associative thickeners, *Carbohydr. Polym.*, 2001, **45**, 1–10.
14. R. Lu, T. Yoshida and T. Miyakoshi, Oriental Lacquer: A Natural Polymer, *Polym. Rev.*, 2013, **53**, 153–191.
15. A. Hofland, Alkyd resins: From down and out to alive and kicking, *Prog. Org. Coat.*, 2012, **73**, 274–282.
16. A. Hofland, Making paint from alkyd emulsions, *ACS Symp. Ser.*, 1997, **663**, 183–195.
17. A. Hofland, Making alkyd emulsion paints, *Pigm. Resin Technol.*, 1994, **23**, 7–11.
18. A. Hofland and F. J. Schaap, Alkyd emulsions for high gloss paint systems: old properties in new particles, *Surface Phenomena and Fine*

Particles in Water-Based Coatings and Printing Technology, [Proc. Fine Part. Soc. Symp.], 1991, pp. 129–144.
19. F. B. Chen and B. G. Bufkin, Crosslinkable emulsion polymers by autoxidation. I. Reactivity ratios, *J. Appl. Polym. Sci.*, 1985, **30**, 4571–4582.
20. F. B. Chen and B. G. Bufkin, Crosslinkable emulsion polymers by autoxidation. II, *J. Appl. Polym. Sci.*, 1985, **30**, 4551–4570.
21. C. L. King, O. W. Smith and S. F. Thames, Acrylic latex properties using CAM monomer, *26th Proceedings of the International Waterborne, High-Solids, and Powder Coatings Symposium*, 1999, pp. 100–112.
22. E. H. Brister, O. W. Smith and S. F. Thames, Castor acrylated monomer (CAM) in vinyl-acrylic latexes, *26th Proceedings of the International Waterborne, High-Solids, and Powder Coatings Symposium*, 1999, pp. 89–98.
23. E. Kaya, S. K. Mendon, D. Delatte, J. W. Rawlins and S. F. Thames, Emulsion Copolymerization of Vegetable Oil Macromonomers Possessing both Acrylic and Allylic Functionalities, *Macromol. Symp.*, 2013, **324**, 95–106.
24. E. Kaya, S. K. Mendon, J. W. Rawlins and S. F. Thames, Emulsion polymerization of a vegetable oil macromonomer based on soybean oil and evaluation of the coating performance, *38th Proceedings of the International Waterborne, High-Solids, and Powder Coatings Symposium*, 2011, pp. 194–204.
25. G. Booth, D. E. Delatte and S. F. Thames, Incorporation of drying oils into emulsion polymers for use in low-VOC architectural coatings, *Ind. Crops Prod.*, 2007, **25**, 257–265.
26. K. L. Diamond, S. N. Shah, O. W. Smith and S. F. Thames, Kinetic investigation of novel vegetable oil derived macromonomers in both bulk and emulsion polymerization, *30th Proceedings of the International Waterborne, High-Solids, and Powder Coatings Symposium*, 2003, pp. 431–443.
27. S. F. Thames, J. W. Rawlins and S. K. Mendon, *Miniemulsion Polymerization Technology*, Scrivener Publishing, 2010, pp. 139–172.
28. C. Quintero, S. K. Mendon, O. W. Smith and S. F. Thames, Miniemulsion polymerization of vegetable oil macromonomers, *Prog. Org. Coat.*, 2006, **57**, 195–201.
29. E. H. Brister, T. Johnston, C. L. King and S. F. Thames, New monomers from vegetable oils, *ACS Symp. Ser.*, 2000, **755**, 159–169.
30. C. L. King, O. W. Smith and S. F. Thames, One-component, low VOC, ambient cure architectural coatings formulated using CAM acrylic latexes, *28th Proceedings of the International Waterborne, High-Solids, and Powder Coatings Symposium*, 2001, pp. 143–157.
31. B. Ghosh and M. W. Urban, Self-Repairing Oxetane-Substituted Chitosan Polyurethane Networks, *Science*, 2009, **323**, 1458–1460.
32. G. Hao, S. K. Mendon, D. E. Delatte, J. W. Rawlins and S. F. Thames, Structure-property relationships of self-crosslinking vegetable oil macromonomer-based latex copolymers, *35th Proceedings of the International Waterborne, High-Solids, and Powder Coatings Symposium*, 2008, pp. 57–68.

33. A. Kohut, Z. Demchuk, K. Kingsley, S. Voronov and A. Voronov, Dual role of methyl-β-cyclodextrin in the emulsion polymerization of highly hydrophobic plant oil-based monomers with various unsaturations, *Eur. Polym. J.*, 2018, **108**, 322–328.
34. K. Kingsley, O. Shevchuk, Z. Demchuk, S. Voronov and A. Voronov, The features of emulsion copolymerization for plant oil-based vinyl monomers and styrene, *Ind. Crops Prod.*, 2017, **109**, 274–280.
35. Z. Demchuk, O. Shevchuk, I. Tarnavchyk, V. Kirianchuk, A. Kohut, S. Voronov and A. Voronov, Free Radical Polymerization Behavior of the Vinyl Monomers from Plant Oil Triglycerides, *ACS Sustainable Chem. Eng.*, 2016, **4**, 6974–6980.
36. Z. Demchuk, O. Shevchuk, I. Tarnavchyk, V. Kirianchuk, M. Lorenson, A. Kohut, S. Voronov and A. Voronov, Free-Radical Copolymerization Behavior of Plant-Oil-Based Vinyl Monomers and Their Feasibility in Latex Synthesis, *ACS Omega*, 2016, **1**, 1374–1382.
37. Z. Demchuk, V. Kirianchuk, K. Kingsley, S. Voronov and A. Voronov, Plasticizing and Hydrophobizing Effect of Plant OilBased Acrylic Monomers in Latex Copolymers with Styrene and Methyl Methacrylate, *Int. J. Theor. Appl. Nanotechnol.*, 2018, **6**, 29–37.
38. Z. Demchuk, W. S. J. Li, H. Eshete, S. Caillol and A. Voronov, Synergistic Effects of Cardanol- and High Oleic Soybean Oil Vinyl Monomers in Miniemulsion Polymers, *ACS Sustainable Chem. Eng.*, 2019, **7**, 9613–9621.
39. H. Kalita, S. Alam, D. Kalita, A. Chernykh, I. Tarnavchyk, J. Bahr, S. Samanta, A. Jayasooriyama, S. Fernando, S. Selvakumar, A. Popadyuk, D. S. Wickramaratne, M. Sibi, A. Voronov, A. Bezbaruah and B. J. Chisholm, Synthesis and characterization of novel soybean oil-based polymers and their application in coatings cured by autoxidation, *ACS Symp. Ser.*, 2014, **1178**, 371–390.
40. I. Tarnavchyk, A. Popadyuk, N. Popadyuk and A. Voronov, Synthesis and free radical copolymerization of a vinyl monomer from soybean oil, *ACS Sustainable Chem. Eng.*, 2015, **3**, 1618–1622.
41. Z. Demchuk, A. Kohut, S. Voronov and A. Voronov, Versatile Platform for Controlling Properties of Plant Oil-Based Latex Polymer Networks, *ACS Sustainable Chem. Eng.*, 2018, **6**, 2780–2786.
42. E. G. Bobalek, Prospects and potential for commercial production and utilization of sucrose fatty acid esters, *ACS Symp. Ser.*, 1977, **41**, 162–175.
43. E. G. Bobalek, A. P. DeMendoza, A. G. Causa, W. J. Coilings and G. Kapo, Preparation and properties of linoleate esters of sucrose, *Ind. Eng. Chem. Prod. Res. Dev.*, 1963, **2**, 9–16.
44. E. G. Bobalek, T. J. Walsh and H.-H. Chiang, Drying oil esters of sucrose, *Off. Dig., Fed. Soc. Paint Technol.*, 1961, **33**, 453–468.
45. J. K. Howie, J. J. Schaefer and J. E. Trout, Synthesis of polyol medium fatty acid polyesters, US6995232, 2006.
46. J. J. Schaefer and J. E. Trout, Synthesis of purified, partially esterified polyester fatty acid compositions, US6887947, 2005.

47. A. Popadyuk, A. Breuer, J. Bahr, I. Tarnavchyk, A. Voronov and B. J. Chisholm, Sucrose octaesters as reactive diluents for alkyd coatings, *J. Coat. Technol. Res.*, 2018, **15**, 481–488.
48. S. Alam, H. Kalita, A. Jayasooriya, S. Samanta, J. Bahr, A. Chernykh, M. Weisz and B. J. Chisholm, 2-(Vinyloxy)ethyl soyate as a versatile platform chemical for coatings: An overview, *Eur. J. Lipid Sci. Technol.*, 2014, **116**, 2–15.
49. S. Alam and B. J. Chisholm, Coatings derived from novel, soybean oil-based polymers produced using carbocationic polymerization, *J. Coat. Technol. Res.*, 2011, **8**, 671–683.
50. A. Chernykh, S. Alam, A. Jayasooriya, J. Bahr and B. J. Chisholm, Living carbocationic polymerization of a vinyl ether monomer derived from soybean oil, 2-(vinyloxy)ethyl soyate, *Green Chem.*, 2013, **15**, 1834–1838.
51. H. Kalita, S. Selvakumar, A. Jayasooriyamu, S. Fernando, S. Samanta, J. Bahr, S. Alam, M. Sibi, J. Vold, C. Ulven and B. J. Chisholm, Bio-based poly(vinyl ether)s and their application as alkyd-type surface coatings, *Green Chem.*, 2014, **16**, 1974–1986.
52. D. J. Kalita, I. Tarnavchyk, M. Sibi, B. R. Moser, D. C. Webster and B. J. Chisholm, Bio-based poly(vinyl ether)s derived from soybean oil, linseed oil, and camelina oil: Synthesis, characterization, and properties of crosslinked networks and surface coatings, *Prog. Org. Coat.*, 2018, **125**, 453–462.
53. H. Kalita, S. Alam, D. Kalita, A. Jayasooriyamu, S. Fernando, S. Samanta, J. Bahr, S. Selvakumar, M. Sibi, J. Vold, C. Ulven and B. J. Chisholm, Novel alkyd-type coating resins produced using cationic polymerization, *J. Coat. Technol. Res.*, 2015, **12**, 633–646.
54. H. Kalita, D. Kalita, S. Alam, A. Chernykh, I. Tarnavchyk, J. Bahr, S. Samanta, A. Jayasooriyama, S. Fernando, S. Selvakumar, D. S. Wickramaratne, M. Sibi and B. J. Chisholm, in *Bio-based and Environmental Benign Coatings*, ed. A. Tiwari, A. Galanis and M. D. Soucek, Scrivener Publishing LLC, 2016, pp. 1–16.
55. J. V. Crivello and R. Narayan, Epoxidized triglycerides as renewable monomers in photoinitiated cationic polymerization, *Chem. Mater.*, 1992, **4**, 692–699.
56. S. Chakrapani and J. V. Crivello, Synthesis and Photoinitiated Cationic Polymerization Of Epoxidized Castor Oil and Its Derivatives, *J. Macromol. Sci., Part A*, 1998, **35**, 1–20.
57. H. Gu, K. Ren, D. Martin, T. Marino and D. C. Neckers, Cationic UV-cured coatings containing epoxidized soybean oil initiated by new onium salts containing tetrakis(pentafluorophenyl)gallate anion, *J. Coat. Technol.*, 2002, **74**, 49–52.
58. Z. Liu and S. Z. Erhan, Ring-Opening Polymerization of Epoxidized Soybean Oil, *J. Am. Oil Chem. Soc.*, 2010, **87**, 437–444.
59. R. Raghavachar, G. Sarnecki, J. Baghdachi and J. Massingill, Cationic, thermally cured coatings using epoxidized soybean oil, *J. Coat. Technol.*, 2000, 72, 125–133.

60. X. Pan, P. Sengupta and D. C. Webster, Novel bio-based epoxy compounds: epoxidized sucrose esters of fatty acids, *Green Chem.*, 2011, **13**, 965–975.
61. X. Pan, P. Sengupta and D. C. Webster, High Bio-based Content Epoxy-Anhydride Thermosets from Epoxidized Sucrose Esters of Fatty Acids, *Biomacromolecules*, 2011, **12**, 2416–2428.
62. P. P. Sengupta, X. Pan, T. J. Nelson, A. Paramarta and D. C. Webster, Cationic UV curing characteristics of epoxidized sucrose esters, *PMSE Prepr.*, 2010, **102**, 888–889.
63. T. J. Nelson, T. P. Galhenage and D. C. Webster, Catalyzed crosslinking of highly functional bio-based epoxy resins, *J. Coat. Technol. Res.*, 2013, **10**, 589–600.
64. S. Ma and D. C. Webster, Naturally Occurring Acids as Cross-Linkers To Yield VOC-Free, High-Performance, Fully Bio-Based, Degradable Thermosets, *Macromolecules*, 2015, **48**, 7127–7137.
65. S. Ma, C. S. Kovash Jr and D. C. Webster, Effect of solvents on the curing and properties of fully bio-based thermosets for coatings, *J. Coat. Technol. Res.*, 2017, **14**, 367–375.
66. S. Ma, D. C. Webster and F. Jabeen, Hard and Flexible, Degradable Thermosets from Renewable Bioresources with the Assistance of Water and Ethanol, *Macromolecules*, 2016, **49**, 3780–3788.
67. H. Pelletier, N. Belgacem and A. Gandini, Acrylated vegetable oils as photocrosslinkable materials, *J. Appl. Polym. Sci.*, 2006, **99**, 3218–3221.
68. H. Liu, W. Lu and S. Liu, Development of acrylated soybean oil-based UV-curable coatings with high impact strength from low viscosity oligomer, *J. Appl. Polym. Sci.*, 2018, **135**, 45698.
69. S. Rengasamy and V. Mannari, Development of soy-based UV-curable acrylate oligomers and study of their film properties, *Prog. Org. Coat.*, 2013, **76**, 78–85.
70. A. Paramarta, X. Pan and C. Webster Dean, Highly functional acrylated bio-based resin system for UV-curable coatings, *RadTech Rep.*, 2013, **27**, 26–32.
71. A. Z. Yu, J. M. Sahouani and D. C. Webster, Highly functional methacrylated bio-based resins for UV-curable coatings, *Prog. Org. Coat.*, 2018, **122**, 219–228.
72. E. Hablot, D. Zheng, M. Bouquey and L. Avérous, Polyurethanes Based on Castor Oil: Kinetics, Chemical, Mechanical and Thermal Properties, *Macromol. Mater. Eng.*, 2008, **293**, 922–929.
73. Z. S. Petrovic, Polyurethanes from Vegetable Oils, *Poly. Rev.*, 2008, **48**, 109–155.
74. Z. S. Petrovic, A. Guo, I. Javni, I. Cvetkovic and D. P. Hong, Polyurethane networks from polyols obtained by hydroformylation of soybean oil, *Polym. Int.*, 2008, **57**, 275–281.
75. J. Argyropoulos, P. Popa, G. Spilman, D. Bhattacharjee and W. Koonce, Seed oil based polyester polyols for coatings, *J. Coat. Technol. Res.*, 2009, **6**, 501–508.

76. X. Pan and D. C. Webster, New Bio-based High Functionality Polyols and Their Use in Polyurethane Coatings, *ChemSusChem*, 2012, **5**, 419–429.
77. T. J. Nelson, L. Bultema, N. Eidenschink and D. C. Webster, Bio-based high functionality polyols and their use in 1K polyurethane coatings, *J. Renewable Mater.*, 2013, **1**, 141–153.
78. T. J. Nelson, B. Masaki, Z. Morseth and D. C. Webster, Highly functional bio-based polyols and their use in melamine-formaldehyde coatings, *J. Coat. Technol. Res.*, 2013, **10**, 757–767.
79. D. Dieterich, Aqueous emulsions, dispersions and solutions of polyurethanes; synthesis and properties, *Prog. Org. Coat.*, 1981, **9**, 281–340.
80. H. Honarkar, Waterborne polyurethanes: A review, *J. Dispersion Sci. Technol.*, 2018, **39**, 507–516.
81. R. Tennebroek, I. van der Hoeven-van Casteren, R. Swaans, S. van der Slot, P. J. M. Stals, B. Tuijtelaars and C. Koning, Water-based polyurethane dispersions, *Polym. Int.*, 2019, **68**, 832–842.
82. D. P. Pfister, Y. Xia and R. C. Larock, Recent Advances in Vegetable Oil-Based Polyurethanes, *ChemSusChem*, 2011, **4**, 703–717.
83. C. Zhang, T. F. Garrison, S. A. Madbouly and M. R. Kessler, Recent advances in vegetable oil-based polymers and their composites, *Prog. Polym. Sci.*, 2017.
84. Y. Lu and R. C. Larock, Soybean-Oil-Based Waterborne Polyurethane Dispersions: Effects of Polyol Functionality and Hard Segment Content on Properties, *Biomacromolecules*, 2008, **9**, 3332–3340.
85. Y. Lu and R. C. Larock, Soybean oil-based, aqueous cationic polyurethane dispersions: Synthesis and properties, *Prog. Org. Coat.*, 2010, **69**, 31–37.
86. Y. Lu and R. C. Larock, Aqueous Cationic Polyurethane Dispersions from Vegetable Oils, *ChemSusChem*, 2010, **3**, 329–333.
87. Y. Lu and R. C. Larock, Synthesis and properties of grafted latices from a soybean oil-based waterborne polyurethane and acrylics, *J. Appl. Polym. Sci.*, 2011, **119**, 3305–3314.
88. Y.-S. Lu, Y. Xia and R. C. Larock, Surfactant-free core-shell hybrid latexes from soybean oil-based waterborne polyurethanes and poly(styrene-butyl acrylate), *Prog. Org. Coat.*, 2011, **71**, 336–342.
89. Y. Xia and R. C. Larock, Castor-oil-based waterborne polyurethane dispersions cured with an aziridine-based crosslinker, *Macromol. Mater. Eng.*, 2011, **296**, 703–709.
90. S. A. Madbouly, Y. Xia and M. R. Kessler, Rheological Behavior of Environmentally Friendly Castor Oil-Based Waterborne Polyurethane Dispersions, *Macromolecules*, 2013, **46**, 4606–4616.
91. Y. Xia, Z. Zhang, M. R. Kessler, B. Brehm-Stecher and R. C. Larock, Antibacterial Soybean-Oil-Based Cationic Polyurethane Coatings Prepared from Different Amino Polyols, *ChemSusChem*, 2012, **5**, 2221–2227.
92. T. F. Garrison, Z. Zhang, H.-J. Kim, D. Mitra, Y. Xia, D. P. Pfister, B. F. Brehm-Stecher, R. C. Larock and M. R. Kessler, Thermo-Mechanical and Antibacterial Properties of Soybean Oil-Based Cationic

Polyurethane Coatings: Effects of Amine Ratio and Degree of Cross-linking, *Macromol. Mater. Eng.*, 2014, **299**, 1042–1051.
93. S. Caillol, Cardanol: A promising building block for bio-based polymers and additives, *Curr. Opin. Green Sustainable Chem.*, 2018, **14**, 26–32.
94. E. Darroman, N. Durand, B. Boutevin and S. Caillol, Improved cardanol derived epoxy coatings, *Prog. Org. Coat.*, 2016, **91**, 9–16.
95. H. Shaghaleh, X. Xu and S. Wang, Current progress in production of biopolymeric materials based on cellulose, cellulose nanofibers, and cellulose derivatives, *RSC Adv.*, 2018, **8**, 825–842.
96. E. Gubbels, J. P. Drijfhout, C. Posthuma-van Tent, L. Jasinska-Walc, B. A. J. Noordover and C. E. Koning, Bio-based semi-aromatic polyesters for coating applications, *Prog. Org. Coat.*, 2014, **77**, 277–284.
97. E. Gubbels, L. Jasinska-Walc, B. A. J. Noordover and C. E. Koning, Linear and branched polyester resins based on dimethyl-2,5-furandicarboxylate for coating applications, *Eur. Polym. J.*, 2013, **49**, 3188–3198.
98. M. N. García González, P. Börjesson, M. Levi and S. Turri, Development and Life Cycle Assessment of Polyester Binders Containing 2,5-Furandicarboxylic Acid and Their Polyurethane Coatings, *J. Polym. Environ.*, 2018, **26**, 3626–3637.
99. M. Lomelí-Rodríguez, R. J. Corpas-Martínez, S. Willis, R. Mulholland and A. J. Lopez-Sanchez, Synthesis and Characterization of Renewable Polyester Coil Coatings from Biomass-Derived Isosorbide, FDCA, 1,5-Pentanediol, Succinic Acid, and 1,3-Propanediol, *Polymers*, 2018, **10**, 600.
100. W. Liu, T. Xie and R. Qiu, Bio-based Thermosets Prepared from Rigid Isosorbide and Flexible Soybean Oil Derivatives, *ACS Sustainable Chem. Eng.*, 2016, **5**, 774–783.
101. Y. Xia and R. C. Larock, Soybean Oil-Isosorbide-Based Waterborne Polyurethane-Urea Dispersions, *ChemSusChem*, 2011, **4**, 386–391.
102. A. Carlson, B. Coggio, K. Lau, C. Mercogliano and J. Millis, Industrial Production of Succinic Acid, *Chemicals and Fuels from Bio-Based Building Blocks*, 2016, pp. 173–190.
103. W. D. Coggio, I. Hevus, C. Webster Dean, A. Schrock, B. Thompson, K. Ulrich and N. Dzadek, Advancing Adhesives: Bio-based succinic acid polyester polyols, ASI Insider, 2015, pp. 36–42.
104. W. D. Coggio, I. Hevus and D. C. Webster, Succinic acid: a bio-based building block in performance driven polyurethane dispersions for coatings, *42nd Proceedings of the International Waterborne, High-Solids, and Powder Coatings Symposium*, 2015, pp. 153–168.
105. G. A. Kraus, Synthetic Methods for the Preparation of 1,3-Propanediol, *Clean: Soil, Air, Water*, 2008, **36**, 648–651.
106. J. V. Kurian, A New Polymer Platform for the Future–Sorona® from Corn Derived 1,3-Propanediol, *J. Poly. Environ.*, 2005, **13**, 159–167.
107. H. Liu, Y. Xu, Z. Zheng and D. Liu, 1,3-Propanediol and its copolymers: Research, development and industrialization, *Biotechnol. J.*, 2010, **5**, 1137–1148.

108. J. Becker, A. Lange, J. Fabarius and C. Wittmann, Top value platform chemicals: bio-based production of organic acids, *Curr. Opin. Biotechnol.*, 2015, **36**, 168–175.
109. T. Robert and S. Friebel, Itaconic acid – a versatile building block for renewable polyesters with enhanced functionality, *Green Chem.*, 2016, **18**, 2922–2934.
110. J. Dai, S. Ma, X. Liu, L. Han, Y. Wu, X. Dai and J. Zhu, Synthesis of bio-based unsaturated polyester resins and their application in waterborne UV-curable coatings, *Prog. Org. Coat.*, 2015, **78**, 49–54.
111. D. M. Patil, G. A. Phalak and S. T. Mhaske, Design and synthesis of bio-based UV curable PU acrylate resin from itaconic acid for coating applications, *Des. Monomers Polym.*, 2017, **20**, 269–282.
112. J. Dai, S. Ma, Y. Wu, J. Zhu and X. Liu, High bio-based content waterborne UV-curable coatings with excellent adhesion and flexibility, *Prog. Org. Coat.*, 2015, **87**, 197–203.
113. P. Li, S. Ma, J. Dai, X. Liu, Y. Jiang, S. Wang, J. Wei, J. Chen and J. Zhu, Itaconic Acid as a Green Alternative to Acrylic Acid for Producing a Soybean Oil-Based Thermoset: Synthesis and Properties, *ACS Sustainable Chem. Eng.*, 2017, **5**, 1228–1236.
114. J. J. Bozell, L. Moens, D. C. Elliott, Y. Wang, G. G. Neuenscwander, S. W. Fitzpatrick, R. J. Bilski and J. L. Jarnefeld, Production of levulinic acid and use as a platform chemical for derived products, *Resour., Conserv. Recycl.*, 2000, **28**, 227–239.
115. A. Maiorana, S. Spinella and R. A. Gross, Bio-Based Alternative to the Diglycidyl Ether of Bisphenol A with Controlled Materials Properties, *Biomacromolecules*, 2015, **16**, 1021–1031.
116. A. Patel, A. Maiorana, L. Yue, R. A. Gross and I. Manas-Zloczower, Curing Kinetics of Bio-based Epoxies for Tailored Applications, *Macromolecules*, 2016, **49**, 5315–5324.
117. C. Leibig, B. Mullen, T. Mullen, L. Rieth and V. Badarinarayana, in *Renewable and Sustainable Polymers*, American Chemical Society, 2011, vol. 1063, pp. 111–116.
118. H. Chung and N. R. Washburn, Chemistry of lignin-based materials, *Green Mater.*, 2013, **1**, 137–160.
119. J. Zakzeski, P. C. A. Bruijnincx, A. L. Jongerius and B. M. Weckhuysen, The Catalytic Valorization of Lignin for the Production of Renewable Chemicals, *Chem. Rev.*, 2010, **110**, 3552–3599.
120. B. M. Upton and A. M. Kasko, Strategies for the Conversion of Lignin to High-Value Polymeric Materials: Review and Perspective, *Chem. Rev.*, 2016, **116**, 2275–2306.
121. E. D. Hernandez, A. W. Bassett, J. M. Sadler, J. J. La Scala and J. F. Stanzione, Synthesis and Characterization of Bio-based Epoxy Resins Derived from Vanillyl Alcohol, *ACS Sustainable Chem. Eng.*, 2016, **4**, 4328–4339.
122. M. Fache, C. Montérémal, B. Boutevin and S. Caillol, Amine hardeners and epoxy cross-linker from aromatic renewable resources, *Eur. Polym. J.*, 2015, **73**, 344–362.

123. M. Fache, A. Viola, R. Auvergne, B. Boutevin and S. Caillol, Bio-based epoxy thermosets from vanillin-derived oligomers, *Eur. Polym. J.*, 2015, **68**, 526–535.
124. M. Fache, B. Boutevin and S. Caillol, Epoxy thermosets from model mixtures of the lignin-to-vanillin process, *Green Chem.*, 2016, **18**, 712–725.
125. M. Fache, R. Auvergne, B. Boutevin and S. Caillol, New vanillin-derived diepoxy monomers for the synthesis of bio-based thermosets, *Eur. Polym. J.*, 2015, **67**, 527–538.
126. M. Fache, B. Boutevin and S. Caillol, Vanillin Production from Lignin and Its Use as a Renewable Chemical, *ACS Sustainable Chem. Eng.*, 2015, **4**, 35–46.
127. M. Fache, B. Boutevin and S. Caillol, Vanillin, a key-intermediate of bio-based polymers, *Eur. Polym. J.*, 2015, **68**, 488–502.
128. M. Fache, E. Darroman, V. Besse, R. Auvergne, S. Caillol and B. Boutevin, Vanillin, a promising bio-based building-block for monomer synthesis, *Green Chem.*, 2014, **16**, 1987–1998.

CHAPTER 11

End-of-life of Plastics/Bioplastics

NEHA MULCHANDANI AND RAMANI NARAYAN*

Department of Chemical Engineering and Materials Science, Michigan State University, East Lansing, MI, 48824, USA
*Email: narayan@msu.edu

11.1 Introduction

Plastics are an indispensable part of human life which have gained global consumer acceptance in the majority of commercial sectors.[1] About 77% of current commercial plastics contain a carbon–carbon backbone in their molecular structure which results in strong resistance to degradation in the environment, resulting in long-lasting materials that persist for decades or longer. Packaging, which has an extremely short use lifetime before becoming waste, accounts for the largest plastic demand by application (44.8% of all polymers). Carbon–carbon backbone polymers [mainly polyethylene (PE), polypropylene (PP), and polystyrene (PS)] comprise 77% of polymers used in packaging.[2] However, this also leads to their persistence in the environment after their end-of-life, where a large amount of plastic is sent to landfills or discarded into the natural environment as litter. According to the global analysis of all the plastics that have ever been made, the total virgin plastics produced until 2017 account for 8.3 billion metric tons (BMT). As of 2015, the total plastic waste generated had been 6.3 BMT, of which only 9% had been recycled, 12% had been incinerated, and 79% entered landfills.[2,3]

The global production of plastics was calculated to be 434 million metric tons (MMT) in the year 2018, of which nearly 77% comprised carbon–carbon backbone polymers[4] such as PE, PP, PS, and polyvinyl chloride (PVC). Ester backbone polymers such as polyethylene terephthalate (PET) accounted for

Green Chemistry Series No. 79
Bioplastics and Biocomposites: A Practical Introduction
Edited by David Grewell

Published by the Royal Society of Chemistry, www.rsc.org

10.2%, urethane backbone polymers such as polyurethane (PU) comprised 8.2%, and 4.9% consisted of other polymers (see Figure 11.1).

If the plastics production, use, and waste management trends continue with the same alarming pace, nearly 12 BMT of plastic waste will have accumulated in the environment by 2050.[2] The accumulation of plastic waste is posing detrimental impacts to the environment, including marine and aquatic life.[5] The presence of plastic waste spans the ocean surface, deep sea sediments, as well as the mid-water region. It has been estimated that nearly 8.8 MMT of plastic waste is discharged into the ocean every year, which is equivalent to dumping a truck of waste plastic every minute into the ocean. With the continuing current practices, the discharge of plastic into the ocean may reach up to 53 MMT per year by 2030.[6] This alarming situation necessitates the need for sustainable alternatives that would reduce the global environmental impact. Bioplastics, therefore, which are bio-based and/or biodegradable in nature, are emerging as alternatives to the conventional carbon–carbon backbone polymers.[7] The global production capacity of bioplastics was accounted to be 2.4 MMT in the year 2021, which is predicted to increase to 7.5 MMT by 2026 (see Figure 11.2).

Globally, plastics have widely been employed in various market segments such as packaging, building and construction, consumer goods, electrical and electronic, and others.[9] Among various market segments, packaging accounted for the largest plastic demand (44.8% of all polymers), which has a very short service life.[4] Nearly 155 MMT of the total produced plastic was used in the packaging sector in 2018, of which 1.2 MMT consisted of bioplastics[8] (see Figure 11.3). This essentially leads to the generation of an enormous amount of plastic waste, which requires strategies to reuse and

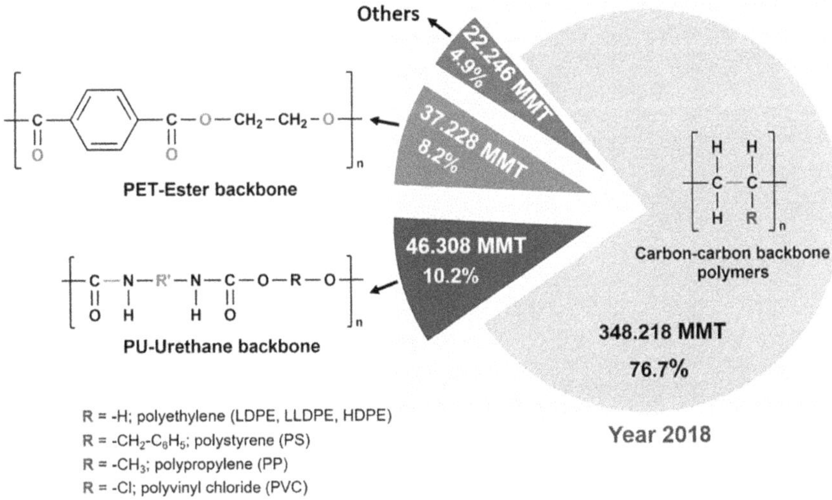

Figure 11.1 Global plastic production|classification of global plastic production according to polymer type. Data from ref. 4.

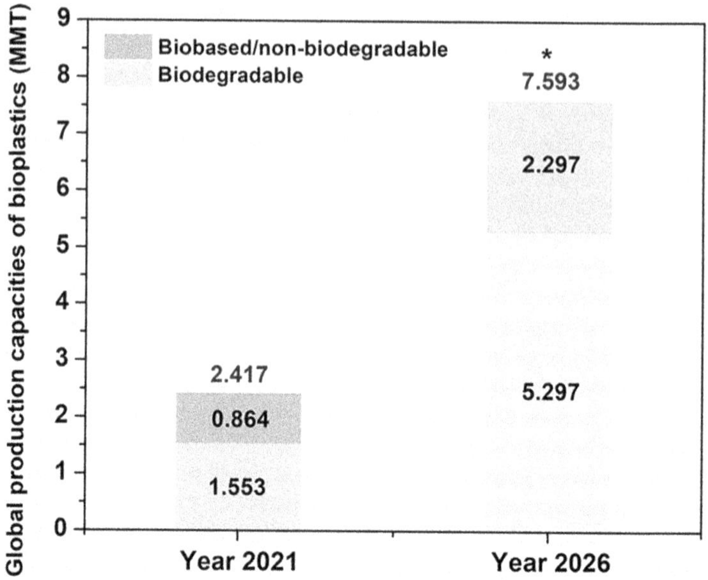

Figure 11.2 Global production capacities of bioplastics in the years 2021 and 2026 (* predicted). Data from ref. 8.

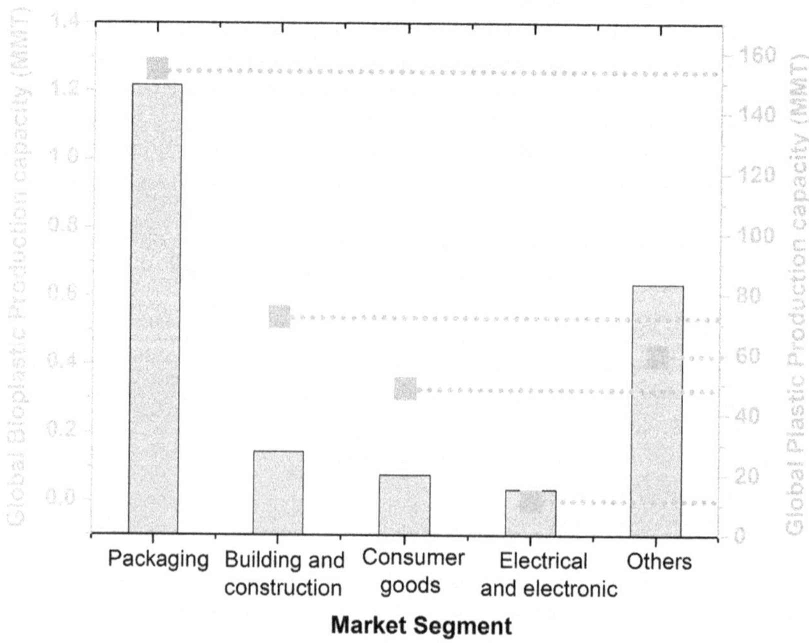

Figure 11.3 Global plastic/bioplastic production capacity in the year 2018 by market segment. The data for global plastic production is from ref. 4 and that for global bioplastic production is from ref. 8.

recycle it in order to explore its complete potential. This chapter thus presents various end-of-life strategies for managing the waste plastic from the perspective of circularity with the aim of reducing global carbon footprint.

11.2 End-of-life Strategies

The rising concerns for managing the enormous amounts of waste plastics has driven the attention of the plastic industry towards their management. Global plastic pollution remains a key challenge which requires multiple strategies for its alleviation and prevention.[10] The strategies for tackling plastic waste adopted by several countries include reducing plastic waste, banning single-use plastics, and implementing regulations on banning waste imports. However, an essential approach towards the global problem of plastic waste management is "circularity."[11] The concept of circular economy encompasses the consumption, waste reduction, and recycling. Circular economy aims to valorize end-of-life plastics in order to eliminate the waste and minimize the resources for new production.[12] Plastic waste recycling, therefore, remains an imperative end-of-life treatment which can mainly be categorized into mechanical recycling, chemical recycling,[13] and biological recycling (see Figure 11.4).

11.2.1 Mechanical Recycling

Mechanical recycling is the prevalent method for recycling where the waste plastic materials are sorted according to their type and properties followed by washing and processing to make secondary raw-materials[14,15] (see Figure 11.5). Packaging materials such as PET and HDPE are largely subjected to mechanical recycling. Mechanical recycling, although being a simple process, encounters the challenge of contamination from residual materials and undesired additives.[16] This often leads to the intricacy of materials which results in the increased waste plastic processing costs. This issue may be addressed by designing materials that are compatible with the existing processes.[17,18]

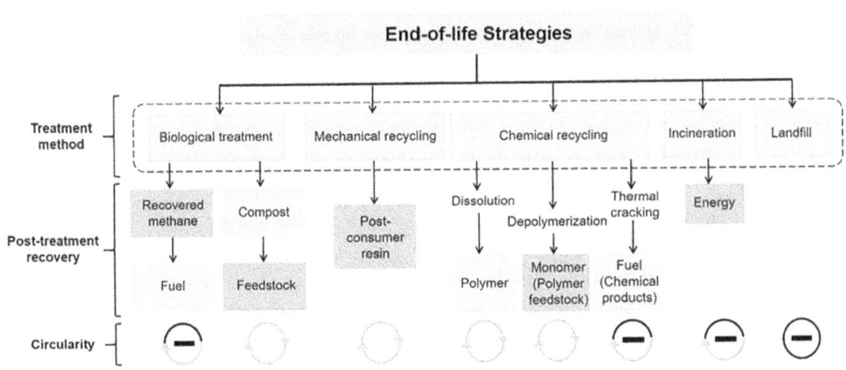

Figure 11.4 End-of-life strategies for managing plastic waste. Adapted from ref. 4 with permission from Springer Nature, Copyright 2021.

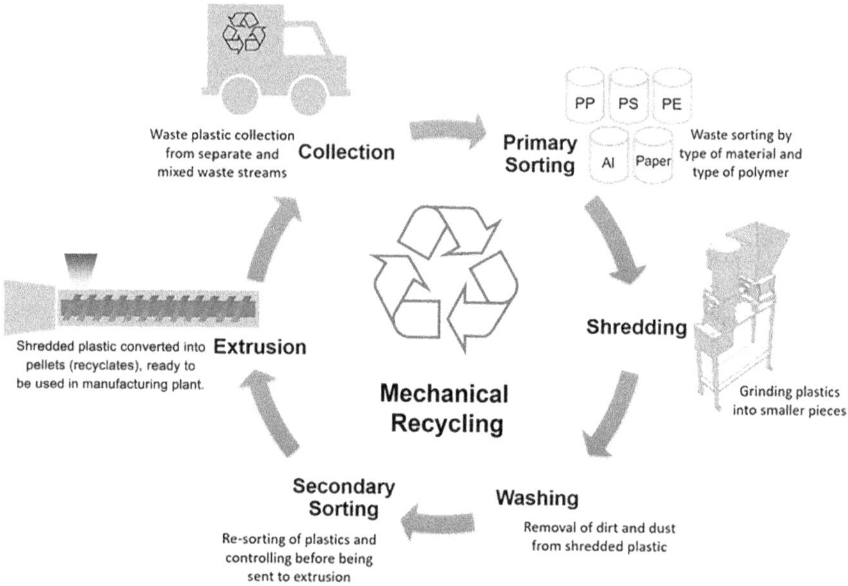

Figure 11.5 Pictorial representation of mechanical recycling of plastic waste in a circular economy.

Furthermore, mechanical recycling is considered to be a downcycling process which results in products of lower-value. According to the National Association for PET Container Resources, approximately 30.1% PET bottles were mechanically recycled in the United States in 2015 where a large amount (>70%) of recycled PET was used for producing lower-value products.[19] This is because of the inevitable polymer degradation occurring throughout the life-cycle of the plastic, including reprocessing. The exposure of plastic to heat, oxygen, and mechanical shear basically alters the polymer properties which results in their usage in lower-value applications.[16] It is essential to realize that recycling, at most, retards the final disposal of plastic waste and does not necessarily prevent it from becoming waste. It is thus essential to develop materials intended for maximum energy recovery over their life-span[20] before being disposed of into landfills.[21] The primary production is, however, influenced by the commercial market forces which may not necessarily be retarded by increased recycling. In some circumstances, the contradictory increase of primary production may be noticed in response to enhanced recycling (or secondary production).[22] It may therefore be perceived that the sole way of preventing plastic waste completely is by minimizing the primary production and augmenting the secondary production.[23,24]

11.2.2 Chemical Recycling

Recycling of waste plastics *via* chemical processes has been adopted as a method complementary to mechanical recycling[25,26] which permits the recycling of waste that is otherwise incinerated or directed to landfills.[27] Chemical recycling may be classified into three approaches: dissolution, depolymerization, and thermal cracking. These approaches differ in the type of materials that can be handled, underlying chemistry, and the resulting product.[28,29]

11.2.2.1 Dissolution

Dissolution is referred to as a physical process which utilizes selected solvents to dissolve the polymer from the stream of mixed plastic waste. This process enables the separation of insoluble contaminants such as pigments and fillers which may be filtered out followed by reprecipitation of the dissolved polymer which may be recovered and reused.[30] The process being physical doesn't affect the composition and structure of the polymer due to which it is considered similar to mechanical recycling rather than chemical recycling. However, in principle, the exposure of waste plastic to thermal and physical stresses during the recovery is much lower than the mechanical recycling process. It is therefore also known as a solvent-based purification method.[31] The selection of solvent plays a crucial role[32] in recovering the target polymer rendering it suitable for the homogeneous streams of plastic waste.[33] This process may ideally be suited for the single-material waste streams containing large amounts of contaminants where their removal would be difficult to achieve *via* mechanical recycling.[34] The ongoing pilot projects incorporating this technology include PureCycle Technologies, US,[35] and Polystyvert, Canada,[28,36] which are targeting PP and PS, respectively. The technology for recovering polyamide and LDPE from multi-layer films is being developed by APK, Germany.[37,38]

11.2.2.2 Depolymerization

Depolymerization is a chemical recycling process which employs heat and preferably a catalyst for decomposing the polymers back to monomers or oligomers via several chemical processes.[39,40] The method therefore is also recognized as "monomer recovery." The obtained depolymerization products may further be subjected to polymerization (see Figure 11.6). From the standpoint of circularity, the aim of chemical recycling is to depolymerize the plastic waste into monomers, which can be used for the production of conventional or new polymers that can withstand several cycles of polymerization and depolymerization.[41] The plastic waste that is sent to the landfills or incineration may be subjected to depolymerization which provides recycled plastic material with properties similar to virgin plastic. However, pre-sorting of the waste stream is necessary as the process requires a consistent polymer composition in order to be subjected to particularly specific processes. This in turn may require substantial energy input. The process of depolymerization is particularly suitable for PET, which is made by polycondensation. The pilot plants that are being developed for

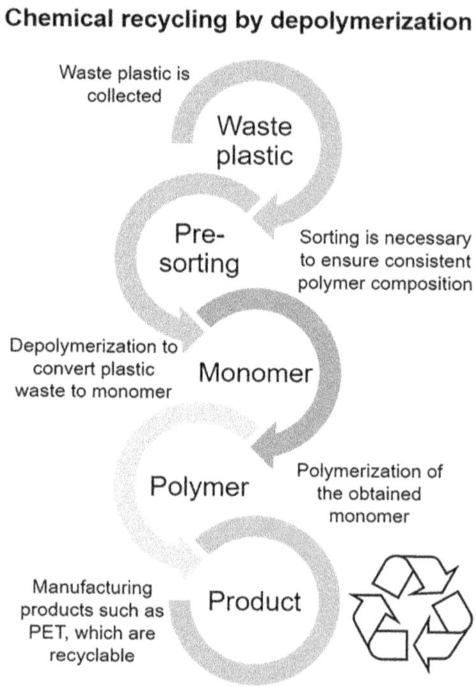

Figure 11.6 Pictorial representation of depolymerization process.

depolymerizing PET include Germany's Rittec,[42] Carbios in France,[43] CuRe Technology,[44] and Ioniqa[45] in the Netherlands, Loop Industries in Canada,[46] and Eastman in USA.[47]

11.2.2.3 Thermal Cracking

Thermal cracking involves pyrolysis and gasification[48] for converting waste plastic to the feedstock components (hydrocarbons and syngas).[49] Pyrolysis typically involves cracking of the polymer chains at elevated temperature in the absence of oxygen.[50] The plastic waste is pre-sorted and shredded before pyrolysis at 400–600 °C. The obtained vapour and gas are purified to generate hydrocarbons such as gas, oils, wax, and char. Polymers with a limited oxygen content such as PP, PE, and PS can be subjected to pyrolysis. It is necessary to ensure the removal of polymers containing higher amounts of oxygen and halogens (such as PVC) from the waste stream. Apart from the oxygen and halogen concerns, the pyrolysis method is capable of handling mixed plastic waste streams that are particularly challenging for all the other recycling methods.[51] Gasification, on the other hand, involves heating the polymer with a limited amount of oxygen. The steps involved in the process include waste pretreatment, gasification, and cleaning of the gas. Gasification is capable of recycling plastic waste and biomass[52] along with polymers containing oxygen and halogens.[53] The end products of the

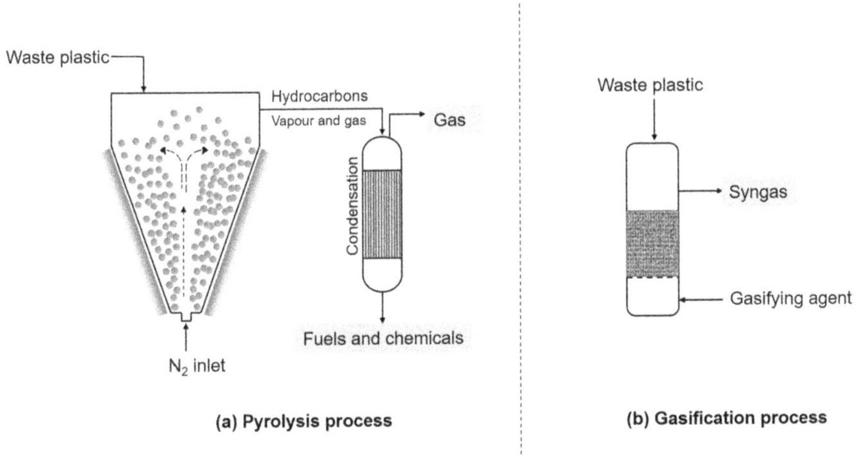

Figure 11.7 Schematic representation of (a) pyrolysis and (b) gasification processes.

pyrolysis and gasification consist of a different mix that may be used in applications varying from fuels to feedstock[54] (see Figure 11.7).

As observed from the above discussion, mechanical and chemical recycling techniques are ideally suitable for the petroleum/fossil-based plastics that are long-lasting materials and cannot be biodegraded under normal environmental conditions and constitute a major market share at present. Additionally, efforts have been focused on developing bioplastics that can replace conventional plastics in various segments including packaging or single-use plastics. As society is moving towards replacing fossil-based plastics with bioplastics in applications having a short-life such as packaging, biological treatment is a promising recycling method for managing the bioplastics after their service life. Bioplastics may be bio-based and/or biodegradable in nature, which have been detailed in the subsequent section.

11.3 Bio-based Plastics

11.3.1 Understanding Bio-based

The term "bio-based" refers to the origin of the carbon in the polymer molecule and indicates whether the carbon is derived from plant-biomass or fossil resources. Bio-based plastics are "plastics in which the (organic) carbon (of the polymer molecule) in part or whole comes from plant-biomass like agricultural crops and residues, marine, and forestry materials, algae, and fungi living in a natural environment in equilibrium with the atmosphere."[55]

Bio-based offers the value proposition of carbon footprint reduction and supports a rural agrarian economy. The bio-based carbon content of a product is the amount of bio-based carbon as a percent of the total organic carbon in the product. The bio-based content is experimentally measured using radiocarbon analysis as codified in ASTM (D6866) & ISO (IS1660 pt2) standards. If inorganic carbon (talc, calcium carbonate) is part of the plastic then one can report biogenic carbon content which is the amount of bio-based carbon (organic + inorganic) as a percent of the total carbon (organic + inorganic) in the product. Figure 11.8 provides an information capsule about bio-based products and radiocarbon analysis for quantifying bio-based (carbon) content.

Plant-biomass (bio-based) resources are photoautotrophs. They convert (remove) CO_2 in the environment to carbohydrates, lipids, and proteins found in plant biomass using water and sunlight energy (photosynthesis). Petro/fossil resources (*e.g.*, oil, coal, natural gas) are formed from plant biomass over millions of years. Therefore, plastics made from plant biomass resources can be credited with removal of CO_2 from the environment and incorporating the carbon into the polymer molecule in a short time (one to ten years). Plastics made from fossil resources cannot be credited with any CO_2 removal from the environment even if one considers a 100 year time scale (the time used in measuring global warming potential, GWP_{100}) because the fossil carbon has formed over millions of years.[56,57]

The term "bioplastics" is used many times in literature and print/e-media to describe "bio-based" and "biodegradable-compostable" plastics without clearly differentiating between them. This has resulted in much confusion in the marketplace and amongst stakeholders and the public. Bio-based refers only to the origins of the carbon in the molecule – the "beginning of life," whether the carbon originated from plant biomass or fossil resources.

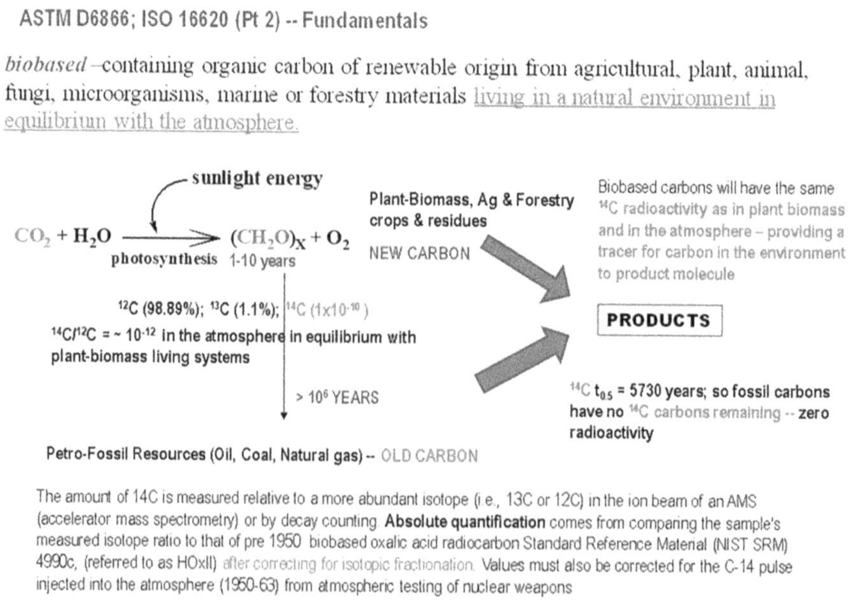

Figure 11.8 Overview of bio-based *vs.* fossil-based product origins, and use of radiocarbon analysis to quantify bio-based carbon content (ASTM D6866; ISO 16620).

Biodegradable or compostable refers to the end-of-life – whether soil or compost microorganisms completely utilize (metabolize) the carbons as their food/energy source. Bio-based plastics are not necessarily biodegradable or compostable and biodegradable-compostable plastics are not necessarily bio-based. Plastics have been sorted according to their origin (fossil or bio-based) and character (persistent or fully biodegradable) in Figure 11.9.

More importantly, bio-based carbon content in a polymer molecule has no connection with its end-of-life behavior. The molecular structure of the polymer, particularly its backbone linkage and physical characteristics (such as crystallinity and morphology), dictate the extent and rate of microbial assimilation (biodegradation) in disposal environments.

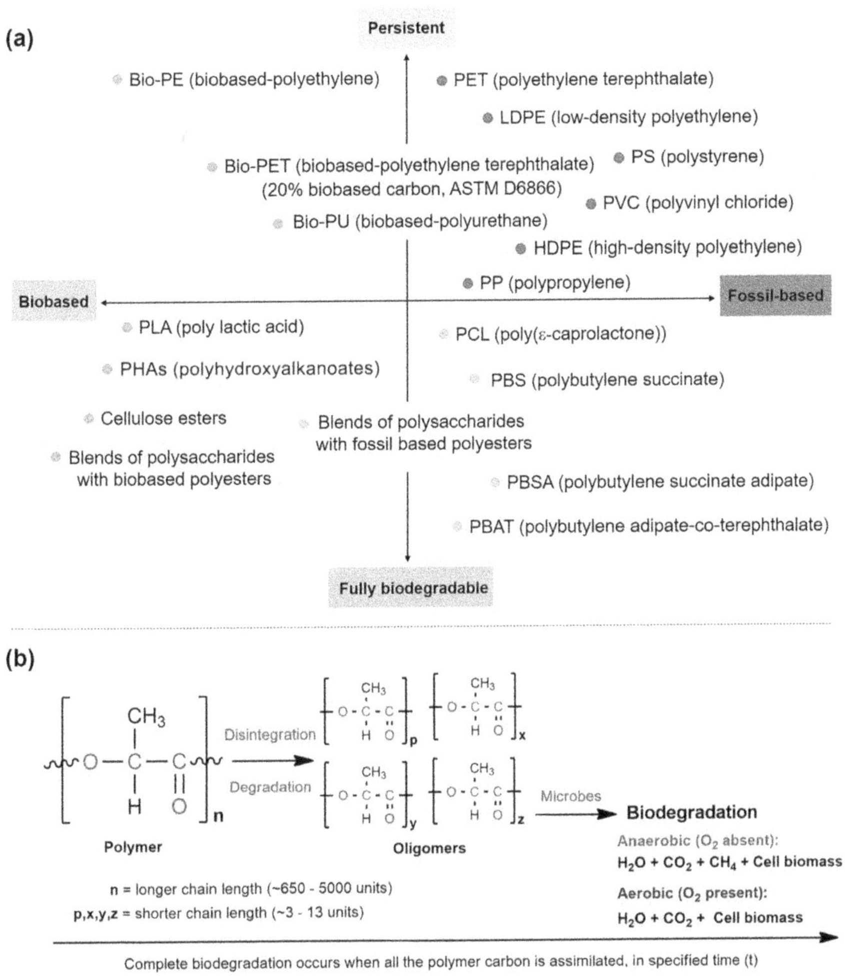

Figure 11.9 Polymers depicted according to carbon feedstock and biodegradability. Adapted from ref. 4 with permission from Springer Nature, Copyright 2021.

11.3.2 Understanding Biodegradability

Biodegradability is an end-of-life process that harnesses soil microorganisms to assimilate/utilize the plastic film's carbon as food/fuel for its life processes in a safe and timely manner. Terms such as "oxo," "hydro," "chemo," or "photo" degradable represent abiotic (non-biological process) mechanisms of degradation and do not constitute "biodegradability." Biodegradability represents the biological process by which microorganisms present in the disposal environment assimilate/utilize carbon substrates as food for their life processes. Figure 11.10 shows this process. Polymer chains with susceptible linkages designed into the polymer backbone degrade by hydrolytic, oxidative (abiotic), or enzymatic (biotic) reactions to smaller molecules or fragments. Microorganisms present in the disposal system (soil, compost) must completely utilize the fragments in a defined time as measured by the evolved carbon dioxide to make a soil biodegradable or compostable claim.

Measuring the rate and extent of carbon dioxide evolved due to microbial metabolism forms the basis of ASTM, European (EN), and International Organization for Standardization (ISO) standards.

Under aerobic conditions, microorganisms biologically oxidize the polymer carbon to CO_2, releasing energy as shown in Scheme 11.1. Microorganisms harness this energy for their life processes in a highly specialized cellular phenomenon that requires the participation of three metabolically interrelated processes: the tricarboxylic acid cycle; electron transport; and oxidative phosphorylation. All of the processes take place inside the cell. Under anaerobic conditions, $CO_2 + CH_4$ (biogas) are produced (see Scheme 11.1).

Thus, a measure of the rate and amount of CO_2 or ($CO_2 + CH_4$) evolved, as a function of total carbon input to the process, is a direct measure of the amount of carbon substrate utilized by the microorganism (*i.e.*, the percent biodegradation).

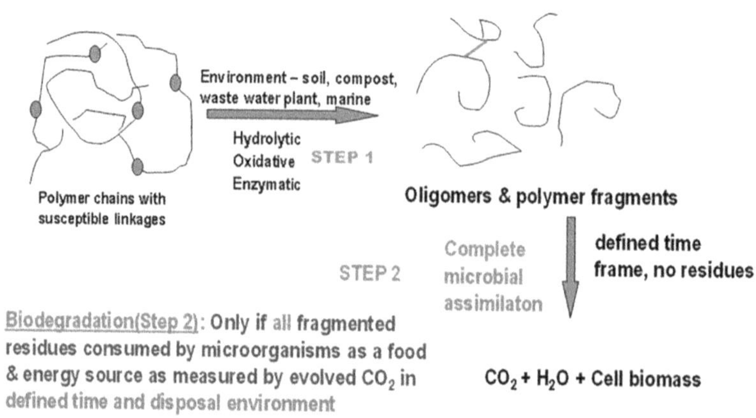

Figure 11.10 Biodegradation process schematic.

Aerobic process

$$C_6H_{12}O_6\text{/C-bioplastic} + 6O_2 \longrightarrow 6CO_2\uparrow + 6H_2O;\ \Delta G^{\circ\prime} = -686\ \text{kcal/mol}$$

Anerobic process

$$C_6H_{12}O_6\text{/C-bioplastic} \longrightarrow 2\ \text{lactate} \longrightarrow 3CO_2\uparrow + 3CH_4\uparrow;\ \Delta G^{\circ\prime} = -47\ \text{kcal/mol}$$

Scheme 11.1 Basic equations describing the microbial utilization of carbon substrates.

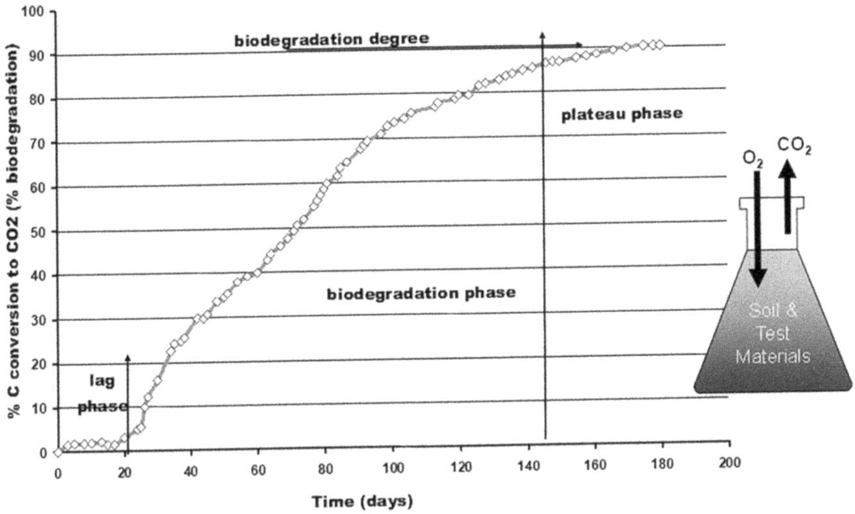

Figure 11.11 Measuring rate and extent of biodegradability using biodegradable polymer as the sole carbon source in compost (ASTM D5338, ISO 14855) or soil (ASTM D5988, ISO 17556).

Thus, one can measure the rate and extent of biodegradation or microbial utilization (mineralization) of the test plastic material by using it as the sole carbon source in a test system containing soil or compost in the presence of air and under suitable temperature conditions (25 °C for soil and 58 °C for compost). Figure 11.11 shows a typical graphical output of the percent carbon converted to CO_2 as a function of time in days. First is a lag phase during which the microbial population adapts to the available test C-substrate. Next is the biodegradation phase during which the adapted microbial population begins to utilize the carbon substrate for its cellular life processes, as measured by the evolved CO_2. When all of the substrate carbon is utilized, the CO_2 output reaches a plateau. Linear or any other form of data extrapolation is strictly prohibited because credible scientific substantiation for the extrapolation model does not exist for such complex biological systems.

Claims of degradable, partially biodegradable, or eventually biodegradable are not acceptable because the degraded fragments can have negative environmental and health impacts.[58] Biodegradable mulch films must demonstrate complete biodegradability in two years or less using ASTM D5988 or ISO 17556 test methods.[59,60]

11.3.3 Applications of Biodegradable, Bio-based Polymers in Agricultural Mulch Films

Conventional plastics are widely used in the agricultural sector as mulch films which have a short service-life. Fossil-based mulching films are not soil biodegradable, resulting in the accumulation of ~462.5 kg of film fragments per hectare per decade considering a film thickness of 20 μm and one cultivation per year. On the other hand, soil biodegradable films reach complete biodegradation in approximately two years under laboratory conditions.[61] Aliphatic and aromatic–aliphatic copolyesters find application as biodegradable mulch films. There are several manufacturers of this class of polymer resins. Using monomers derived from plant-biomass feedstocks such as succinic or adipic acids or butanediol or other diols in the synthesis of copolyesters creates bio-based content in the copolyester molecule. An example is polybutylene adipate terephthalate (PBAT),[62] which is completely biodegradable (see Scheme 11.2) and a potential alternative to PE.

Blending with other bio-based polyesters such as polylactic acid (PLA) or polyhydroxyalkanoates is another approach to incorporating bio-based content. Blends of PBAT–PLA[63] are commercially available and used in agricultural mulch film applications as well.

11.4 Conclusions

The current chapter underlines the existing challenges and strategies for managing plastic waste with the aim of mitigating global environmental pollution. The carbon–carbon backbone of commercial fossil-based polymers increases their persistence in the environment, which occupies a massive region of landfills. In this regard, mechanical and chemical recycling methods are proposed for recycling waste plastics. An emphasis is placed on reducing the primary production and intensifying the secondary production in order to cease the accumulation of plastic waste in the environment. Further, bioplastics and their role in addressing the problem of

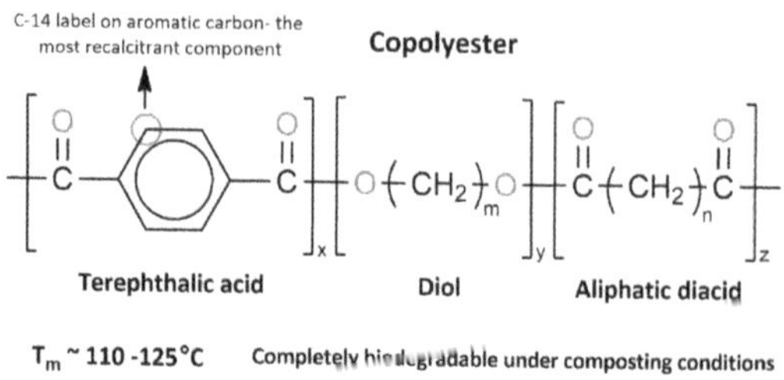

Scheme 11.2 General structure of biodegradable PBAT copolyesters.

waste plastic have been discussed. The concept of circular economy has been recognized as a tool to support the goals of sustainable development.

Abbreviations

BMT	Billion metric tons
MMT	Million metric tons
PBAT	Polybutylene adipate terephthalate
PE	Polyethylene
PET	Polyethylene terephthalate
PLA	Polylactic acid
PP	Polypropylene
PS	Polystyrene
PU	Polyurethane
PVC	Polyvinyl chloride

References

1. R. Palos, A. Gutiérrez, F. J. Vela, M. Olazar, J. M. Arandes and J. Bilbao, *Energy Fuels*, 2021, **35**, 3529–3557.
2. R. Geyer, J. R. Jambeck and K. L. Law, *Sci. Adv.*, 2017, **3**, e1700782.
3. A. Gupta, A. K. Mohanty and M. Misra, *Composites, Part A*, 2022, **154**, 106759.
4. K. L. Law and R. Narayan, *Nat. Rev. Mater.*, 2022, **7**, 104–116.
5. J. R. Jambeck, R. Geyer, C. Wilcox, T. R. Siegler, M. Perryman, A. Andrady, R. Narayan and K. L. Law, *Science*, 2015, **347**, 768–771.
6. The National Academies of Science Engineering Medicine, Reckoning with the U.S. Role in Global Ocean Plastic Waste. Available at https://www.nap.edu/resource/other/dels/plastics-in-the-ocean/, (accessed January 16, 2022).
7. N. Kawashima, T. Yagi and K. Kojima, *Macromol. Mater. Eng.*, 2019, **304**, 1900383.
8. European Bioplastics, nova-Institute, Global production capacities of bioplastics, 2021. Available at https://www.european-bioplastics.org/market/, (accessed January 14, 2022).
9. H. Millet, P. Vangheluwe, C. Block, A. Sevenster, L. Garcia and R. Antonopoulos, in *Plastics and the Environment*, The Royal Society of Chemistry, 2019, pp. 1–20.
10. L. Lebreton and A. Andrady, *Palgrave Commun.*, 2019, **5**, 6.
11. S. Billiet and S. R. Trenor, *ACS Macro Lett.*, 2020, **9**, 1376–1390.
12. Z. O. G. Schyns and M. P. Shaver, *Macromol. Rapid Commun.*, 2021, **42**, 2000415.
13. J. M. Garcia and M. L. Robertson, *Science*, 2017, **358**, 870–872.
14. S. M. Al-Salem, P. Lettieri and J. Baeyens, *Waste Manage.*, 2009, **29**, 2625–2643.

15. I. A. Ignatyev, W. Thielemans and B. Vander Beke, *ChemSusChem*, 2014, **7**, 1579–1593.
16. K. Ragaert, L. Delva and K. Van Geem, *Waste Manage.*, 2017, **69**, 24–58.
17. M. J. Rogoff and D. E. Ross, *Waste Manage. Res.*, 2016, **34**, 181–183.
18. I. Janajreh, M. Alshrah and S. Zamzam, *Sustainable Cities Soc.*, 2015, **18**, 13–20.
19. X. Zhang, M. Fevre, G. O. Jones and R. M. Waymouth, *Chem. Rev.*, 2018, **118**, 839–885.
20. G. Dodbiba, K. Takahashi, J. Sadaki and T. Fujita, *J. Cleaner Prod.*, 2008, **16**, 458–470.
21. T. Zink and R. Geyer, *J. Ind. Ecol.*, 2019, **23**, 541–548.
22. T. Zink and R. Geyer, *J. Ind. Ecol.*, 2017, **21**, 593–602.
23. P. T. Anastas and J. B. Zimmerman, *Environ. Sci. Technol.*, 2003, **37**, 94A–101A.
24. Y. Sunaga, S. Ogoe, K. Aoki, H. Ito and Y. Teramoto, *ACS Sustainable Chem. Eng.*, 2018, **6**, 12110–12118.
25. Chemical Recycling Is A Complementary Solution To Tackle Plastic Waste, *European Chemical Industry Council*, https://cefic.org/media-corner/newsroom/chemical-recycling-is-a-complementary-solution-to-tackle-plastic-waste/, (accessed January 15, 2022).
26. R. P. Lee, M. Tschoepe and R. Voss, *Waste Manage.*, 2021, **125**, 280–292.
27. M. Häußler, M. Eck, D. Rothauer and S. Mecking, *Nature*, 2021, **590**, 423–427.
28. T. Thiounn and R. C. Smith, *J. Polym. Sci.*, 2020, **58**, 1347–1364.
29. M. Pohjakallio, T. Vuorinen and A. Oasmaa, in *Plastic Waste and Recycling*, ed. T. M. Letcher, Academic Press, 2020, pp. 359–384.
30. L. T. J. Korley, T. H. Epps, B. A. Helms and A. J. Ryan, *Science*, 2021, **373**, 66–69.
31. D. S. Achilias, A. Giannoulis and G. Z. Papageorgiou, *Polym. Bull.*, 2009, **63**, 449–465.
32. C. Gutiérrez, M. T. García, I. Gracia, A. de Lucas and J. F. Rodríguez, *Waste Biomass Valorization*, 2013, **4**, 29–36.
33. J. G. Poulakis and C. D. Papaspyrides, *Resour., Conserv. Recycl.*, 1997, **20**, 31–41.
34. I. Georgiopoulou, G. D. Pappa, S. N. Vouyiouka and K. Magoulas, *Resour., Conserv. Recycl.*, 2021, **165**, 105268.
35. Polypropylene recycling venture launches $120 million project, Purecycle, https://purecycle.com/2017/07/polypropylene-recycling-venture-launches-120-million-project/, (accessed January 15, 2022).
36. Government of Canada supports leading-edge company specializing in polystyrene recycling that helps protect the environment, https://www.canada.ca/en/innovation-science-economic-development/news/2021/11/government-of-canada-supports-leading-edge-company-specializing-in-polystyrene-recycling-that-helps-protect-the-environment.html, (accessed January 15, 2022).

37. T. W. Walker, N. Frelka, Z. Shen, A. K. Chew, J. Banick, S. Grey, M. S. Kim, J. A. Dumesic, R. C. V. Lehn and G. W. Huber, *Sci. Adv.*, 2020, **6**, eaba7599.
38. J. Carey, *Proc. Natl. Acad. Sci. U. S. A.*, 2017, **114**, 612–616.
39. B. Whajah, N. da Silva Moura, J. Blanchard, S. Wicker, K. Gandar, J. A. Dorman and K. M. Dooley, *Ind. Eng. Chem. Res.*, 2021, **60**, 15141–15150.
40. C. A. Navarro, Y. Ma, K. H. Michael, H. M. Breunig, S. R. Nutt and T. J. Williams, *Green Chem.*, 2021, **23**, 6356–6360.
41. G. W. Coates and Y. D. Y. L. Getzler, *Nat. Rev. Mater.*, 2020, **5**, 501–516.
42. Rittec Umwelttechnik, Revolpet® A cycle for plastic. Available at: https://www.rittec.eu/solutions/revolpet-r.html, (accessed January 15, 2022).
43. V. Tournier, C. M. Topham, A. Gilles, B. David, C. Folgoas, E. Moya-Leclair, E. Kamionka, M. L. Desrousseaux, H. Texier, S. Gavalda, M. Cot, E. Guémard, M. Dalibey, J. Nomme, G. Cioci, S. Barbe, M. Chateau, I. André, S. Duquesne and A. Marty, *Nature*, 2020, **580**, 216–219.
44. CuRe Polyester Rejuvination, Targeted PET/Polyester applications. Available at: https://curetechnology.com/applications/, (accessed January 15, 2022).
45. Ioniqa, Ioniqa's Circular Solution, 2019. Available at https://ioniqa.com/applications/, (accessed January 15, 2022).
46. Loop Industries, Revolutionary Technology, 2022. Available at https://www.loopindustries.com/en, (accessed January 15, 2022).
47. Eastman, Eastman offers innvovative recycling technology for polyesters. Available at https://www.eastman.com/Company/News_Center/2019/Pages/Eastman-offers-innovative-recycling-technology-for-polyesters.aspx, (accessed January 15, 2022).
48. A. Antelava, N. Jablonska, A. Constantinou, G. Manos, S. A. Salaudeen, A. Dutta and S. M. Al-Salem, *Energy Fuels*, 2021, **35**, 3558–3571.
49. T. Malkow, *Waste Manage.*, 2004, **24**, 53–79.
50. C. Wu and P. T. Williams, *Fuel*, 2010, **89**, 3022–3032.
51. N. Déparrois, P. Singh, K. G. Burra and A. K. Gupta, *Appl. Energy*, 2019, **246**, 1–10.
52. C. Xu, M. Nasrollahzadeh, M. Selva, Z. Issaabadi and R. Luque, *Chem. Soc. Rev.*, 2019, **48**, 4791–4822.
53. C. Wang, Z. Jiang, Q. Song, M. Liao, J. Weng, R. Gao, M. Zhao, Y. Chen and G. Chen, *Energy*, 2021, **232**, 121005.
54. O. Dogu, M. Pelucchi, R. Van de Vijver, P. H. M. Van Steenberge, D. R. D'Hooge, A. Cuoci, M. Mehl, A. Frassoldati, T. Faravelli and K. M. Van Geem, *Prog. Energy Combust. Sci.*, 2021, **84**, 100901.
55. J. R. Pires, V. G. Souza, P. Fuciños, L. Pastrana and A. L. Fernando, *Polymers*, 2022, **14**(7), 1359.
56. R. Narayan, in *Degradable Polymers and Materials*, American Chemical Society, 2006, vol. 939, ch. 18, pp. 282–306.
57. R. Narayan, *MRS Bull.*, 2011, **36**, 716–721.
58. R. C. Thompson, C. J. Moore, F. S. Vom Saal and S. H. Swan, *Philos. Trans. R. Soc., B*, 2009, **364**, 2153–2166.

59. R. Narayan, in *Handbook of Biodegradable Polymers*, ed. B. Catia, De Gruyter, 2020, pp. 455–474.
60. R. Narayan, in *Soil Degradable Bioplastics for a Sustainable Modern Agriculture*, ed. M. Malinconico, Springer Berlin Heidelberg, Berlin, Heidelberg, 2017, pp. 23–34.
61. OWS, Expert statement: (Bio)degradable mulching films. Available at https://www.ows.be/news/expert-statement-biodegradable-mulching-films/, (accessed January 16, 2022).
62. M. T. Zumstein, A. Schintlmeister, T. F. Nelson, R. Baumgartner, D. Woebken, M. Wagner, H.-P. E. Kohler, K. McNeill and M. Sander, *Sci. Adv.*, 2018, **4**, eaas9024.
63. H. Khan, S. Kaur, T. C. Baldwin, I. Radecka, G. Jiang, I. Bretz, K. Duale, G. Adamus and M. Kowalczuk, *ACS Sustainable Chem. Eng.*, 2020, **8**, 5360–5370.

Subject Index

www.ingramcontent.com/pod-product-compliance
Lightning Source LLC
Chambersburg PA
CBHW052236030425
24539CB00008B/274